Planning for Place and Plexus

"A lively, engaging book ... which uses 'neoclassical' economic principles ... in a digestible format. The authors go so far as to draw from the film *Thelma and Louise* to show how game theory can be applied in predicting whether someone will drive or take public transit. This provocative, highly relevant book deserves to be on the bookshelf of everyone concerned with urban planning and transportation."

Robert Cervero, Professor and Chair
Department of City and Regional Planning
University of California, Berkeley

Congestion is worse than ever and land development continues unabated into the countryside. Recent accounts suggest that unprecedented conditions are ahead; coordinated land use and transport planning in metropolitan areas is both targeted as a solution and the subject of increased scrutiny.

Planning for Place and Plexus provides a fresh and unique perspective on metropolitan land use and transport networks, challenging current planning strategies and offering frameworks to understand and evaluate policy.

The book suggests actions for the future urban growth of metropolitan areas and includes current and cutting edge theory, findings, and recommendations which are cleverly illustrated throughout using international examples. *Planning for Place and Plexus* is a valuable resource for students, researchers, practitioners, and policy advisors working across transport, land use, and planning.

David M. Levinson holds the RP Braun/CTS Chair in Transportation Engineering and is the Director of the Networks, Economics and Urban Systems (Nexus) Research Group at the University of Minnesota.

Kevin J. Krizek is Associate Professor of Planning, Design, and Civil Engineering at the University of Colorado and Director of the Active Communities/Transportation (ACT) Research Group.

Planning for Place and Plexus

Metropolitan land use
and transport

David M. Levinson and
Kevin J. Krizek

Routledge
Taylor & Francis Group
NEW YORK AND LONDON

First published 2008 in the USA and Canada
by Routledge
270 Madison Avenue, New York, NY 10016

Simultaneously published in the UK
by Routledge
2 Park Square, Milton Park, Abingdon, Oxon OX14 4RN

*Routledge is an imprint of the Taylor and Francis Group,
an informa business*

© 2008 David M. Levinson and Kevin J. Krizek

Typeset in Sabon by
Florence Production Ltd, Stoodleigh, Devon
Printed and bound in Great Britain by
TJ International Ltd, Padstow, Cornwall

All rights reserved. No part of this book may be reprinted or reproduced or utilized in any form or by any electronic, mechanical, or other means, now known or hereafter, invented including photocopying and recording, or in any information storage or retrieval, system, without permission in writing from the publishers.

Every effort has been made to contact and acknowledge copyright owners, but the editor and publishers would be pleased to have any errors or omissions brought to their attention so that corrections may be published at a later printing.

British Library Cataloguing in Publication Data
A Catalogue record for this book is available
from the British Library

Library of Congress Cataloging in Publication Data
Levinson, David M., 1967–
  Planning for place and plexus: metropolitan land use and transport/David Levinson and Kevin J. Krizek.
    p. cm.
  1. City Planning. 2. Land use, Urban. 3. City Traffic.
  I. Krizek, Kevin J. II. Title.
  HT166.L47 2008
  307.1′216—dc22   2007025388

ISBN10 0–415–77490–X (hbk)
ISBN10 0–415–77491–8 (pbk)
ISBN10 0–203–93539–X (ebk)

ISBN13 978–0–415–77490–1 (hbk)
ISBN13 978–0–415–77491–8 (pbk)
ISBN13 978–0–203–93539–2 (ebk)

For KJK: To my parents, Claudia and Ray, who taught me to always finish what I started.

For DML: To my children Benjamin and Olivia, who have yet to start.

Contents

List of illustrations	viii
Preface	xii
Acknowledgments	xiv
1 At a crossroads. Again.	1
2 Diamond of Action	17
3 Homebuying	39
4 Jobseeking	70
5 Traveling	94
6 Scheduling	119
7 Diamond of Exchange	138
8 Siting	153
9 Selling	167
10 Diamond of Evaluation	190
11 Designing	218
12 Assembling	251
13 Operating	278
14 Drawing the curtain	306
Index	327

Illustrations

Figures

1.1	Conceptual framework for understanding land use-transportation interactions	10
1.2	Overview of the structure of place and plexus	11
2.1	Washington, DC, including site of Reagan assassination attempt, local hospitals, and the reach that can be achieved on a grid-based transportation network	18
2.2	Prospect theory of value versus gains	22
2.3	The Diamond of Action	24
2.4	Trends in household income, 1967–2002	27
2.5	Theorized parabolic effect of others on quality of experience	29
2.6	Residential tenure for location decisions	34
2.7	Average hours per day spent in primary activities	35
3.1	Von Thünen's isolated state	42
3.2	Bid-rent curves and distance from city center	44
3.3	"Landlord's Game" board (1904)	46
3.4	"Landlord's Game" board (1910)	47
3.5	Example networks	47
3.6	Law of the network, showing increasing or decreasing returns	49
3.7	Accessibility matrix showing examples of modes and preferences	54
3.8	Schelling's model in initial (a) and final (b) states	57
3.9	Average size of new homes in the United States	65
4.1	Illustration of a gravity model	78
4.2	Gravity model of relationship between willingness to travel and available opportunities	78
4.3	Relationship of travel time and city size	80
4.4	Social network connecting the Reagan, Bush, and Hinckley families	85
4.5	A simple example of an individual's social network	86
5.1	The Mohring Effect	100
5.2	A positive feedback loop operating on bus service	101
5.3	Road costs: complementarity and competition	103
5.4	Modal competition model	108

5.5	Relationship of bus ridership to wait time	109
5.6	Bicyclists in Shanghai	112
5.7	Poster encouraging car sharing, Second World War	115
6.1	Mean minutes spent in various activities, Minneapolis-St. Paul and Washington, DC	122
6.2	Comparison of time spent in various activities versus required travel time	123
6.3	Example travel diary form	125
6.4	A space-time prism	128
6.5	Average trip distances, by trip purpose	129
6.6	Frequency of travel, by trip purpose	129
6.7	Temporal distribution of trips, by trip purpose	131
6.8	The positive utility of travel	134
7.1	The Diamond of Exchange	148
7.2	Network model of the economy	149
8.1	Locational triangle	156
8.2	Varignon frame	157
9.1	Leasable retail area of US shopping centers (1970–2003)	174
9.2	Average size of supermarket food stores in the US	175
9.3	Number of community retail pharmacy outlets, by type of store	176
10.1	Location of current Stillwater Bridge	192
10.2	The Diamond of Evaluation	194
10.3	Decision process options for NEPA	207
10.4	The product development cycle	214
11.1	The Fulton Mall	220
11.2	The Diamond of Design	223
11.3	Christaller's central place theory	224
11.4	Hierarchy of roads	229
11.5	Grid-based morphology shaped by the Land Ordinance of 1785	234
11.6	A woonerf in Utrecht, the Netherlands	243
11.7	Signs indicating the start and end of a woonerf in Utrecht, the Netherlands	244
12.1	Examples of political rhetoric surrounding the assembly of infrastructure investments	253
12.2	Modes of distinction: the Routemaster double-decker red bus in London	254
12.3	Change in consumers' surplus due to a shift in (a) the supply curve, (b) the demand curve	257
12.4	Primary travel routes between the Mall of America and Downtown Minneapolis	258
12.5	The Diamond of Assembly	262
12.6	Examples of elasticity curves	265
12.7	Construction of revealed demand (fulfilled expectations) curve with positive network externalities	272
12.8	Mount Transit and Mount Auto	274

13.1	The Diamond of Operation	280
13.2	Input-output diagram	281
13.3	Bottleneck	283
13.4	The relationship between temporal equity and mobility (travel delay)	286
13.5	Optimal congestion toll, and welfare loss without toll	287
13.6	Parking demand, by time of day	291
13.7	Traditional versus Georgist taxing proposals	297
14.1	Miles of public roads in the United States	314
14.2	Year of construction of existing residential properties	315
14.3	Year of construction of existing commerical properties	316
14.4	Size of networks as a proportion of their maximum extent	321

Tables

2.1	Framework for Part I	35
3.1	Hedonic regression results	63
4.1	Illustrative trip table	77
4.2	Summary statistics and regression results to predict mean commuting time	83
4.3	Social capital scores, by state	87
4.4	Definition of "low-income" households, 1995 NPTS	90
5.1	Sources of information for commute mode share to downtown Minneapolis	97
5.2	Regional travel survey estimates of transit mode share of inbound travelers broken out by time of day	97
6.1	Example records in a trip file	127
6.2	Classifying trip tours	135
9.1	A matrix of exchange	170
9.2	Evolution of the retailing hierarchy	172
9.3	Profile of retail in the United States	173
9.4	Shopping center classifications	175
10.1	Measures of efficiency	196
10.2	Equity impact statement checklist	204
11.1	Layers of place, telecommunications, and plexus	238
11.2	Data structures representing the transportation network	239
11.3	Medalists in the town planning event of the international Olympic Games	247

Boxes

1.1	The American Dream	6
2.1	Cooperation amongst consumers (collective action without government)	31
3.1	Monopoly and anti-monopoly	45
3.2	Computing accessibility	51

3.3 Self-selection	59
3.4 Hedonic regression analysis	62
3.5 Should affordable housing be new?	64
4.1 History of gravity models	74
4.2 Performing gravity model calculations	76
4.3 Modeling aggregate home to work travel times in the aggregate	81
4.4 Social networks, Bush, and Hinckley	85
4.5 Do the rich travel more?	89
5.1 Understanding (and questioning) mode split values	96
5.2 Grand Central Station	110
5.3 Rebranding transit	114
6.1 More than 24 hours in a day?	124
6.2 Travel surveys	124
6.3 Language of travel behavior	126
6.4 Does travel have positive utility?	133
7.1 Firm behavior as increasingly open systems	143
7.2 Cooperation amongst competitors	144
7.3 Defining complementarity and competition	147
7.4 Networks in the economy	151
8.1 The "new" economic geography	159
8.2 Spatial mismatch and the banlieue	164
9.1 The abandoned mall blues	168
9.2 Ground floor retail everywhere	178
9.3 Starbucks at the end of the universe	187
10.1 Manhattan or Manitoba: access versus mobility	200
10.2 Alternative evaluation paradigms	211
11.1 Systems of cities	226
11.2 Flexible design	236
11.3 Japan's building-line system: attaching plexus to place	240
12.1 Consumers' surplus	259
12.2 Elasticity	264
12.3 Causality	268
13.1 Ramp meters	285
13.2 Righteous lanes	288
13.3 Growth management programs	293
13.4 Henry George's single (land) tax	296
13.5 Commuter tax	297
13.6 History of travel demand management	299

Preface

Our motivation for this work was straightforward. We were slated to jointly teach a course at the University of Minnesota on land use-transportation, combining perspectives from urban planning and civil engineering. Like most instructors, we looked for a suitable text.

This is where "straightforward" morphed into "complex." In pursuit of an appropriate text, we inventoried syllabi from other instructors who taught similar courses. We surveyed the instructors; we studied their syllabi; we digested the information. We subsequently published an article describing our content analysis of this work in the *Journal of Planning Education and Research* (2005; vol. 24, 3:304–316). Among other things, we concluded—as did many instructors—that no single text satisfactorily covered all important land use-transportation topics from a pedagogical perspective. We therefore endeavored to write our own.

Given the "hot" topics of land use and networks, especially transportation, we thought we could do more than just write a textbook. We desired a piece of work with more zip than a book whose life would be limited to the classroom. We wanted to write a book that high-level policy advisors could also sink their teeth into and one that stood up to the rigor of academe. This is the point where "complex" morphed to "evasive."

We realized we were trying to satisfy three goals. We endeavored to write a text that would: (1) merge two sub-disciplines (land use and transport) in a straightforward and coherent, but also compelling manner; (2) be useful for graduate-level education in urban planning, civil engineering, geography, regional science, urban studies, and other allied disciplines; and (3) be interesting and engaging enough for other "professional" citizens, high-level policy advisors, or even politicians to want to wade through. It became apparent to us that there is good reason why no single book stands out in terms of satisfying these demands. Were we aiming for the impossible?

We produced what we thought to be a good outline and structure. We then scrapped it. We tried another outline. We scrapped it again. We reworked the structure. This happened over and over. To say our efforts were iterative would be an understatement. After countless iterations to the overall structure, we achieved equilibrium. It was a long time coming. Did we satisfy all three of our goals? We think so. Ultimately, the readers (and reviewers) will let us know.

We divided the book into three parts; the subject of each part represents a class of agent that acts over the landscapes of metropolitan areas in terms of patterns of land use and transport: individuals; businesses (firms); and governments. Separately examining the behavior of these three agents provided an effective strategy to better dissect and understand their actions. Is this the best structure? We thought so; it makes sense to us. We hope it makes sense to you, the reader, as well.

We kept the discussion at a relatively high level—being sure to explain concepts in sufficient terms—but by no means strictly serving as a trade manual. We thought broadly, being sure not to equate urban transportation with congestion. We were always conscious of the larger planning, sociological, and economic context and most importantly how individuals, businesses, and governments interact with one another and with urban landscapes over networks. Those looking to this book for recipes on how to conduct traffic impact analyses, write zoning codes, or run four-step travel demand models will come away disappointed. We feel such skills, although important, are specifics best left to practice. The more challenging part is understanding the larger transportation-land use system and its components, articulating reasons for change, and prescribing workable solutions to vexing urban problems.

The 14 chapters are intended to mesh well with the timing of the semester system on many university campuses—a chapter for each week. Instructors will undoubtedly want to season to their own taste, particularly in terms of scheduling with assignments and exams and supplementing the text with other materials. In our course we make extensive use of student-run case studies to help bring theory to practice.

We offer this edition to help today's students of cities, be they enrolled in school or experience life, think about how both place (areas of space with definite or indefinite boundaries) and plexus (the combination of networks, including both social networks and physical infrastructure) function. Despite identifying flaws in existing systems, as a result of past decisions and dysfunctional decision-making processes, we think these systems can be made better. If we did not have hope there would have been no point in the penning of *Planning for Place and Plexus: Metropolitan Land Use and Transport*.

David M. Levinson
Minneapolis, Minnesota

Kevin J. Krizek
Boulder, Colorado

Acknowledgments

Both in and outside of academe, many of us realize the degree to which our own perspectives are shaped by teachers and colleagues who happened to affect us in ways that profoundly changed our thinking on different matters. These people—with whom we have had contact either directly (in the classroom, in hallways, or at professional meetings) or virtually (through email, the Internet or reading their work)—are too many to mention. We thank you all, particularly those who we have had the pleasure of working with or studying under; you know who you are. Of course, this work could not come to fruition without the support of countless individuals. We thank the dozens of graduate students at the University of Minnesota who endured draft copies of the manuscript while completing our class on Land Use and Transportation. We greatly appreciate their patience in wading through the typos and providing suggestions for improvement. Ahmed El-Geneidy meticulously prepared the maps presented herein. Robin McWaters diligently proofed and strengthened earlier versions of the chapters. Douglas Benson crafted the hand-drawn sketches that appear; thank you, Doug, for enduring so many iterations. Thanks to the various publishers who provided copyright permission to use the epigraphs found herein. The cadre of reviewers that Routledge commissioned shared valuable comments at both the micro- and macro-level. We trust we honored their suggestions.

Kevin is fortunate to have had the opportunity to closely collaborate with one of the most versatile minds in the field of transportation, David Levinson. David's ability to uncover unique and inter-disciplinary perspectives in addressing old problems and his clarity in approaching urban systems is admirable.

Finally, the support of the Center for Transportation Studies at the University of Minnesota was invaluable to this project. In particular, we thank Peter Park Nelson for providing editorial assistance and preparing numerous graphics. At Routledge, the support of Eleanor Rivers was unwavering. We thank her for her help in jumping through all of the various hoops involved in producing any publication.

Chapter 1

At a crossroads. Again.

"One needs to have a lot of money to sleep in this town. The vehicles moving down the narrow, winding streets, the quarrelsome crowd refusing to move on. The rich man, when called away on business, will have himself borne through the crowd, which opens to make way for him; he will make swift progress over everyone's head in his vast Liburnian litter. As he goes, he will read, write, sleep within. And for all that he will arrive before us. In my case, the human tide in front of me prevents me from hurrying; the hastening throng behind me is thrusting into my back. Someone shoves an elbow into me; another man gives me a nasty jolt with a long beam. Here's a fellow also set on giving my head a whack with his joist and yet another with his mighty cask. A wagon is coming forward with a great bulk of timber swaying about on it; a second is loaded with a pine trunk. These are threatening the crowd as they swing in the air."

Juvenal describing congested conditions in ancient Rome
as translated in Roman Roads [1]

Urban areas are at a crossroads. Again. Traffic congestion is worse than ever. Again. The urban crisis is getting more severe. Again. Environmental catastrophe awaits us. Again.

We title this book *Planning for Place and Plexus*. Most readers will easily recognize that *Place* refers to the land use pattern and the distribution of activities across space. The less familiar word, *Plexus*, refers to the com*plex* of networks that connect people and places. These networks include transportation, but also communication and information, other infrastructure, and perhaps most neglected, the social networks that serve to glue people together.

2 At a crossroads. Again.

Modern society has unconsciously created a Place and a Plexus that disaffects many. In the developed world, places are increasingly made up of dispersed, low-density developments; the plexus is dominated by an automobile-highway system that connects but simultaneously disconnects us. Both place and plexus are also subject to new information and communication technologies that, ironically, enable citizens to cocoon rather than to consociate. The current "default setting" for urban growth relies on the automobile to reach an ever-widening set of destinations.

In contrast, the conventional wisdom in turn-of-the-millennium urban planning urges tightly knitting land use and transportation together, preferably in compact developments containing diverse uses, which make it easier to walk, bicycle or take transit and discourage driving. Scores of practitioners, politicians, and professors claim that designing communities which more closely resemble built environments of centuries ago will allow households to live simpler, easier, higher quality, and altogether copasetic lives. The attention and policy focus devoted to these issues is real. The desires to contain development and control traffic are passionately felt by many and these matters are receiving increasing attention, both in the press and in policy.

Identifying urban traffic problems and devising strategies to remedy them, however, is not a novel pursuit. Julius Caesar, for example, fashioned a system that involved banning unnecessary vehicles from the streets during daylight hours:

> ... no one shall drive a wagon along the streets ... where there is continuous housing after sunrise or before the tenth hour of the day, except whatever will be proper for the transportation ... of material for ... public works, or for removing from the city rubbish. [2]

Imagine the economic consequences of implementing this edict in modern cities. Influential urban writers of the mid-twentieth century therefore touted a more holistic solution to congestion that relies on land use to address transportation problems.

> In a nation that is both motorized and urbanized, there will have to be a closer relation between transportation and urban development. We will have to use transportation resources to achieve better communities and community planning techniques to achieve better transportation. The combination could launch a revolutionary attack on urban congestion that is long overdue. [3]

> If the problem of urban transportation is ever to be solved, it will be on the basis of bringing a larger number of institutions and facilities within walking distance of the home; since the efficiency of even the private motorcar varies inversely with the density of population and the amount of wheeled traffic it generates. [4]

However, these ideas—while powerful—are now over a half-century old. Thus, if the problems and proposed strategies to solve them are not new, then what is? The problems of urban growth—particularly efforts to mitigate relentless expansion and increases in traffic congestion—have been on political radar screens for years. It seems as if modern society should be in an opportune position to pointedly address these issues. It seems.

Previous research and debate prompt us to re-evaluate whether society really is at the crossroads we implied at the outset. The problems are more than vaguely familiar. Has civilization merely been spinning its wheels for half a century (or for two millennia)? The opening epigraph suggests that congestion was problematic even in Roman times. Congestion was also reported in Renaissance Paris, in Victorian times (where novelists describe hordes of workers crowding bridges into European cities following the Industrial Revolution), and in major US cities at the turn of two centuries ago (caused by horse-drawn carriages). Even a 1958 essay by William Whyte warned Americans that their penchant for using five acres to do the work of one was not only "bad aesthetics" but "bad economics." [5] If the issues of congestion and concern over dispersed land uses are not new—and many of the proposed strategies have been around for some time—then what is? Can a more contemporary spin on urban planning and policy satisfactorily address the transportation-land use problem?

Efforts to harness the automobile and its associated relatively expansive land use practices has grown in fits and starts following cycles loosely correlated with the economy and government administrations. The issues and solutions introduced once before are re-emerging for several reasons. Although the nature of the problem has not changed, its breadth, scale, and intensity have increased. At the same time, matters of place and plexus have been increasingly under the microscope. Despite highly visible calls put forth decades ago, development on the expanding fringe of metropolitan areas has blistered ahead, riding roughshod over common values and principles espoused in many professional fields concerned with the built environment. For example, environmentalists cringe with the filling of each wetland. Architects despise "garagescape" housing. Urban designers demonize seas of parking lots. Civil engineers seek context-sensitive design. Planners lament dispersed land uses as the root of many problems. Even economists are frustrated by the subsidies provided for expensive exurban development requiring new infrastructure. The list goes on. Perhaps only now is society wealthy enough to deal with the problems of wealth.

Current situation

The modern world's relentless march to develop the pristine countryside is driven by numerous factors, not the least of which is population. To illustrate, the US population is growing faster (in an absolute sense) than at any time in history, adding some 2.5 million people every year—equivalent to a new city the size of San Diego. Many urban planning concerns result from an increase in population. The Census Bureau expects a total US population of 392 million by 2050; an increase of about 100 million people over 50 years. [6] But answers to population growth—war, famine, disease, space exploration, emigration (to where?), rapture, or draconian birth regulations—are either undesirable or unlikely. A growing population needs somewhere to live and Americans are consuming more land per capita for living space than ever before.

Land consumption, however, needs to be placed in the context of other related phenomena. For example, reliable and long-term sources of energy remain uncertain. The transportation system, though cleaner than decades ago, still pollutes the air and threatens to change the climate. The heralded hydrogen economy holds out hope of eviscerating the energy and environmental enigmas, but gives no guarantees. Other hopefuls turn their attention to the possible "renewable" power from wind and solar energy.

Addressing the problems of expensive imported energy and environmental emissions may be possible through technological solutions such as hybrid-powered or hydrogen-fueled vehicles. The issue of congestion, however, requires something else. People are driving more while transit use continues to hold steady. Even in the auto-friendly United States more than one in four of all dollars (both federal and state) spent on surface transportation has gone to transit over the past 25 years ($29,000 million for transit vs. $95,500 million for highways in 1999).[a] Can society build its way out of congestion? Congestion is not merely an inconvenience—it costs money and time: time that could be better spent doing almost anything else, time that in today's hurried lifestyle is scarcer and scarcer.

There is little to suggest that trends of migration outward from central cities will subside in absolute terms. This trend will continue to leave, as it has for decades now, a wake of disadvantaged people with declining municipal services. [7] A transportation network that adequately connects a landscape of spatially separated activities only with automobiles inadequately serves the poor, the physically and mentally disabled, immigrants, the elderly, and children. Children call on their parents to act as chauffeurs, while other groups require special services, suffer unreasonably long trips, or simply remain sequestered in their homes.

Residents care most about what they confront daily: traffic congestion and the relatively ambiguous concept known as *sprawl*. In fact, these two issues, congestion and sprawl, are now beginning to trump other, highly visible and long belabored matters such as crime and education in public opinion polls. [8] Issues associated with urban growth are now among the most important concerns facing public officials, business interests, and citizens. In fact, one would be hard pressed to identify planning efforts from any growing community *not* trying to control sprawl (however it is defined).

The path to the present

The present situation cannot be traced back to a single event, policy, or invention. Rather it is the product of a variety of factors. For example, urban scholars have identified the top ten influences on the US Metropolis over the past half-century:

1. the 1956 Interstate Highway Act and the dominance of the automobile;
2. federal Housing Administration mortgage financing and subdivision regulation;
3. de-industrialization of central cities;
4. urban renewal: downtown redevelopment and public housing projects (1949 Housing Act);
5. Levittown (the mass-produced suburban tract house);
6. racial segregation and job discrimination in cities and suburbs;
7. enclosed shopping malls;
8. sunbelt-style sprawl;
9. air conditioning;
10. urban riots of the 1960s. [9]

But even the product of fifty years of dispersed settlement results from more than ten influences. Missing from that list are motives. Box 1.1 considers the American Dream as a motive for individuals to shape the landscape.

At the local level, zoning regulations were developed in the early twentieth century in part to isolate noxious uses (and improve public health) and to avoid lawsuits over public nuisances. Zoning's purview has been significantly expanded since then. Separating land uses—in the manner referred to as Euclidean zoning due to a legal case concerning Euclid, Ohio—is alive and well in communities today. Separation implies distance, and large distances create space for free parking and encourage reliance on the automobile, which in turn has prompted not only road building, but also minimum parking requirements and excess pavement. And on top of this, one cannot look past the role of exclusionary development regulations: minimum lot sizes and the like. The regulatory "stick" surely shapes contemporary cities.

That which cannot be conquered with the stick of regulation is seduced with the carrot of subsidy. Over the past half-century, an array of state, local and federal programs have, through incentives, been built into the development process. The biggest contribution lies in the billions of dollars spent by all levels of government on building new roads, thereby enabling and directing future development into corridors formerly considered appropriate only for agriculture. On the state and local levels, the "corporate enticement game" played by everyone from county supervisors to state governors encourages footloose commercial development most often located on the urban outskirts. Corporations have become increasingly skilled at pitting communities against one another to wrest perks from their governments. The result is often isolated

Box 1.1 The American Dream

Policy factors have shaped the landscape, but so have individuals. Individuals in the US desire to partake of the "American Dream," and engage in two mechanisms to satisfy that desire, for which we use the terms rugged individualism and collective action.

One cannot underestimate the power of people's desire for the American Dream. In 1928, US presidential candidate Herbert Hoover ran on a platform that called for "a chicken in every pot and a car in every garage." Today, citizens demand more from their politicians: expecting multiple cars in two- and three-car garages—all in an area devoid of congestion, pollution, urban ills, crime, or signs of classes lower than yourself, paid for with a minimum of taxes. These desires have prompted demands for the idyllic existence that many believe can only be had in suburban environments.

Following on the heels of striving for the American Dream is the pursuit of "rugged individualism." Again, Herbert Hoover coined this term as he extolled free, private enterprise and initiative as the foundation of America's "unparalleled greatness." Government entry into commercial business, he argued, would destroy political equality, increase corruption, stifle initiative, undermine the development of leadership, extinguish opportunity, and "dry up the spirit of liberty and progress."[a] Rugged individualism implies a spirit of discovery, entrepreneurship, and innovation. The same entrepreneurial spirit that opened up the wilderness applies to outward migration within cities, where new residents of the suburban fringe consider themselves pioneers. Collective action for the public good may take place with or without government, but increasingly with. Another of Hoover's great quotes was, "it is just as important that business keep out of government as that government keep out of business." The irony, however, is that despite Hoover's advocacy of rugged individualism, individuals could not express their ruggedness without government support, even before Hoover was shuttled out in favor of an even more activist government under Franklin Delano Roosevelt's New Deal. Government intervention began with the use of the military to seize territory held by indigenous populations, and has included massive land giveaways (in the US to the railroads, to homeowners via the Homestead Act, and to public universities via the Morrill Act, among others). In the twentieth century, government involvement continued with rural electrification and telephone subsidies, road building, irrigation and hydroelectric projects, farm subsidies, and farm foreclosure loans. In fact, "it would be hard to find a Western family today or at any time in the past whose land rights, transportation options, economic existence, and even access to water were not dependent on federal funds." [10]

Is it too simple to trace the current situation back to the American Dream and attempts to achieve it through both rugged individualism and collective action? Probably.

The American Dream, however it is manifested today (possibly in the form of continued auto reliance and wide, expansive lots) is alive and strong and railroading public policy. US Presidential candidate Adlai Stevenson may have been alluding to similar sentiments in his 1952 campaign by questioning the relentless nature of the American Dream:

> Our people have had more happiness and prosperity, over a wider area, for a longer time than men have ever had since they began to live in ordered societies 4,000 years ago. Since we have come so far, who shall be rash enough to set limits on our future progress? Who shall say that since we have gone so far, we can go no farther? Who shall say that the American Dream is ended? [11]

Anthony Downs, echoing the American Dream, provides a straight-forward account to explain, from a land use-transportation perspective, the bulk of household decision-making in US metropolitan environments. [12] He suggests that there are five overriding preferences informing the location choices of most Americans: owning a detached single-family home on a spacious lot; relying on private automobiles for movement; working in attractively landscaped low-rise places; residing in small communities with responsive and localized government; and living free from the signs of poverty. He further explains how most local governments have done an exemplary job through the course of their planning and development in responding to the vision created and shared by their constituents. Few would argue that the factors identified by Downs are not dear to many. The power of preferences provides a succinct explanation of the reasons for the existing built environment.

a Herbert Hoover, "Rugged Individualism Speech" (October 22, 1928) www.pinzler.com/ushistory/ruggedsupp.html accessed June 6, 2004.

---

business parks or big-box retailers at the metropolitan edge, surrounded by acres of asphalt.

But many subsidies are also built into the development process. The majority of new residential development costs government more to build and service than it generates in taxes and fees. [13] New residential or commercial development requires roads, sewer systems and water lines, and eventually schools and emergency services. On-site development costs (e.g., sidewalks, sewer laterals, cul-de-sacs) are often passed on to buyers by developers as part of the price of a home, but off-site costs (e.g., trunk sewers, water mains, schools, fire stations, wastewater treatment plants, arterial streets) are not. Some governments charge impact fees to developers to connect to existing community infrastructure. [14] But it is frequently the case that the costs of off-site infrastructure are averaged across the entire population, and are thus mostly paid by people who don't benefit from it, leading to an over-consumption of public services and straining the budgets of local and state governments.

This book is not so much about understanding the whys, whens, and hows of twentieth century urban form as it is about knowing how the legacy of this form will impact the future. George Santayana asserted that "Those who do not remember the past are condemned to repeat it." [15] Accounts from planning history and closely aligned fields have provided a good understanding of the whys, whens, and hows (despite Santayana also asserting that "history is a pack of lies about events that never happened told by people who weren't there").

From a transportation perspective, several explicit and implicit policy decisions have spawned generations who rely almost exclusively on auto travel; they expect to travel at free-flow speeds for only the cost of their car and their gasoline. By this logic, there must be something wrong with non-drivers—for what kind of person would willingly rely on a slow, unreliable transit system that lacks amenities and is often touted as a welfare service? Many advantages result from the policies that have transpired. Free-market economists continually remind us that, in the aggregate, our quality of life has never been higher. There is undoubtedly going to be a ripple or two of contention here or there, but in the developed world, people's freedom and flexibility of choice is indeed unprecedented.

Perspective

Municipal policies, programs or initiatives that fall under the rubric of land use and transportation issues have been in place for years—sometimes decades. We have subsequently been privy to decades of research examining the efficacy of policies and initiatives. We question, however, the degree to which research has effectively informed or guided current policy.

Economist Anthony Downs contended in 2004 that "a region can reduce its peak hour traffic congestion—or at least slow such congestion's rate of increase by relying on the principle of one hundred small cuts." [16] Just as a woodsman with a small axe can only chop down a large tree with many small blows struck over a long time, this incrementalist theory suggests that a region needs to apply many different tactics simultaneously in a coordinated manner. Having seen over one hundred small cuts (in the form of different policies and initiatives), we contend that our rate of progress in harnessing congestion, or sprawl for that matter, has been minuscule to date.

If society is indeed at a new crossroads, the situation may be related to changes in the composition of the typical American household, a changing real estate market in which the demand for traditional suburban housing is waning, or the anticipated large growth in non-residential space and new residential construction over the next 20 years. [17] Although, it may be due to nothing more than society having recently turned the corner into a new century and a new millennium, providing the opportunity to critique and evaluate the efficacy of past—and current—transportation and land use policy regimes.

Now may be the time to question, and perhaps to abandon, an incrementalist paradigm suggested by the death of one hundred cuts. Can real solutions emerge from weak or in some cases even misguided policy? We contend that most cities have reached a point that requires a more comprehensive approach to refabricate land use-transportation outcomes that will result in a superior place and a sounder plexus. A comprehensive approach, of course, takes time: decades if not half-centuries (major change does not come overnight). Our approach, described and documented in the following pages, stresses the need to recognize the focuses of *Planning for Place and Plexus*.

Place (Metropolitan issues, not intercity issues)—We focus discussion on issues that affect metropolitan areas. These issues traverse urban, suburban, and what has been referred to as exurban areas. In so doing, we do not focus exclusively on the inner city or on exurbia, but rather focus our attention on the metropolitan issues of land use and transportation. We cannot adequately consider rural land uses or intercity transportation, and so do not.

Problem (Accessibility, not congestion)—Considerable attention of late has focused on the worsening nature of traffic congestion in metro areas. [18] We question the degree to which congestion is the disease or merely a symptom. Some claim the root cause of congestion is dispersed land uses that trigger automobile travel. [19] The refutation of that argument lies in the evidence that there was congestion before the automobile (as noted in the quotation from Juvenal at the outset of this chapter), and there appears to be more localized congestion in dense areas. Some even see additional congestion as a solution to auto reliance, begging the question of "what is the problem?" Congestion is the result of many factors, including socio-demographic forces, technology, and economic development. To the extent that travel is undertaken so individuals can engage in activities in other places—work, recreation, shopping, health services—it is important to understand how these activities are distributed throughout a metropolitan area. The success of land use in influencing travel behavior depends in large part on the opportunities that land use provides. Our attention therefore turns to examining the relation between these activities and the manner in which people travel to them.

Perspective (Activity, not travel)—Finally, our analysis does not limit itself to the set of activities that occur once an individual leaves home. As mentioned above, it is important to consider the broader aspects that influence the demand for travel. For example, many shopping tasks can be satisfied via the Internet without leaving the home or the workplace, minimizing the need for personal travel (replacing shopping with shipping). Building on a foundational principle of current activity-based transportation modeling efforts, we consider travel relative to the larger influences that govern one's daily interactions.

Overview and approach

A comprehensive examination of Place and Plexus requires holistic thinking. Figuring out creative approaches to move people across networks over time and space, all in keeping with the goals of a community, requires us to move beyond our disciplinary upbringings. Although we, the authors, have been trained as transportation planners, transportation policy analysts, transportation engineers, and transportation economists, we are interested here in the inter-relationship of transportation and location or land use. The study of this system is sufficiently interdisciplinary to warrant a discipline of its own. We therefore think of ourselves as transportationists. [20] This means we are interested in understanding the transportation system holistically. However, we are also trained in land use planning and consider ourselves locationists because we seek to understand land use and location holistically. Turning the emphasis from the methodological area (policy, planning, engineering, economics) to the substantive (transportation and location) this book provides unique insights into important questions. The traditional methodologies are often reductionist in approach. Even the field of city planning, despite often classifying itself as an interdisciplinary profession, falls into this trap. Although at times a reductionist approach is needed in order to understand specifics, integration is also required—seeing the whole from the parts.

This book uses several techniques to approach these important questions, considering both theory and observation, and allowing them to inform each other. We argue both that theories destroy data and that data destroy theories. By the first statement, we imply that a simple, clear theory, model, or world-view is worth thousands of observations and anecdotes in shaping understanding and ultimately decisions. By the second, we mean that solid, well-founded, and replicable observations that contradict theories (especially so-called common sense theories, which may be common but seldom make sense) destroy those theories as valid world-views.

We divide the book into three parts; the subject of each part represents a class of agent that flexes its muscle over the landscapes of metropolitan areas: (1) individuals, (2) businesses (firms), and (3) governments. The behavior of each

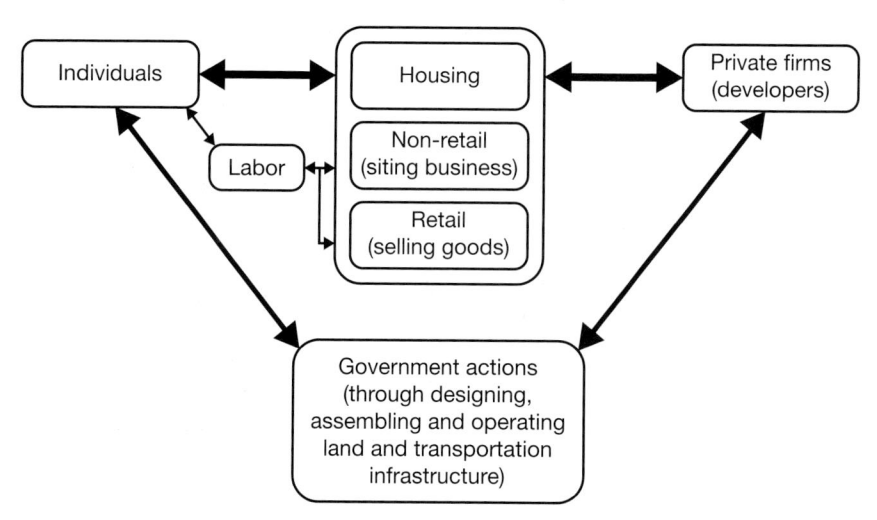

Figure 1.1 Conceptual framework for understanding land use-transportation interactions

class of agent is inextricably linked to the behaviors of the others, as shown in Figure 1.1. Individuals (and their households) consume land (where they live) and space on transportation infrastructure (provided, in large part, by the government). Individuals also consume housing that is (or, at one time, was) provided by developers as well as goods that retailers sell. Developers respond to firms' preferences for land uses and locations, and are influenced by the property rights and transportation infrastructure that government provides. Firms come primarily in two forms: retail-based (those that sell goods and services to consumers) and non-retail based (those that produce goods and services).

Separately examining the behavior of these three agents provides an effective strategy to better dissect and understand their actions. The behavior, and subsequent action, of each of these three agents, we claim, is largely influenced by different series of factors we call "diamonds" because of the way we array their relationships. At the beginning of each part of the book, we therefore describe the Diamond of Action, the Diamond of Exchange, and the Diamond of Evaluation, respectively. Figure 1.2 describes this structure.

Part 1 begins by focusing on factors that affect individual behavior. We rely on four factors—Chances, Constraints, Complementors, and Competitors— to explain individual decision-making using the Diamond of Action. We find it useful to decompose individual actions into four different decision frameworks, each representing choices of individuals. For example, when exploring long-term individual behavior, we describe residential location decisions. We discuss notions of accessibility, which measures the opportunities provided by the transportation location system, and the ease of reaching places from other places. We also examine aversion principles to explain the spatial separation of

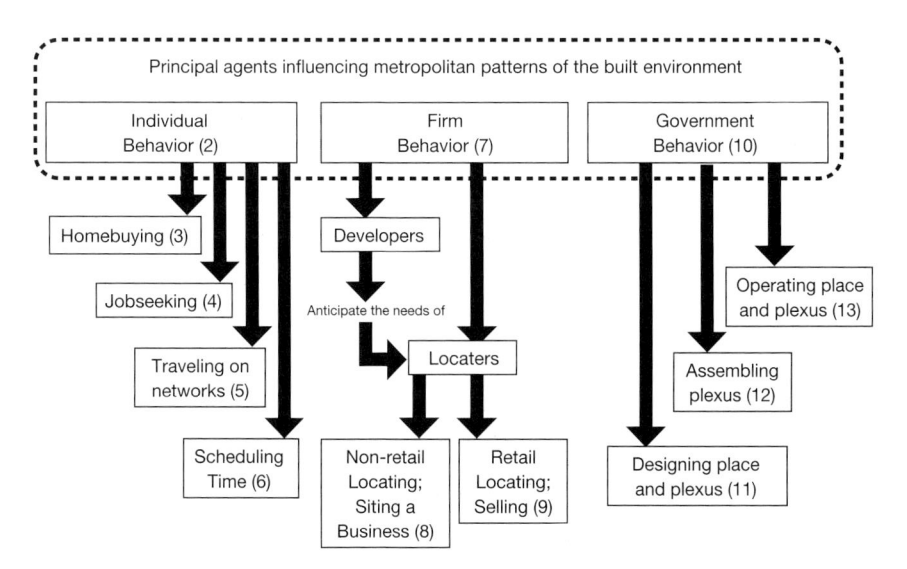

Figure 1.2 Overview of the structure of place and plexus

racial, socio-cultural, and ethnic groups that can be observed around the world. Although the legal barriers enforcing segregation have been eliminated in many areas, the informal social barriers, though often deplored, remain. We introduce Thomas Schelling's segregation model, which proposes a mechanism by which individuals having different comfort levels with people of other races will lead to self-segregation according to race. [21]

Next we consider aspects of the most frequent single destination for most people (besides home)—their workplace—and how workplace locations relate to land use-transportation systems. A key tenet in this discussion is association, the matching of origins and destinations. How do workers choose jobs? This choice is in many ways a product of social networks and "the strength of weak ties" as suggested by Mark Granovetter. [22] But it is also the product of physical networks, and gravity models demonstrate that travel time still matters.

This discussion is followed by examining the behaviors associated with mode choice and vehicle ownership decisions. Game-theoretic models (e.g., the Prisoner's Dilemma and Arms Race scenarios) apply to a variety of circumstances, but are particularly useful in understanding the difference between a user equilibrium, wherein individual users seek to minimize their own cost and time regardless of the effects on others, and a social optimum, in which everyone acts for the collective good. This difference explains why people prefer automobiles to transit for personal travel yet continue to advocate for additional transit service, why bicycles are being driven off the road in Beijing, and why sport utility vehicles (SUVs) are getting taller and taller.

Once residents have decided on where to live and work and how to travel, people have to schedule their daily activities and make decisions about when to travel, taking into account the congestion that is a product of longer-term decisions. These factors affect which destinations people frequent, when they travel, and how they combine different types of trips. We explain how a set of constraints helps to visualize what choices are feasible and uncover what we refer to as the language of travel.

Part 2 focuses on firms, which come in various shapes and sizes. One breakdown differentiates between developers (responsible for making decisions such as where to build, what to build, and how much floor space to build) and locators (responsible for deciding where to place their businesses to receive maximum return on investment). Developers, however, respond to and anticipate the needs of locators (occupants who would locate in the buildings that are developed). Often, the importance of both developers and locators has been underestimated in studies of how cities are formed.

As with the Diamond of Action, the Diamond of Exchange considers competitors and complementors. However, we apply a supply chain approach to examine suppliers and customers as well, allowing us to analyze the economy as a network phenomenon in the Diamond of Exchange. The issues for firms arise from agglomeration economies—which provide the spark for the formation of place and plexus in the beginning. Agglomeration explains why firms cluster. The necessary consequence is that some places have many more jobs than houses (and most areas have more houses than jobs). This arrangement of jobs and houses is a key element in many plans. Agglomeration is counterpoised against job–worker balance, as a trade-off that markets and regulators must face.

Firms locate in relation to both their suppliers and their customers. Selling and retailing are major shapers of cities, and of people's opinions of cities. Again, agglomeration is important, though alternative communication, information, and media networks are reshaping the traditional shopping experience.

Part 3 considers place and plexus in terms of design, provision and regulation—tasks generally associated with the public sector. This part begins with the Diamond of Evaluation, focusing on four planning objectives: Efficiency, Equity, Environment, and Experience, mediated by the politics of Expediency. By and large, government actions rely on a combination of approaches: designing place and plexus, assembling transportation infrastructure, and then allocating use of space and facilities.

In Chapter 11 we focus on design, because proper design of systems of place and plexus ultimately sets the stage for all that follows. Our framework is built around four themes, suggesting that a hierarchy of roads and transit routes needs to be synchronized with a hierarchy of places. The design approach embodies a morphology that applies pattern and structure to the environment and needs to be compatible with the community's history and its goals for the future. To accomplish this, good design is built with different layers and is sensitive to its architectural content.

We follow with Chapter 12 describing the impacts of assembling different types of infrastructure. Governments may build infrastructure to expand supply (e.g., roads and transit systems). However, this additional supply changes demand patterns.

The penultimate Chapter 13, Operating, responds by describing strategies to appropriately allocate capacity of the land use-transportation system, highlighting the important role played by pricing. That is, governments attempt to influence the use of these systems or regulate land development to affect demand. Congestion arises because of allocation problems: a scarce resource (road space) is under-priced and over-consumed. Given demand and supply, how does congestion actually work? Queues form and discharge. Understanding this engineering concept enables the transportationist to fully comprehend traffic. Understanding the causes of congestion in terms of supply and demand leads us to some conventional solutions, and to some radical solutions.

The final chapter emphasizes how "mature" land use-transportation systems are subject to the same conditions as developing systems. Incremental improvements have limited power to make dramatic changes. We also suggest that technologies such as specific transportation networks advance over time. This evolution and revolution in technologies follows from innovation. Not all innovations are viable. The lifecycle model explains the deployment of networks, but it is important to consider how new networks are launched. We explain these ideas in the form of suggestions to better position and "straighten" many of the policies and actions debated in land use-transportation discussions.

Throughout each of the chapters a series of boxes reinforce, illustrate, or challenge a particular theme presented in the text. They highlight some key points or concepts that support the reader's understanding of the rest of the book, and of transportation-land use interactions as a whole. These boxes fall in one of five forms because they:

1   reveal further technical detail about particular topics;
2   describe the efficacy of a particular policy, program, or initiative and comments on its potential to serve its stated aims;
3   shine additional light on a particular behavior that is important to understand and consider;
4   provide additional interesting information (sometimes interesting trivia) that provides context around a particular issue; or
5   challenge some longstanding dogma.

Together, the boxes provide a refreshing diversion from the linear narrative of the text, while conveying important information related to land use-transportation discussions.

Cumulatively, the three parts of this book examine the theories underlying transportation and location interactions. Collectively, the chapters explain how cities work: a bold claim perhaps, but one which we strive to live up to. Adapting Schelling, we believe identifiable structural forces explain the travel behavior and activity patterns that we see. These structures, although created by millions of individual actions, are only marginally affected by any one individual. Yet these structures shape what each individual, firm, or agency can and will do. The macro-structures we observe all have micro-foundations, the actions of agents. Yet the micro-structures (individual patterns of behavior) are shaped by macro-foundations (the set of choices and constraints that others impose). These chapters present a series of inter-related models that relate transportation and land use. Throughout this book, we strive to minimize the use of equations, except where they are necessary to clearly understand the theories under discussion.

The land use and transportation system in many metropolitan areas is often thought of as the problem leading to growth, gridlock, and sprawl. We disagree. These are mere symptoms. The problem is how to satisfy people's needs and desires for where and how to live. We do not mean just members of the middle and upper classes, but all people. The problems of congestion and unconstrained growth are the problems of the fortunate. Collective problems do not necessarily require legislative solutions, but the search for solutions should not preclude public policy. An understanding of the past and a vision of the future sheds light on the decisions required to travel the path ahead. We conclude by suggesting strategies for enabling the transportation-land use system to drive out of the rut in which it has been spinning its wheels. A new millennium argues for new approaches, new ways of thinking, and new ways of acting. Although it has not started especially well, we still have well over nine centuries to get things right and see little reason to discover the familiar crossroads, again.

Notes

a   Bureau of Transportation Statistics (2003) *National Transportation Statistics*, Table 3-29a: Transportation Expenditures by Mode and Level of Government from Own Funds, Fiscal Year (Current $ millions) www.bts.gov/publications/national_transportation_statistics/2002/html/table_03_29a.html (accessed June 6, 2004).

References

[1] Juvenal, 'Satire III', in R. Chevallier (ed.), *Roman Roads*, Berkeley, CA: University of California Press, 1976, pp. 67–70.
[2] Hamblin, D.J. and Grunsfeld, M.J., *The Appian Way: A Journey*, New York: Random House, 1974.
[3] Owen, W., *The Metropolitan Transportation Problem*, Washington, DC: Brookings Institution Press, 1966.
[4] Mumford, L., 'Neighborhood and Neighborhood Unit', in L. Mumford (ed.), *The Urban Prospect*, New York: Harcourt, Brace and World, 1968, pp. 56–78.
[5] Whyte, W.H., Jr. , 'Urban Sprawl', in W.H. Whyte, Jr. (ed.), *The Exploding Metropolis*, Berkeley, CA: University of California Press, 1993, pp. 133–156.
[6] United States Census Bureau, *Population Projections of the United States by Age, Sex, Race, and Hispanic Origin: 1995 to 2050*, US Census Bureau, 2002.
[7] Orfield, M., *Metropolitics: A Regional Agenda for Community and Stability*, Washington, DC: Brookings Institution Press, 1997.
[8] Pew Center for Civic Journalism, *Straight Talk from Americans*, 2000. Available at: www.pewcenter.org/doingcj/research/index.html (accessed July 23, 2007).
[9] Fishman, R., 'The American Metropolis at Century's End: Past and Future Influences', *Housing Facts and Findings*, vol. 1 (4), 1999.
[10] Coontz, S., *The Way We Never Were: American Families and the Nostalgia Trap*, New York: Basic Books, 1992.
[11] Stevenson, A.E., *Major Campaign Speeches of Adlai E. Stevenson, 1952*, New York: Random House, 1953.
[12] Downs, A., *New Visions for Metropolitan America*, Washington, DC: Brookings Institution Press, 1994.
[13] Altshuler, A.A. and Gómez-Ibáñez, J.A., *Regulation for Revenue: The Political Economy of Land Use Exactions*, Washington, DC: Brookings Institution Press, 1993.
[14] Nelson, A.C. (ed.), *Development Impact Fees*, Chicago, IL: American Planning Association, 1998.

[15] Santayana, G., *The Life of Reason*, 1905. Available at: www.gutenberg.org/etext/15000 (accessed May 1, 2007).
[16] Downs, A., *Still Stuck in Traffic: Coping with Peak-Hour Traffic Congestion*, Washington, DC: Brookings Institution Press, 2004.
[17] Nelson, A., 'Leadership in a New Era', *Journal of the American Planning Association*, 2006, vol. 72 (4): 393–407.
[18] Texas Transportation Institute, *Annual Urban Mobility Report*, College Station, TX: Texas A & M University, 2005.
[19] Levinson, D.M., Krizek, K.J., and Gillen, D., 'The Machine for Access', in D.M. Levinson and K.J. Krizek (eds.), *Access to Destinations*, San Diego, CA: Elsevier, 2005, pp. 133–156.
[20] Garrison, W.L. and Levinson, D.M., *The Transportation Experience*, New York: Oxford University Press, 2006.
[21] Schelling, T., *Micromotives and Macrobehavior*, New York: Norton, 1978.
[22] Granovetter, M., *Getting a Job*, Cambridge, MA: Harvard University Press, 1974.

Chapter 2

Diamond of Action

Angie: "What do you feel like doing tonight?"
Marty: "I don't know, Ange. What do you feel like doing?"
Angie: "We're back to that, huh? I say to you, 'What do you feel like doing tonight?' And you say back to me, 'I dunno. What do you feel like doing tonight?' Then we wind up sitting around your house with a couple of cans of beer watching the Hit Parade on television."
    from the 1955 movie *Marty*
    by Paddy Cheyefsky

The Washington Hilton Hotel is known to transportation professionals as the site of the annual meeting of the Transportation Research Board, the world's largest gathering on transportation issues, which attracts nearly 10,000 participants every year. Though unmarked by a plaque, the hotel is also the site where, on March 30, 1981, John Hinckley shot President Ronald Reagan, White House Press Secretary James Brady, Secret Service agent Timothy McCarthy, and District of Columbia police officer Thomas Delahanty. Within moments, Secret Service bodyguards had hustled Reagan—still unaware that he had been hit—into a limousine and were speeding away from the scene and toward the White House, following Secret Service protocol. But the discovery that Reagan had been injured by a bullet and not, as he initially believed, by the force of being pushed into the limousine, forced Secret Service agent Jerry Parr to immediately order the car re-routed to George Washington University Hospital, the closest hospital to the scene of the shooting.[a]

With the life of the President of the United States hanging in the balance, Jerry Parr had to quickly choose between several hospitals. The President is usually treated at the National Naval Medical Center in Bethesda, Maryland, northwest of the Hilton; this location was relatively far away. Another hospital roughly the same distance from the Hilton but due east was Howard University Hospital (Figure 2.1), and a little farther east of that were the Washington Hospital Center, the Veterans' Hospital, and Children's Hospital, which are clustered together; in the opposite direction (roughly due west from the Hilton) sits Georgetown University Hospital. That Reagan was routed to George Washington University Hospital makes sense given the local geography, the time urgency, and the fact that the car had started out southbound toward the White House.

Several opportunities (referred to in this chapter as *chances*) were assumed to have been evaluated by Jerry Parr: George Washington University Hospital, Bethesda Naval, Georgetown University Hospital, Howard University Hospital, Washington Hospital Center. All of these are highly regarded hospitals, though they have different specialties. The quality of care at several of these hospitals would probably have been sufficient to save the President's life.

One *constraint* in this particular scenario was time: the President was actively bleeding. Agent Parr had an obligation to minimize travel time, subject to another *constraint* that the quality of care at the chosen destination would be sufficient. The travel time depended on several factors. There was other

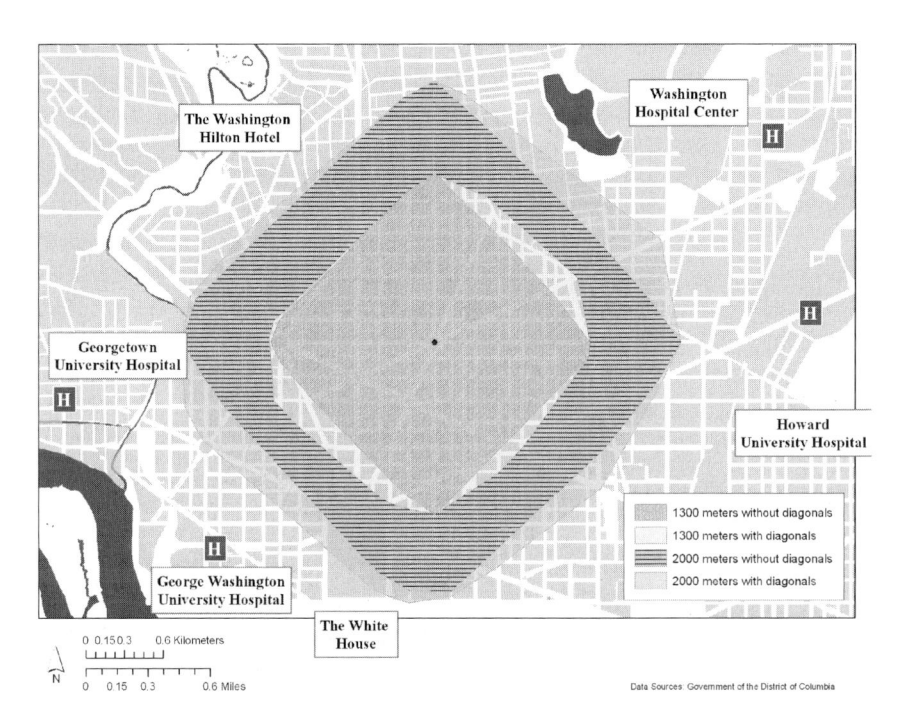

Figure 2.1 Washington, DC, including site of Reagan assassination attempt, local hospitals, and the reach that can be achieved on a grid-based transportation network

traffic on the road, and unlike a normal Presidential excursion, there was no prepared route or advance escort service to clear the way. The road network itself was a *constraint*, as the presidential limousine could only make time driving on existing streets, with buildings, Washington's famous traffic circles, and other objects posed as barriers.[b]

The other vehicles on the road were clearly *competitors* in this life or death journey. Once at the hospital, other *competitors* appeared in the form of patients who placed their own demands on the doctors Reagan needed (in this particular case, however, the doctors had been in the hospital at a meeting, and so were readily available).

The existence of a road between the hotel and the hospital, of the hospital itself, and of a staff of skilled physicians all depended on the presence of a city full of people who would use those facilities on a daily basis, making them available when needed by the President. These factors acted as *complementors*. The availability of alternative routes and alternative hospitals is testament to the large number of *complementors* that could be drawn on. The hospital would not exist but for George Washington University, which though a private university, grew out of a Congressional charter signed by President James Monroe for "The Columbian College."[c]

Predicting decisions

The behavioral and social sciences are long on theories aiming to model and predict human decision-making. The goal of almost all theory is to help explain some outcome (behavior) as a function of some inputs (variables), positing some relationship between them. Some theories are relatively abstract, while others are more applied; some theories are empirically tested, while others are more useful as frameworks. Depending on the context, some hold up pretty well in explaining what they are supposed to explain. However, all have more than a healthy dose of error—an often unaccounted for dimension—and the manner in which this error is incorporated into the prediction process helps set apart one of the theories most often relied on in transportation-oriented work.

Imagine that you are presented with two objects. You must try to guess their weights and select the heavier one. Sometimes you will be right, sometimes wrong. The likelihood of being right depends on the difference in their weights as well as your particular skills. In 1929, Louis Leon Thurstone proposed a "Law of Comparative Judgment," which, greatly simplified, says that perceived weight is $w = v + e$, where v is the true weight and e is a random error with an expected (or average) value of zero ($E(e) = 0$). [1] The finding from this experiment is that the greater the difference in weight, the greater the probability of choosing correctly. This experiment is analogous to predicting decisions, such as which mode of travel individuals decide to use. Adapting this approach to study how people make choices, economists have developed a method for explicitly accounting for the actual choice as well as the error term. They do so wrestling with a concept likely familiar to emerging transportationists known as utility (where the observed utility = predicted utility + random term).

Economists developed the concept of utility based on the general proposition that "people make decisions to advance their self-interest." [2] People prefer alternatives with the higher utility—an unobservable characteristic assessed by examining choices. One of the conditions of utility theory is non-satiation: more is preferred to less of any normal good (and if less is preferred to more, we just define the "good" to be less of something). However, utility theory does not assume that two of something is necessarily twice as good as one of something; it allows for diminishing (or increasing) returns. According to utility theory, the demand for different goods takes into account the prices of all goods, income, and tastes and is subject to budget constraints. In a similar manner to elementary consumer economic theory, consumers take advantage of a good when the utility of consuming the good is higher than the disutility of its cost (including the opportunity cost of alternatives). The demand for different goods or services (defined as everything from a community, a car, or a cell phone) depends on prices of all goods, household income, and personal tastes.

Travel behavior theory differs from consumer choice theory in that transportation choices tend to be discrete (e.g., where to go, when to go, which mode to use) rather than continuous (e.g., how much to buy). [3] I drive because, for me, it is safer, more flexible, or cheaper than taking the bus. Alternatively, the demand for a transit trip may be viewed as a function of both the benefits of the trip and its costs: time (access and egress time, wait time, travel time), money (transit fare), and uncertainty (schedule adherence, safety). This thought process allows us to identify a basic demand function that incorporates the relationship between the cost (or price) of a good and the level of demand. As long as the cost of consuming a good is lower than an individual's willingness to pay, the good is consumed (or a mode of travel is chosen).

Mode choice models thus have roots in Thurstone's psychological theory and in Kelvin Lancaster's consumer behavior theory, [4] as well as in modern statistical methods. Stanley Warner first applied concepts of utility theory to disaggregate travel in 1962. Using data from the Chicago Area Transportation Study (CATS), Warner investigated classification techniques using models from biology and psychology. Building on the work of Warner and other early investigators, disaggregate demand models emerged.[d] This theory was formalized by Daniel McFadden, who is largely credited with introducing the utility-maximizing framework derived from economics and psychology to travel behavior research [5]; in 2002 he received the Nobel Prize in Economics or his efforts. Utility theory concepts have been further advanced by Moshe Ben-Akiva [6] and other transportation modelers who developed the form that has become most widely applied, the multinomial logit model (MNL). In practice, utility maximization has been used in travel behavior research for decades and has become closely associated with decisions to minimize cost, broadly defined. These include both monetary ("out of pocket") costs and non-monetary costs such as time and inconvenience.

Although the utility maximization framework has proven useful in efforts to forecast choice of travel mode (or other relatively defined discrete choices), its advantage is less evident in understanding the many choices that households make over space and time. For example, what if the choice set is not clearly defined? What if there is little guidance about the array of factors that influence a choice? Fortunately, other social sciences (particularly psychology) and other models have come to the rescue with approaches firmly grounded in cognitive processes. A main advantage of these other models is that they tend to be more explicit about the specific variables that explain behavior. However, they tend to be less explicit about the mechanism by which these variables act on behavior.

For example, social learning theory (largely credited to Albert Bandura) posits that by "observing others, one forms an idea of how new behaviors are performed, and on later occasions this coded information serves as a guide for action." [7]. Although the theory was devised to apply to criminology, its assertions—that outcomes occur as the result of an individual's behavior and that individuals then expect similar outcomes to occur in the future in response to the same actions—transfer well to other applications. Social learning theory aims to explain human behavior in terms of continuous reciprocal interaction between the characteristics of a person, the behavior of a person, and the environment in which the behavior is performed. [8, 9]

A different approach, the theory of planned behavior, [10–12] focuses on the role of different types of beliefs in explaining behavior. Behavioral beliefs ("What will result?") contribute to people's perceptions of possible outcomes weighted by an evaluation of those outcomes. Normative beliefs ("What would other people think?") consider the reactions of referent individuals weighted by an individual's motivation to comply with those referent individuals. Control beliefs ("What else would facilitate or constrain this behavior?") suggest the user considers an array of factors that may advance or inhibit the behavior and these are weighted by the perceived power of each factor.

These cognitively oriented theories, however, give little play to the role of the physical environment (as opposed to the social environment). [13] Therefore, social ecological models suggest that there are a variety of contexts —individual, interpersonal, organizational and community—that operate at multiple levels to influence individual action. In addition to intra-individual factors, ecological models say that human behavior is shaped by higher-level factors including organizational, policy, social and physical environments, as well as dynamic interactions across multiple domains. [14–16] In this case, the community context usually refers to the physical environment, from the micro-scale (e.g., the home), to the meso-scale (e.g., the neighborhood), to the macro-scale (e.g., the region and beyond).

Each of the above approaches (particularly utility theory), however, assumes that individuals usually act in apparently rational ways. What if this is not the case, or what if their rationality is difficult to interpret? Yet another alternative, prospect theory, developed by Daniel Kahneman and Amos Tversky, explains why people are seemingly irrational when analyzed through the prism of utility theory. [17] Whereas formal utility theory assumes people only care about final outcomes, prospect theory suggests that decisions depend on how the alternatives are presented. The theory suggests that people are risk-averse when seeking potential gains (they prefer a certain $100 to a 50 percent chance at $200, and 50 percent chance of no gain); they are also risk-seeking when addressing potential losses (they prefer a 50 percent loss of $200 and 50 percent chance of no loss to a certain loss of $100). The value function that represents this relationship is shown in Figure 2.2. Prospect theory is consistent with psychological research into happiness, and suggests that utility is reference-based, rather than an absolute. Individuals frame decisions differently depending on how they are presented, and use heuristics rather than rigorous evaluation to decide.[e]

Understanding human behavior is not as simple as applying one of the above theories. There are countless explanations for why individuals make the decisions they do. Take, for instance, a residential location decision: households strive to balance the values of various attributes: new construction, schools, size, quality, crime, race, and accessibility. This decision can be thought of as an optimization problem. It is not that utility theory is wrong; of course people maximize their utility, what else are they going to do? And, if they do something seemingly irrational, it is just that we need to redefine their utility.

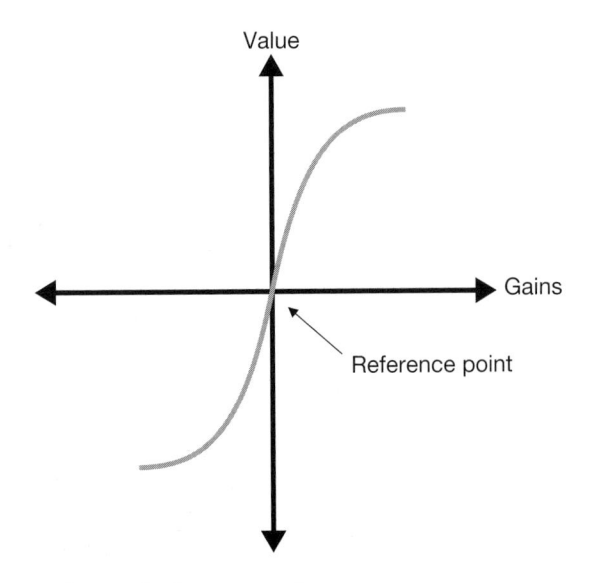

Figure 2.2 Prospect theory of value versus gains

Utility remains impossible for the analyst to accurately measure: if individuals do not even know what they want, how can an economist or psychologist know? The analyst chalks up the unexplained portion of behavior to error. But if the error far exceeds what is known, is the model still useful? Perhaps, in the aggregate—but analysts must acknowledge its limitations.

Most importantly, the reason the above theories are incomplete is that people's decisions depend upon the behavior of others, and the actions of others depend upon individual decisions. Utility can only examine decisions, given the state of the world. No single theory can explain all behavior. It would be misleading to say otherwise. But in this book, we aim to learn the state of the world, but we also want to know what produces the state of the world, and what decisions that world produces.

Diamond of Action

To better understand how individuals or households make decisions—or take action—in land use-transportation contexts requires a fuller theoretical framework. Our approach examines observed behaviors and infers which bundles of activities are preferred, given a willingness to engage in travel or seek destinations as a function of attributes associated with individuals, their families, and their physical and social environments, subject to budget and time constraints. Decisions (e.g., where to live, work or travel, and what activities to perform) are subject to the actions of other individuals. An individual's actions are both enabled by and constrained by the actions of others (see Figure 2.3). In the sections that follow, we provide a more detailed explanation of how four "Cs" of the Diamond of Action manifest themselves and ultimately affect the fifth C—the Choice.

Constraints

We begin by imagining the array of opportunities available to any individual (or set of individuals), faced with decisions about where to live, where to work, how to get around, and in what types of activities they will participate. Constraints are matters that—voluntarily or involuntarily, explicitly or implicitly—set bounds on the daily, weekly, annual, or longer-term decisions that a household makes; they limit the range of opportunities available to any one person and demarcate the frontier that an individual cannot or will not cross. Primary constraints include time (e.g., 24 hours in a day), space (e.g., geography), finances (e.g., income), and responsibility (e.g., care of children or parents), and can be thought of as representing the four sides of the Diamond of Action. These primary constraints range on the continuum from being relatively fixed to more flexible. This set of opportunities varies by individual.

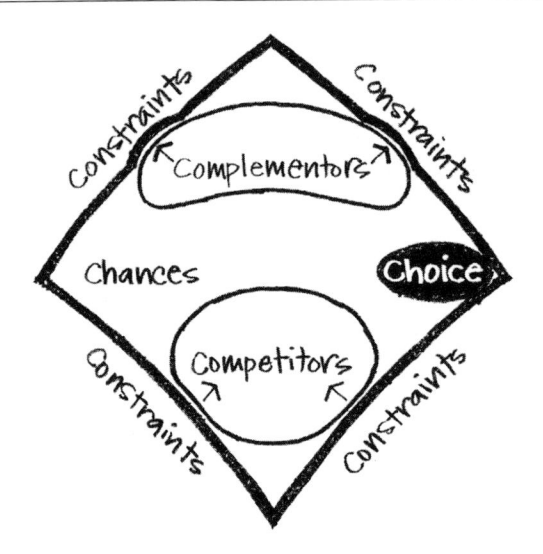

Figure 2.3 The Diamond of Action

The anecdote about the attempted assassination of President Reagan demonstrates how a particular action (i.e., which hospital) played out relative to four different factors: time, other hospitals, traffic congestion, and facilities.

Pierre L'Enfant created the city plan of Washington, DC with intersecting diagonal avenues superimposed over a grid system—but imagine the nation's capital without its diagonal avenues: a strict grid-based network with equal speeds on each link, as found in many other US cities. The ultimate decision—the choice—plays out in a manner influenced by four C's: constraints, chances, competitors, and complementors. The city blocks that can be reached within a fixed amount of travel from a point on the grid (such as the Washington Hilton, or Logan Circle) form a diamond. Figure 2.1 illustrates that phenomenon (showing how far someone can drive in five minutes and ten minutes in the area surrounding Logan Circle). The time constraint is expressed as a boundary in the shape of a diamond. The opportunities (chances) that someone has in five minutes is the dark shaded area, the opportunities within ten minutes is the light shaded area. Competitors, by congesting roads and making travel slower, reduce the space that can be traveled, contracting the diamond. Complementors, by making travel faster (for instance, by justifying exclusive transit service in a corridor instead of sitting in traffic), stretch the diamond outward, meaning a larger area could be traversed in less time.

The above is just an illustration. Real networks, as in Washington, DC, are seldom perfect grids; the space of opportunities will always be distorted from the ideal diamond shape suggested above. However, the metaphor gives meaning to a framework we apply to individual (household) decision-making. In order to better understand how and why individuals and households react to the set of policies in the ways they do—and the ultimate *choices* they make, it is important to understand how the *chances* that are available are influenced by the *constraints* imposed as well as how *competitors* reduce chances and how *complementors* expand the chances.

Time

Fixed constraints are structural elements that set hard bounds on the range of decisions that can be made and activities that can be pursued. The most immediate is time: people are inherently constrained by the time available in a day – 24 hours (1,440 minutes). Few people travel anywhere near this amount. Starting with work or school (say, 8 hours) and sleeping (8 hours), leaves us

8 hours a day for meals, family time, errands, travel, and everything else that makes up our existence. One quickly sees how activities leave relatively limited time for travel. But despite advances in transportation technology over the centuries, in the aggregate, we have seen little change in the time people spend commuting to and from work: most commute times remain on the order 60 minutes round trip per day. [18, 19] Time spent traveling by workers for other (non-work) travel is typically less than 30 minutes, suggesting that individuals spend a fixed amount of time per day (on the order of 90 minutes) engaging in travel from one activity to another. The phenomenon—dubbed the travel time budget [20]—has enjoyed a level of attention that is something of a paradox. The concept has shown a stubborn persistence in the literature (discussed in greater detail in Chapter 6), despite the fact that the more closely it is examined, the more elusive it becomes. [21] For our purposes, the issue is less about whether a travel budget exists and more about recognizing that inherent bounds exist—some physical (e.g., 1,440 minutes), others logistical (balancing the competing demands in a day).

Given time constraints and the need to complete other activities, people are inherently limited in how much time they spend traveling. They seek ways to maximize the efficiency of their chores. Budgetary constraints, long referenced in the field of land use and transportation, elicit tradeoffs between home location and transportation costs.

Geography

While behavioral preferences and activity patterns shape the time budget, the distance that can be traversed in that time (geographically) is affected by the available technologies. As the dominant mode in cities has evolved from walking, to the streetcar, to the automobile, the distance that can be covered in the same time period has increased as well. New forms of information and communications technology continue to emerge as forces influencing people's activities and residential location patterns. More than a century ago, rural residents in particular were able to purchase products through catalogues or wish-books (e.g., Sears or Montgomery Ward) and have them delivered to their outlying locations because of the US Postal Service's rural free delivery. Today, the Internet is a catalogue of nearly everything you could ever want, with markets such as eBay and stores such as Amazon shipping via carriers such as Federal Express and United Parcel Service. Services can be found as well, including job boards for those seeking employment and relationship-finding services for lonely hearts. Communication technologies are burgeoning, prompting many to question the degree to which geography remains relevant. People (and businesses) demonstrate considerable freedom and flexibility in where they locate (either farther outside the city center in urban fringe areas or in remote locations); fax, email, and video instant messaging facilitate conversations between even the most remote locations.

Money

Another set of constraints is the financial resources available per individual and household to make travel, residential, and other decisions. Budgetary constraints, long referenced in discussions of place and plexus, force households to make trade-offs between home location and overall transportation costs (though, most of the time, these discussions refer only to the work commute). Figure 2.4 presents income trends for US households from 1967 to 2002 showing how household incomes have risen in constant 2002 dollars. Coupled with simultaneous significant reductions in household size, this suggests that financial constraints are generally relaxing.

Incomes across the country are increasing for most people, implying higher disposable incomes and more resources available. Considering observed increases in geographic mobility, the rising quality of goods and services, and the expanding floor space of the average single-family home, it seems clear that as a whole, Americans are becoming wealthier. But at the same time, a variety of demands are being placed on that wealth. A recent study shows average expenditures per household across 11 categories. [22] Dollars spent on shelter are the largest slice of the pie; funds devoted to transportation are a close second, comprising, on average, at least 28 percent of total household expenditures.

Responsibility

Other constraints, however, may not be relaxing in the same manner. Declining household size may mean fewer household responsibilities for some; but it also means there are fewer adults to share the responsibilities incurred by the household, and there may be fewer adults per child in many households with the rise of the single-parent family. Constraints of responsibility affect the demand for travel. Non-work trips are tied to some very basic and necessary human activities such as shopping, performing errands and socializing. Although these activities must be undertaken, there is a great deal of flexibility in when and how they are undertaken. A large body of research relates trip-making and activity patterns to demographic and socio-economic conditions, and ties trip generation to variations in land use patterns and metropolitan size. However, these activity patterns vary even more significantly across some fundamental structures: the natural and cultural cycles reflected in the calendar and the clock. Consider, for instance, the role of household life cycles. Progressing through different stages of employment and income brackets has ramifications for one's quality of life. Progressing through relationships, marriage, the presence of young children, adolescents, and then an empty nest, for example, gives rise to a series of different self-imposed demands and constraints on one's decisions. There are certain basic considerations—lifestyle, quality of life, household size, and household structure—that impose responsibility and thus constraints, but do allow flexibility in how those responsibilities are satisfied. In the end, travel and residential location depend on a variety

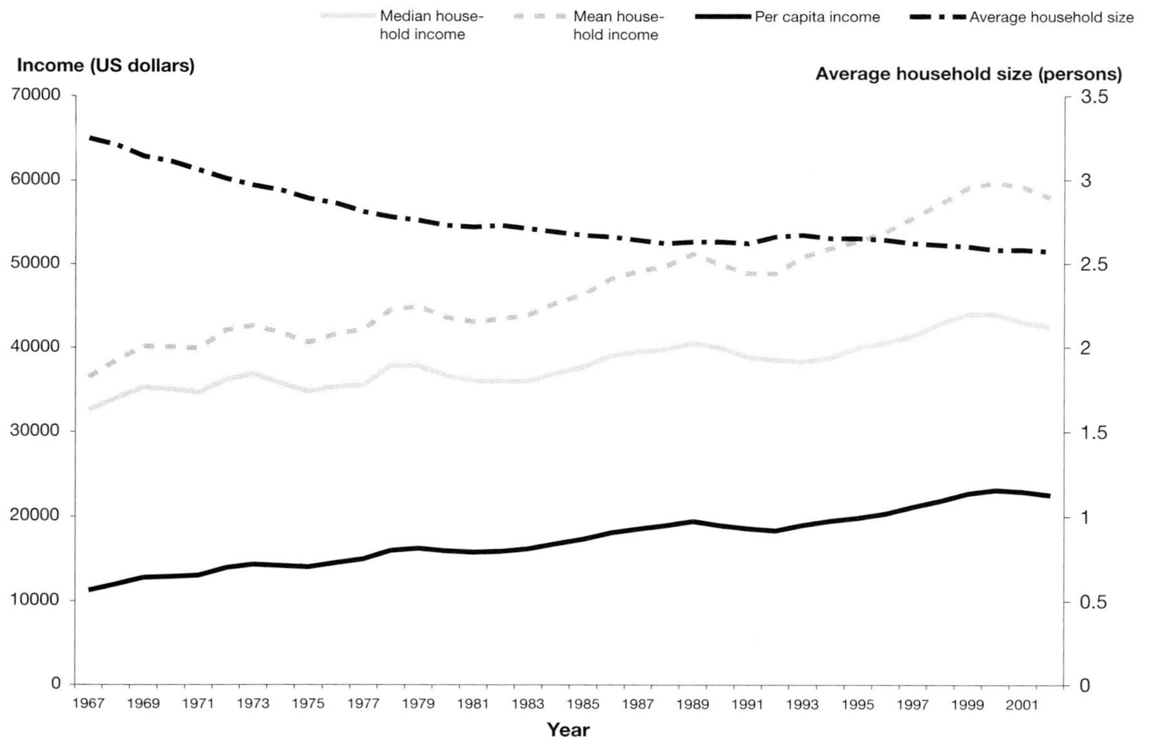

Income (US dollars)

Average household size (persons)

— Median household income   — Per capita income
— — Mean household income   — · — Average household size

Figure 2.4 Trends in household income, 1967–2002

Source: US Census Bureau, Current Population Survey, 1967 to 2002 Annual Social and Economic Supplements. Income reported in 2002 adjusted dollars.

of factors, including constraints inherent in the regional and local opportunities for access, workforce participation (retirement, at-home parenting, etc.), life cycle, and employment location.

Complementors and Competitors

Our friends are our enemies and vice versa. The array of possible activities, and the likelihood we will engage in them, reflects that they are available. That services are available is enabled by our companions in the first place, through economies of scale. If there were not a critical mass of neighbors desiring coffee, the neighborhood café would vanish. Without demand for people to get from point A to B, there might not be a roadway.

However, one must be aware of the consequences of decisions in a market economy. Demand is inherently endemic to attractive places, goods or services. It has to do with others competing for the same good. Most of us desire more and better alternatives: new places to go, better things to do, faster ways to get there. To the extent that we can have more opportunities that are costless (or nearly so, in the case of a simple rearrangement of existing resources), these opportunities are usually seen as beneficial. But when alternatives cost money or time, we face constraints and make trade-offs. This is true in both the personal and political realms. It is therefore important to consider the behavior of those competing for the same good. This may be as personal as competing for the same life partner, or it may mean striving against another individual who desires the same job. In the land use-transportation context, it may mean competing for the same desirable home in the right neighborhood, or even for a stretch of highway during rush hour. Such factors can conceivably be measured, estimated, and quantified—but this is a task considerably easier said than done.

The more attractive destinations engender higher demand and greater competition, which plays out on several levels. For example, commuters battle for throughput on a major arterial at rush hour. Trying to out-guess other travelers in terms of selecting a route is part of the game. In areas with thriving residential markets, those looking to purchase a house compete against other potential bidders. Employment searches result in competition against other businesses, locations, or individuals.

Parabolic effects

The goods or services people desire are enabled because of others demanding similar goods or services; they have similar preferences. The irony, however, is that access to these goods and services are tempered by competition with others. In many respects, this principle applies to everything from bakery goods at the corner store to limited capacity on the road system. We illustrate the relationship as parabolic in nature (Figure 2.5) showing how the quality of a good or service increases to a certain extent, at which point it may decline due to excessive demand.

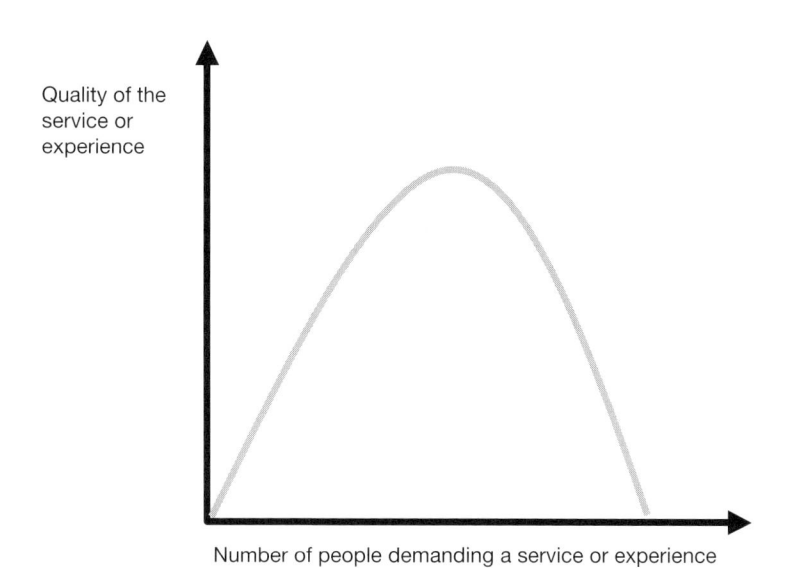

Figure 2.5 Theorized parabolic effect of others on quality of experience

Zero-sum and non-zero-sum games

Complementors and competitors may even be the same people simply playing different roles. In brief, complementors reduce constraints and increase chances; competitors increase constraints and reduce chances. Other individuals come into play as competitors, who reduce chances by increasing the constraints on actions. But complementors, by creating markets and ensuring that networks are in place to serve them, push out the constraints and create more chances (opportunity, variety, alternatives). Complementors cannot directly create more time, money, or space, but they do increase what can be done with the time, money, and space available.

In game theory terms, you are playing zero-sum games with competitors and non-zero-sum games with complementors. A zero-sum game is so called because the payoff of the game is fixed and split between the players, often with one player getting the entire payoff, and the other player losing an equivalent amount. [23] In business terms, these are win-lose situations. An example is a football game: if my team wins, your team must lose. In a non-zero-sum game, on the other hand, there are gains from cooperation (either conscious or unconscious). In business terms, these are win-win situations. An example is gains from trade: to illustrate with an old Reese's Peanut Butter Cups commercial, one guy is walking down the street with a chocolate bar, the other (strangely) has an open jar of peanut butter, and when they collide the result is chocolate mixed with peanut butter. At first they are angry— "You got chocolate in my peanut butter!" "You got peanut butter in my chocolate!"—but in the end the two clumsy pedestrians conclude that the combination is tastier than either chocolate or peanut butter alone. They were involved in a non-zero-sum game, though unaware of it. Box 2.1 examines the emergence of the consumer cooperative movement, an attempt to consciously achieve the benefits of non-zero-sum games.

Chances

The array of chances comprises the opportunities remaining after constraints rule out other possibilities (i.e., a choice is selected from the available opportunities). Also thought of as a choice set, the array of chances is represented by the shaded area inside the Diamond of Action. The area may have a geographical significance, representing nearby opportunities unconstrained by lack of access. Utility theory teaches that consuming more is better than consuming less. It also implies that having more chances or opportunities is better than having fewer, since you are no worse off with additional opportunities or possibilities. That implication would be true if search and decision-making were costless. Economically, search and decision-making are not free, although that fact is often ignored in analysis. However, psychologists argue that decision-making can be prohibitively costly, and people with more opportunities (beyond a certain point) are not necessarily happier than those with less. [24] Although more chances are not necessarily better or worse, chances are vital in analyzing what people actually choose.

There are many things to do with one's time, but we can only pursue a few. Although I, like Marty and Angie at the beginning of this chapter, may sometimes feel there is nothing to do, this is because I have already eliminated many implausible possibilities (such as visiting the moon for a holiday). Even if technologically feasible, these implausible opportunities may be outside my time budget, if I am subject to the constraints of other activities, or they may be beyond my money budget (or both, in the case of the lunar vacation).

Choices

Finally, we arrive at the bundle of decisions an individual makes—their choices. One's choice is probably at the edge of their constraints. Given additional time, money, space, or less responsibility, an individual might do something differently. From a prediction standpoint, we tend to assume that people have defined preferences that they apply to different bundles of goods and that their choices remain stable. The difficult matter, however, is that segments of the population weigh their time, money, activities, and even personal image differently. Furthermore, their preferences and how they weigh such considerations are not stable over time but vary by geographical context, life cycle or even time of day.

Some preferences are well acknowledged, such as the "love affair with the car." The most obvious reason for using the car is the freedom, flexibility,

Box 2.1 Cooperation amongst consumers (collective action without government)

A cooperative is an autonomous association of persons united voluntarily to meet their common economic, social, and cultural needs and aspirations through a jointly owned and democratically controlled enterprise. (Source: ICA Statement on the Cooperative Identity)

Although much of the cooperation described in this chapter, and in this book as a whole, is implicit, explicit cooperation between people has a long history. The ability to coordinate human actions for the common good, be it hunting a mastodon or organizing fire insurance as Benjamin Franklin did in 1752 with the Philadelphia Contributorship for the Insurance of Homes from Loss by Fire. Agricultural cooperatives (such as organizations of dairy farmers) began in the early nineteenth century. [25] The modern consumer cooperative movement is generally credited as beginning in 1844 in Rochdale, England, with the Rochdale Equitable Pioneers Society, when textile workers opened up a rival to the company store. This was not the first cooperative store, but it was the first successful one. The consumers owned the store, and so the store would only charge its costs, but not seek to make profits (and especially not the exorbitant profits that a monopolistic company store would charge).

The Rochdale Pioneers established seven principles of cooperation, which largely remain at the heart of modern cooperatives:

> (1) open membership to all who will cooperate in good faith, without restrictions regarding race, color, or creed; (2) one member has only one vote; (3) no proxy voting; (4) limited return on capital; (5) net savings distributed on the basis of patronage; (6) trading on a cash basis only; (7) audited accounts made available to members; and (8) regular membership meetings in support of cooperative education. [26]

The idea spread quickly, and by 1845 a cooperative opened in Boston, Massachusetts founded on the Rochdale Principles. Although most cooperatives in the end did not succeed, by 1920 there were 2,600 consumer cooperatives in the United States and the Great Depression saw the formation of more. Another wave of cooperatives began in the late 1960s and early 1970s.

The risk to the cooperative is that democratically run business may spend too much time on politics and not enough on business. The limits to return on capital greatly constrain expansion, as gains are returned to members, but ensure that ownership remains local and the co-ops remain focused. The distribution of reward requires a distribution of responsibility that flies in the face of economic theory as it has been practiced since Adam Smith posited that increasing the division of labor was at the core of markets and the wealth of nations. Craig Cox [27] details the chaos that emerged in many Twin Cities natural food co-ops during the politically turbulent 1970s, as member-owners contended (sometimes violently) for the control of the distribution warehouse and disputed about the "one true way" to govern these stores. Despite the chaos, these stores remain a part of the Minnesota food distribution economy.

and speed it provides. It has been this way for the better part of a century and there is little to indicate automobile use will subside—a point that is still grudgingly acknowledged by even the hardiest of transit supporters. But other, less direct issues influence modal decisions as well. These include status (driving a new and flashy automobile, not wanting to take transit or bike to maintain appearances), privacy and solitude (desiring personal time even while crawling through stop and go traffic during rush hour), and convenience (going when you want, not when the bus company wants).

Timing

In considering one's ultimate choice, timing is an underlying and critical dimension. The timing of a decision may differ over one's lifecycle. Given similar constraints and changes, one may pursue a radically different choice at age 25 than age 45. Preferences change.

But there is also a natural sequence to the timing (or frequency) associated with how any individual or household decides about matters. For example, the high cost of housing and other real estate has given rise to financial markets for long-term financing of mortgages, and to rental markets for housing. This adds a dimension to housing demand that includes housing tenure, and the dual roles of homeownership as a means of investment as well as of consumption. [28] Such factors require households to consider longer-term matters such as future income and real estate price appreciation. The longer time frame of a year or several years may see households changing in composition, acquiring a vehicle, moving to another home, or having a member change jobs or even join or depart from the labor force. Contrast that with deciding when to run errands, what route to choose and accompanying mode choice decisions.

Recognizing that choices vary by time frame is by no means a new pronouncement. [29] Furthermore, the literature is increasingly acknowledging that long-term choices serve to condition short-term choices (e.g., activity participation or travel behavior); it is just that we do not know a lot about how decisions within different time frames interact. The interdependence between longer-term and shorter-term choices is poorly understood and rarely addressed in modeling applications.

The utility-maximizing framework assumes, for example, that a residential choice is influenced by access, an economic measure of the transport user's benefit that combines information on travel accessibility and the attractiveness of opportunities at destinations. Alternatively, it predicts mode of travel as a function of the accessibility provided at a given location. The influence of longer-term choices concerning the housing and labor market, household composition, socio-economic status, and stage of life cycle are commonly ignored. Longer-term choices might simply be assumed to be exogenous (given from outside) in travel demand models. Or, if they are considered to be interdependent, the longer-term and shorter-term choices have typically been addressed operationally by assuming that shorter- and longer-term choices are made simultaneously. Researchers have made progress by employing more sophisticated modeling approaches to consider how longer-term choices condition shorter-term choices [30] and this will likely be a fruitful area of research in upcoming years. We do not claim to offer the final word or even a more elaborate modeling approach to consider such temporal interdependencies. Our task in this volume is to consider the major decisions households experience, think about their temporal horizons, and recognize that many choices, in effect, serve to condition other choices.

Framework for Part I

The timing of important household decisions related to place and plexus serves to organize the first part of this book. The next four chapters focus on four different decision-making processes—where to live, where to work, how to travel, and how to schedule activities. Each decision process is critically important, and the decisions are interdependent. However, these processes operate on different time horizons, and we present each roughly according to the frequency with which most individuals make such types of choices. For example, Figure 2.6 shows the average tenure for residential location decisions and the breakdown by three different age groups. The dramatic drop-off after the 1–5 year category reveals that almost half of the population moves less frequently than every six years. Where in the metropolitan area a household decides to live is a critically important factor for land use-transportation; however, it is a decision that a household embarks upon relatively infrequently.

Contrast the timing of a home relocation with choices about the duration of daily activities. Figure 2.7 shows the number of hours spent in one of a dozen categories inventoried by the Bureau of Labor Statistics (travel is not broken out). Other than personal care activities (e.g., sleeping), no single category sums to more than five hours in an average weekday. Activity related decisions—including what to do, where to go, and when to do it—form the other bookend for the time horizon covered in Part 1 (Chapter 6).

Temporally flanked by decisions about where to live and what to do, we discuss two other matters. The first relates to where individuals work and the importance of job location strategies (Chapter 4). Given that the median tenure of wage and salary workers with employers is 4.0 years, [31] this represents a long-to-medium-term decision that more closely mirrors the change in one's home. The topic near and dear to the hearts of those interested in land use-transportation issues, choice of travel mode (Chapter 5), is habitual in nature [32] and is therefore considered a medium-to-short term decision.

The following four chapters discuss the underlying theories and importance for each of the primary groups of choices. We present dominant theories, document salient trends, review relevant literature, and illuminate knowledge that could shed new light on decision-making patterns. We rely on Table 2.1

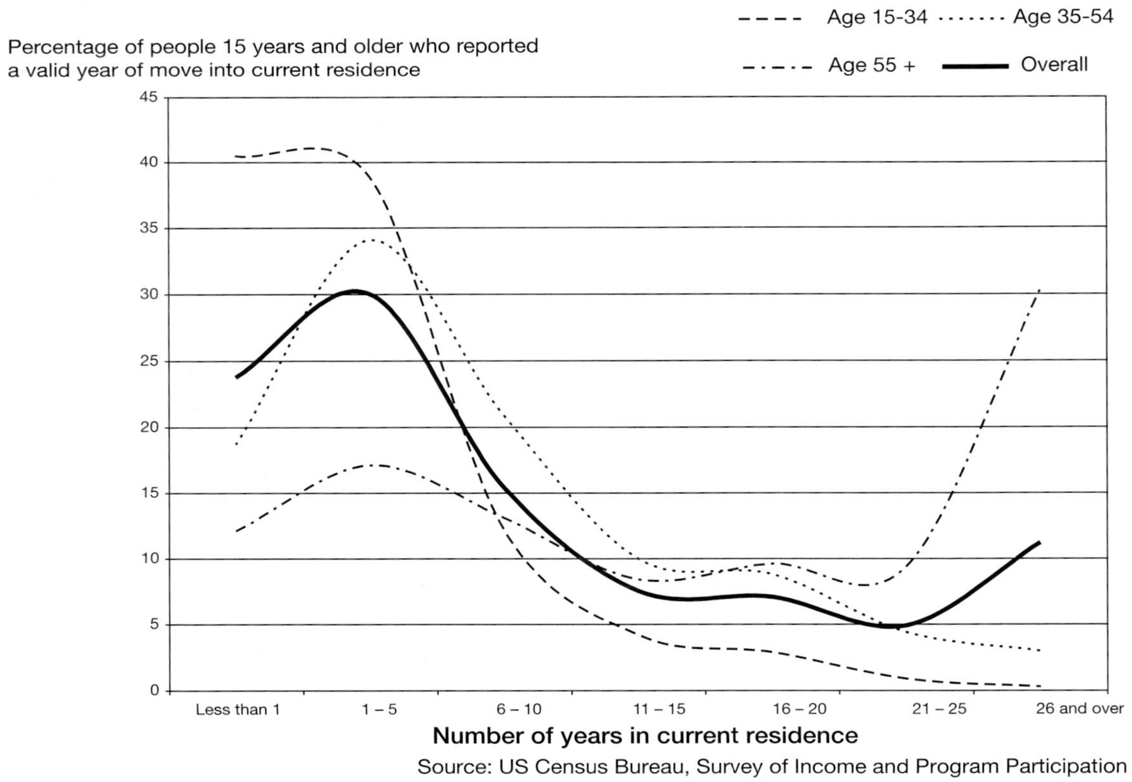

Percentage of people 15 years and older who reported a valid year of move into current residence

– – – – Age 15-34 ⋯⋯ Age 35-54
– · – · Age 55 + —— Overall

Number of years in current residence

Source: US Census Bureau, Survey of Income and Program Participation

Figure 2.6 Residential tenure for location decisions

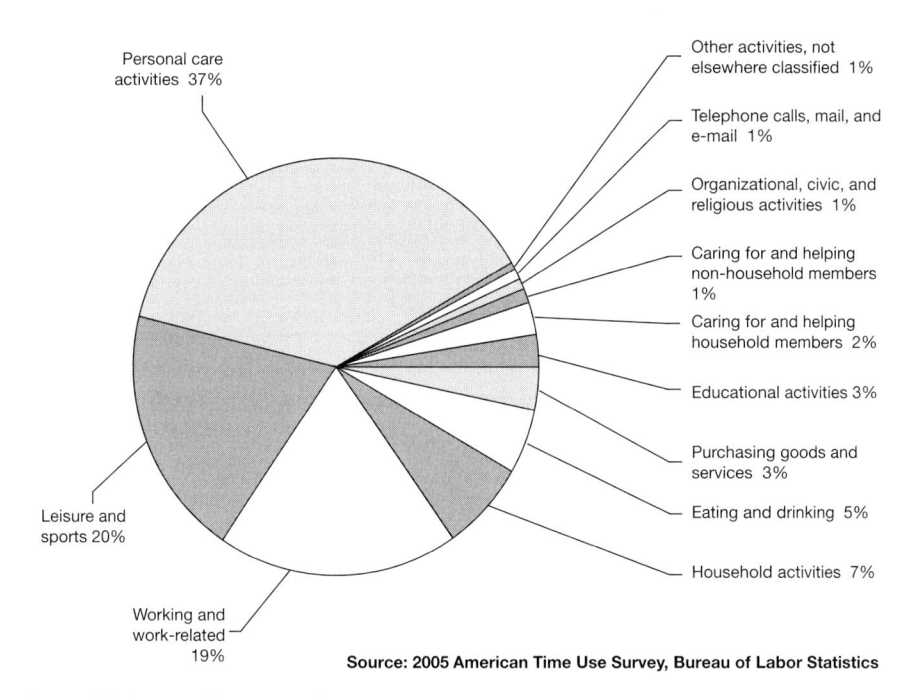

Figure 2.7 Average hours per day spent in primary activities

- Personal care activities 37%
- Other activities, not elsewhere classified 1%
- Telephone calls, mail, and e-mail 1%
- Organizational, civic, and religious activities 1%
- Caring for and helping non-household members 1%
- Caring for and helping household members 2%
- Educational activities 3%
- Purchasing goods and services 3%
- Eating and drinking 5%
- Household activities 7%
- Working and work-related 19%
- Leisure and sports 20%

Source: 2005 American Time Use Survey, Bureau of Labor Statistics

as a framework to present components that influence each outcome. We discuss a series of "pull" factors that predominantly rely on theories about the networks that serve to connect people to each other physically, socially or through information. We also discuss a series of "push" factors that are loosely based on the foundations of game theory. This allows us to consider the interactions that occur as individuals compete for resources. Game theory is broader than just "push" and network analysis is more than just "pull," but the framework allows us to consider a number of important phenomena and their interactions. Ultimately, the outcome (choice) is presented as the solution of an optimization problem. The set of decisions is the optimization set of conditions that result from combining each of the above two approaches.

Table 2.1 Framework for Part I

		Temporal horizon				*Guiding models and theory*
		Years			*Minutes*	
		Home	Job	Mode	Activity	
Influence component	Pull factors					*Networks*
	Push factors					*Game*
Outcome	Product					*Optimization*

Notes

a At the hospital, Reagan was operated on by a team of specialists. Reagan reportedly asked Joseph Giordano, the lead surgeon "Please tell me you're all Republicans." Giordano, a Democrat, replied, "We are all Republicans today."
b When the President is normally traveling for medical care, he takes the Marine One helicopter from the White House. However, on March 30, 1981, time was not available to use the fastest mode.
c The university was also the recipient of a grant of 50 shares of the Potomac Company bequeathed by George Washington himself. The university took its present name in 1904, and is now the largest private landholder in Washington, DC. The Potomac Company, founded by Washington in 1772, aimed to improve the navigability of the Potomac River, through such enhancements as locks, side canals, and channel dredging. The company's rights to improve the river were later transferred to the Chesapeake and Ohio Canal Company, which broke ground for a new canal between the District of Columbia and Cumberland, Maryland on July 4, 1828—the same day the Baltimore and Ohio Railroad broke ground. Use of the C&O Canal peaked in the 1870s, but the railroad proved the more successful enterprise, and it eventually acquired the Canal in 1899. In 1924, the canal was abandoned in the wake of severe damage due to flooding. In 1938, the federal government acquired the right-of-way, and after considering the land for use as a highway, ultimately decided to create the Chesapeake and Ohio Canal National Historic Park. Taking a long view, the canal was one of many complementors that affected the choice of where to take the wounded President.
d The analysis is disaggregate in that individuals are the basic units of observation, yet aggregate because models yield a single set of parameters describing the choice behavior of the population.
e This heuristic decision process may have arisen to prevent individuals from continuing to search for better (or the best) alternatives rather than deciding on a satisfactory alternative. Search costs are not free, and can become burdensome.

References

[1] Thurstone, L.L., 'A Law of Comparative Judgement', *Psychological Review*, 1927, vol. 34: 278–286.
[2] McFadden, D., 'The Path to Discrete Choice Models', *Access*, 2002, vol. 20 (Spring): 2–7.
[3] Handy, S., 'Critical Assessment of the Literature, on the Relationships Among Transportation, Land Use, and Physical Activity', in *TRB Special Report 282: Does the Built Environment Influence Physical Activity? Examining the Evidence*, Report to the House Committee on Physical Activity, Transportation, and Land Use, Washington, DC: Transportation Research Board and the Institute of Medicine, 2004.
[4] Lancaster, K.J., 'A New Approach to Consumer Theory', *Journal of Political Economy*, 1966, vol. 74 (2): 132–157.
[5] McFadden, D. and Domencich, T., *Urban Travel Demand: A Behavioral Analysis*, Amsterdam: North-Holland, 1975.
[6] Ben-Akiva, M. and Lerman, S., *Discrete Choice Analysis: Theory and Application to Travel Demand*, Cambridge, MA: MIT Press, 1985.
[7] Bandura, A., *Social Learning Theory*, New York: General Learning Press, 1977.

[8] Bandura, A., *Social Foundations of Thought and Action: A Social Cognitive Theory*, Englewood Cliffs, NJ: Prentice-Hall, 1986.
[9] Baranowski, T. and Perry, C.L., 'How Individuals, Environments, and Health Behavior Interact: Social Cognitive Theory', in K. Glanz, B.K. Rimer, and F.M. Lewis (eds.), *Health Behavior and Health Education: Theory, Research, and Practice*, San Francisco, CA: Jossey-Bass, 2002.
[10] Ajzen, I., *Attitudes, Personality, and Behavior*, Milton Keynes: Open University Press, 1988.
[11] Ajzen, I., 'The Theory of Planned Behavior', *Organizational Behavior and Human Decision Processes*, 1991, vol. 50 (2): 179–211.
[12] Montano, D.E. and Kasprzyk, D., 'The Theory of Reasoned Action and the Theory of Planned Behavior', in K. Glanz, B.K. Rimer, and F.M. Lewis (eds.), *Health Behavior and Health Education: Theory, Research, and Practice*, San Francisco, CA: Jossey-Bass, 2002.
[13] King, A.C. and Stokols, D., 'Theoretical Approaches to the Promotion of Physical Activity: Forging a Transdisciplinary Paradigm', *American Journal of Preventive Medicine*, 2002, vol. 23 (2): 15–25.
[14] McLeroy, K.R., Bibeau, D., Steckler, A., and Glanz, K., 'An Ecological Perspective on Health Promotion Programs', *Health Education Quarterly*, 1988, vol. 15 (4): 351–377.
[15] Sallis, J.F. and Owen, N., 'Ecological Models', in K. Glanz, F.M. Lewis, and B.K. Rimer (eds.), *Health Behavior and Health Education: Theory, Research, and Practice*, San Francisco, CA: Jossey-Bass, 1997.
[16] Stokols, D., 'Establishing and Maintaining Health Environments: Toward a Social Ecology of Health Promotion', *American Psychologist*, 1992, vol. 47 (1): 6–22.
[17] Kahneman, D. and Tversky, A., 'Prospect Theory: An Analysis of Decision under Risk', *Econometrica*, 1979, vol. 47 (2): 263–291.
[18] Schafer, A. and Victor, D.G., 'The Future Mobility of the World Population', *Transportation Research Part A*, 2000, vol. 34 (3): 171–205.
[19] Schafer, A. and Victor, D., 'The Past and Future of Global Mobility', *Scientific American*, 1997, vol. 277 (4): 58–61.
[20] Zahavi, Y., *Travel Time Budget and Mobility in Urban Areas*, Washington, DC: United States Department of Transportation, 1974.
[21] Mokhtarian, P. and Chen, C., 'TTB or Not TTB, That is the Question: A Review and Analysis of the Empirical Literature on Travel Time (and Money) Budgets', *Transportation Research Part A-Policy and Practice*, 2004, vol. 38 (9–10): 643–675. Washington DC: Transportation Research Board, 2003.
[22] Surface Transportation Policy Project and Center for Neighborhood Technology, *Driven to Spend: A Transportation and Quality of Life Publication*, Surface Transportation Policy Project, 2000.
[23] Wright, R., *Nonzero: The Logic of Human Destiny*, New York: Pantheon Books, 2000.
[24] Schwartz, B., *The Paradox of Choice: Why More is Less*, New York: Ecco, 2004.
[25] Cropp, B. and Graf, T., *The History and Role of Dairy Cooperatives*, 2001. Available at: www.wisc.edu/uwcc/info/dairy/history.pdf (accessed June 30, 2006).
[26] Duft, K., 'A Popularist View of Cooperative Principles', *Agribusiness Extension Newsletters: Cooperatives*, 1991. Available at: www.agribusiness-mgmt.wsu.edu/ExtensionNewsletters/coop/PopViewCoopPrin.pdf (accessed May 15, 2007).

[27] Cox, C., *Storefront Revolution: Food Co-ops and the Counterculture*, New Brunswick, NJ: Rutgers University Press, 1994.
[28] Waddell, P., 'Towards a Behavioral Integration of Land Use and Transportation Modeling', in *9th International Association for Travel Behavior Research*, Queensland, Australia, 2001.
[29] Cullen, I.G., 'The Treatment of Time in the Explanation of Spatial Behavior', in P. Carlstein (ed.), *Human Activity and Time Geography*, London: Edward Arnold, 1978, pp. 27–38.
[30] Ben-Akiva, M. and Bowman, J.L., 'Integration of an Activity-based Model System and a Residential Location Model', *Urban Studies*, 1998, vol. 35 (7): 1131–1153.
[31] United States Department of Labor, *Employee Tenure Summary*, Washington, DC: Bureau of Labor Statistics, 2004.
[32] Garling, T. and Axhausen, K.W., 'Introduction: Habitual Travel Choice', *Transportation*, 2003, vol. 30 (1): 1–11.

Chapter 3

Homebuying

> "If you lived here, you would be home now."
>
> Motto of Bridgeville, Delaware

The self-proclaimed, "[m]ost perfectly planned community in America," Levittown, Pennsylvania is a suburb of Philadelphia. Its development was initiated four years after its older sibling, Levittown, New York, emerged on Long Island in 1947, and several years before a third Levittown, in Willingboro, New Jersey, broke ground in 1959. During this twelve year period, it is estimated that Abraham Levitt and Sons, the developer of these communities, built one in eight new houses in the United States. [1] These homes were located not only in the eponymous "Levittowns," but also in smaller developments across the country. The influence of the Levittowns was widespread. Levittown-like communities opened up vast new areas for development, thereby fully cementing the suburban trend that had begun with the streetcar. Some 20 million Americans moved to the suburbs in the housing boom following World War II, and less than 200,000 (or 1 percent) of them lived in a Levittown proper.

Levitt and Sons was founded in 1929 as a custom homebuilder. The company's big break came during World War II, when it was contracted by the US government to construct housing for defense workers in Norfolk, Virginia; it was here that Levitt and Sons pioneered the mass production methods that would come to prominence just two years after the War, in 1947, on 1,200 acres of potato fields of Long Island, with the launch of the first Levittown.

The original Levittown accomplished several goals. It provided affordable housing for an emerging middle class, especially war veterans. It employed the technologies of mass production and standardization, which had been used up to this point for everything from guns to automobiles, taking an almost assembly-line approach to housing construction. Although the houses could not move, the construction workers could. Workers would arrive on an empty street; the first group would lay slab foundations on lot after lot (the houses lacked basements); the next group would add the second component (the frame), house by house, and so on, until all the houses on the street were complete. Prices ranged from $8,000 to $10,000—inexpensive for the era. By keeping costs down, Levitt and Sons was able to offer more residents a chance to partake in the American Dream: a single-family home in the suburbs.

But Levittown, and the suburbia it symbolized, was also reviled by critics. Lewis Mumford derided it as "a one-class community on a great scale, too congested for effective variety and too spread out for social relationships ... Mechanically, it is admirably done. Socially, the design is backward." [2] The comment implied several critiques. First, these communities only served a single income class, thereby only serving residents similar to one another. Second, residents of single-family homes on small lots do not have the same number of interactions as people living in (presumably) denser cities. The third criticism claimed that these developments were "too congested for effective variety." This critique is perplexing, as it is congestion that creates variety. (Adam Smith wrote that the "[d]ivision of labor is limited by the extent of the market." [3]) It is easy for one group to criticize how another lives, but one must live as they do to really understand.

Sociologist Hebert Gans lived in Levittown to do just that. In his book, *Levittowners* [4] the urbanologist detailed the complexity of social interaction, showing that despite the model houses looking the same on the outside, people's lives were as normal as elsewhere. Before condemning those who selected Levittown, one should be aware of the situation they were in before. As William Levitt commented "[w]hat would you call the places our homeowners left to move out here? We give them something better and something they can pay for." [5] Levittown, in other words, provided additional opportunities by relaxing the money constraint that kept people living in places they otherwise would prefer to leave. Although this benefit may or may not have come at the cost of eroding social interaction (and any move from one place to another requires severing some social ties), it was a price that those who moved were willing to make for what they believed would be a better life—a life they, and not Lewis Mumford, would have to live.

Today Levittown, Pennsylvania, houses nearly 54,000 people; Levittown, New York, has about 53,000; Willingboro Township, New Jersey, contains 33,000 people. There is even a Levittown in Puerto Rico with 30,000 residents (and houses made of concrete instead of wood to weather the hurricane season and termites), and the firm built 9,000 houses in the City of Bowie, Maryland.

Many more people live in Levittown-like planned suburban communities of greater or lesser scale. The firm Levitt and Sons was sold to the multinational conglomerate ITT Industries in the 1960s, and has since been folded.

One of the consequences of thousands of households moving to Levittown-like communities in the middle of the twentieth century was the need for people to reach such outlying locations. Transit service to the suburbs was inadequate. Concomitant with the widespread availability of the automobile, people decided to rely on cars. Thus, moving to Levittown and using the automobile were joint choices. It was difficult to live in these communities without a car.

Using Levittown as an example demonstrates the spatial importance of one's home location as well as the travel dimensions. In discussing these matters, the transportation component is the one most often considered and the issue most scrutinized. As described in Chapter 1, congestion is the most visible dimension of the land use-transportation problems—a condition that is *typically* thought of as a result of people's daily mode, route, and scheduling decisions (e.g., car versus bus; surface streets versus interstate highway; 8:00 A.M. versus 10:30 A.M.). These shorter-term decisions prompt many planners, most policy makers, and the general public to diagnose the travel aspects as the root of many people's dissatisfaction with most metropolitan areas, not the dimensions related to where to live.

However, based on the pulse of recent discussions on land use-transportation, an intellectual breakthrough is surfacing. Researchers, practitioners, and even politicians are realizing that the act of travel is more complicated than how people travel from origin "O" to destination "D." Just as important are the characteristics of how people choose "Os" and "Ds." That is, the locational and spatial dimensions of where to live, work, shop, and other are equally important; location decisions (where the trip starts and ends) and travel decisions (how to go) are often a byproduct of each other.

Von Thünen's isolated state

Formally acknowledging that these concepts are interrelated is not new; rather, these thoughts might be surfacing again today because of a more contemporary understanding of cities (or a more visible outcome: congestion). Almost two centuries ago, a German landowner and economist named Johann Heinrich von Thünen [6] suggested that a farmer's profit at a given location depended on two factors: (1) how much people in the city were willing to pay for different crops (legend has it his example was based on tomatoes); and (2), how much it would cost to produce and transport tomatoes to market. The two key variables were the market price of the good and the cost of transport.

Assuming a city has a single center, von Thünen argued that activities would be arranged based on their travel cost (including both temporal and monetary costs). The most time-sensitive activities would be closest to the center of the city. Immediately surrounding the central city would be farms that produce perishable commodities like fruits, vegetables, and milk. In the ring surrounding would be forest products, which at the time were burned for heat. Wood was heavy, and so to minimize transport costs, would be located in the next ring out. Proceeding outward, this would be followed by field crops like grains and, finally, an outermost ring devoted to ranching and animal products, as shown in Figure 3.1.

Von Thünen understood that this was a model, and that the assumptions underlying it were to provide understanding, not to make predictions. The first assumption was that farmers sought to maximize profit (a safe bet, even today). Second, the city was isolated, surrounded by an undifferentiated plane of unoccupied wilderness (with no messy geographical features like mountains, rivers, or oceans). Third, farmers travel directly to the city center via oxcart (there were no roads to navigate).

To illustrate using an example adapted from Nelson, [7] Figure 3.1 shows four points: *a*, *b*, *c*, *d*. Line *a-c* is the farm price of dairy, while line *b-d* is the price of fruits and vegetables. In the city, point *a* is higher than point *b*, so the price of dairy is higher than the price of fruits and vegetables. Furthermore, the cost of transporting dairy products (especially in the days before refrigeration)

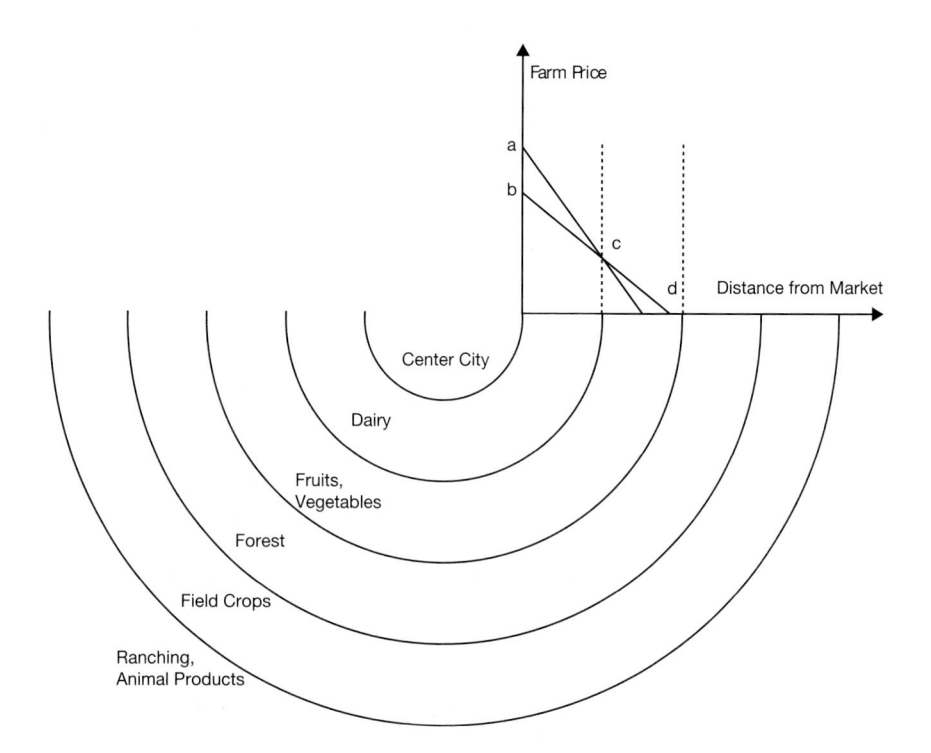

Figure 3.1 Von Thünen's isolated state

is higher than the cost of moving fruits and vegetables. So the dairy price drops more quickly than the fruit and vegetable price as you move out from the city, since dairy from too far away will likely spoil if used for milk (thus, giving rise to the cheese industry). Beyond point c, the farm price of fruits and vegetables is greater than the farm price of dairy. This mandates a different but better and higher use of the land in that ring.

Von Thünen's model was developed before the widespread deployment of modern industry or railroads. Still, the idea that rent would be highest in the city center, and diminish with distance, has become a cornerstone of modern urban economics. This is reflected when those who would benefit most from proximity to the center (e.g., fruit, vegetable, and dairy producers) would pay higher rent than forest owners, and so on. It can be seen today applied to urban real estate markets. Proud of his economic analysis to the end, on his tombstone von Thünen had engraved his famous equation for the "natural wage" (now called the marginal product of labor): $w = \sqrt{ap}$, where the natural wage, w, is the square root of (ap), where p is the worker's product and a is his subsistence requirements.

Alonso's bid-rent function

Economist William Alonso extended Von Thünen's model to a more contemporary urban setting in his 1960 dissertation. He determined land use, rent, and density by type (population, employment) as a function of distance to the central business district by solving the equilibrium problem of land allocation. In brief, people trade-off space for time and money. The bid-rent function shows the trade-off for a given level of satisfaction or utility. A locator-traveler can enjoy constant utility by opting for lower travel costs and less space at a given level of rent, or for higher travel costs but more space. Different locator-travelers display different levels of preference for space and for time, which is what allows markets to form and differentiate the housing stock by location. The bid-rent is the maximum anyone is willing to bid for a given piece of land (given all of the other bidders (competitors), and all of the other pieces of land (chances) on the market).

Figure 3.2 shows multiple bid-rent curves (U_1, U_2, U_3) for a single household, along with the actual gradient of rents (the market price) R (where U_3 is preferred to U_2, which is preferred to U_1). Note they are all downward-sloping: the price people are willing to pay decreases as they get farther from the city center, reflecting the traveler's aversion to commuting. This comports with most empirical evidence (including the hedonic model presented later in the chapter), though the world is certainly more complicated when there are multiple centers, rather than a single center. It is well-recognized today that most major metropolitan areas are now polycentric in nature. [8]

Note also that lower bid-rent curves have greater utility: the less that is paid in rent, the more other commodities can be consumed. Furthermore, note

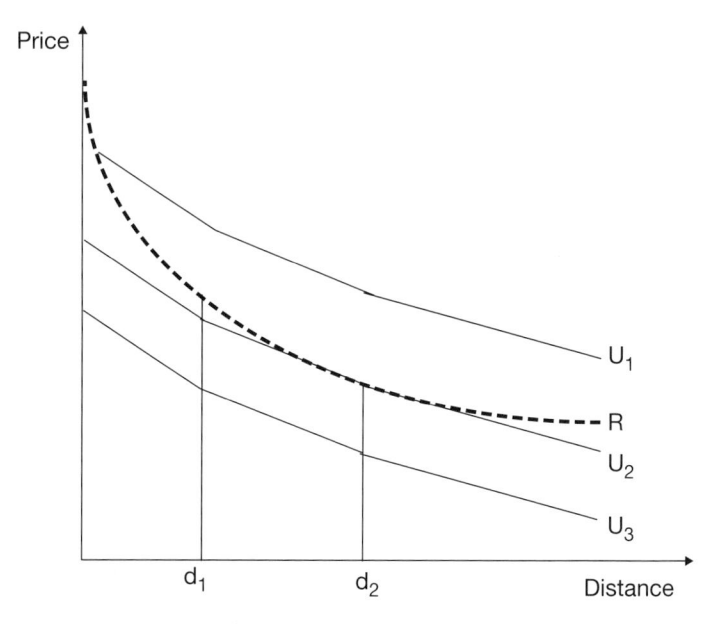

Figure 3.2 Bid-rent curves and distance from city center

that for a given distance from the city center, there is only one level of rent, and this rent curve is tangent to a single utility curve (in this case U_2) for each household.

The idea of rent varying linearly with location plays out in the game of Monopoly (see Box 3.1). Rents increase as a player goes around the board, in the same way that rents increase as one moves towards the center in the monocentric model. However, in Monopoly, the "center"—the block with the highest rents—is actually the edge of Atlantic City, Boardwalk.

Accessibility (attraction)

The core rationales presented above are instrumental in explaining the spatial form of metropolitan areas even in the twenty-first century. People will pay a premium for more opportunities, explaining why rent downtown is more expensive than at locales further from the center. The reason people pay a premium to live closer to attractions is because they enjoy heightened levels of accessibility. Past writings have suggested accessibility is "perhaps the most important concept in defining and explaining regional form and function." [9]

Accessibility is a term with a history of being cavalierly thrown about and even misused in the worlds of land use and transportation (for more discussion, see Chapter 10). In this work, we define accessibility as the ease of reaching land use (place) given the transportation system (plexus). [10–13] In other

Box 3.1 Monopoly and anti-monopoly

Widely recognized as the most popular board game, Monopoly at its core is about place, plexus and accessibility. Each player is an actor, who both collects rents and pays them out. The objective is to maximize this difference through acquiring properties and improving them. One could think of many ways to make the game more realistic, though that would add complexity. For instance, the rents in the game are fixed and built in (i.e., Broadway and Park Place in the US version have the highest rents). Yet rents are determined by accessibility. Accessibility in the game could be a function of other properties that have been developed, or it could be a function of where one is most likely to land (statistically, Illinois Avenue is the property most frequently landed on). One has to assume that the rents in the game of Monopoly are exogenous, perhaps associated with off-board activities like offices. However, advanced players, knowing the accessibility of various locations, will pay a premium for lots that are frequently landed on.

Although the game is obviously about place, it is also about plexus, though this is modeled less accurately. Movement around the game board is carried out with tokens (in the Parker Brothers 1937 version, they included flatiron, purse, lantern, car, thimble, shoe, top hat (representing Mr. Monopoly), and the rocking horse; later the horse and rider, Mr. Monopoly's dog, Scotty, and the wheelbarrow replaced the lantern, purse and rocking horse). The more recent tokens are more transportation-oriented than, say a lantern or a purse. But the more prominent transportation orientation are the four railroad squares, which indicate the profits that can be obtained by controlling inter-city transportation, particularly if there is a monopoly on such movement.

The Parker Brothers did not invent the game Monopoly. Nor was it invented by Charles Darrow, who sold his patent rights to Parker Brothers. Hasbro, current owner of the game, still insists that Charles Darrow first brought the game to Parker Brothers, but the legal history indicates otherwise. The game was first patented as "The Landlord Game" in 1904 by Lizzie Maggie, a Quaker follower of Henry George and the single tax, and was designed to illustrate the concepts therein. The 1904 and 1910 versions are illustrated in Figures 3.3 and 3.4. Parker Brothers paid off several other competing game developers at about the time they bought Darrow's version. [14] The game had been played on various college campuses, and was known colloquially as Monopoly. (Notice that the "Free parking" square was divided between the "Poorhouse" or "Central Park Free" in the 1910 version).

A game called "Anti-Monopoly" was developed to illustrate the breaking of monopolies. The game starts with a bunch of monopolies on the board and players compete as trustbusters to make it a competitive market by bringing indictments. In 1974, General Mills Fun Group, the company that then owned Parker Brothers and the trademark on the game, sued Ralph Anspach, who along with Iowa State University mathematics professor Irvin Hentzel, developed Anti-Monopoly. The issue was trademark, since "Anti-Monopoly" included the trademarked "Monopoly" as part of its title. The owners of Monopoly went further, and signed contracts with distributors that prohibited them from selling competing games. Judge Spencer Williams ruled that Anti-Monopoly did violate the Monopoly trademark and ordered 40,000 Anti-Monopoly games buried in a Minnesota dump. Yet in the end, Williams' ruling was overturned, with the United States Court of Appeals accepting the argument that no consumer would confuse Monopoly with its opposite "Anti-Monopoly." Hasbro purchased the name "Anti-Monopoly," and now licenses it back to Anspach.

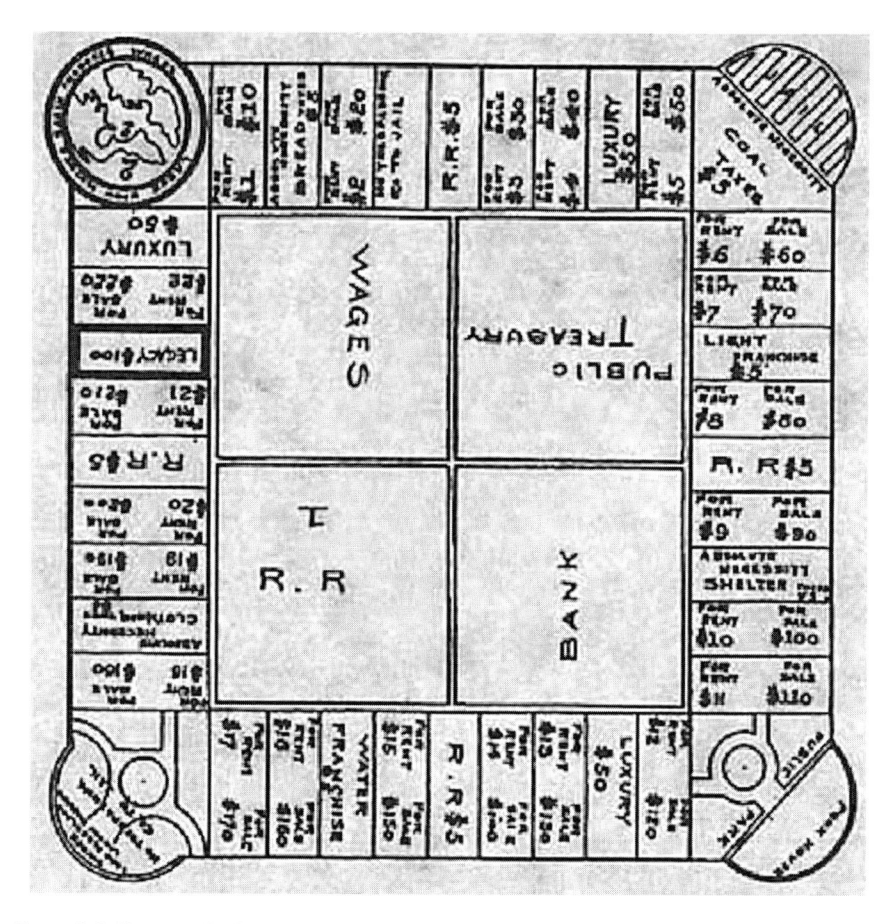

Figure 3.3 "Landlord's Game" board (1904)

words, accessibility is a measure of the glue holding place and plexus together, and of the ease with which one can get where one wants to go; it is what ties land use to transportation.

Accessibility as a network concept

Accessibility can also be thought of as a network concept. For example, imagine there are two cities (or nodes), City A and City B, as shown on the top of Figure 3.5. The two cities create two travel markets: A-B and B-A. The middle case adds one city, and one link, but greatly increases the number of travel markets: A-B and B-A remain, but A-C, C-A, B-C and C-B are added (we increased by four markets to a total of six). The addition of one link tripled the number of Origin-Destination (O-D) pairs served. The bottom case adds

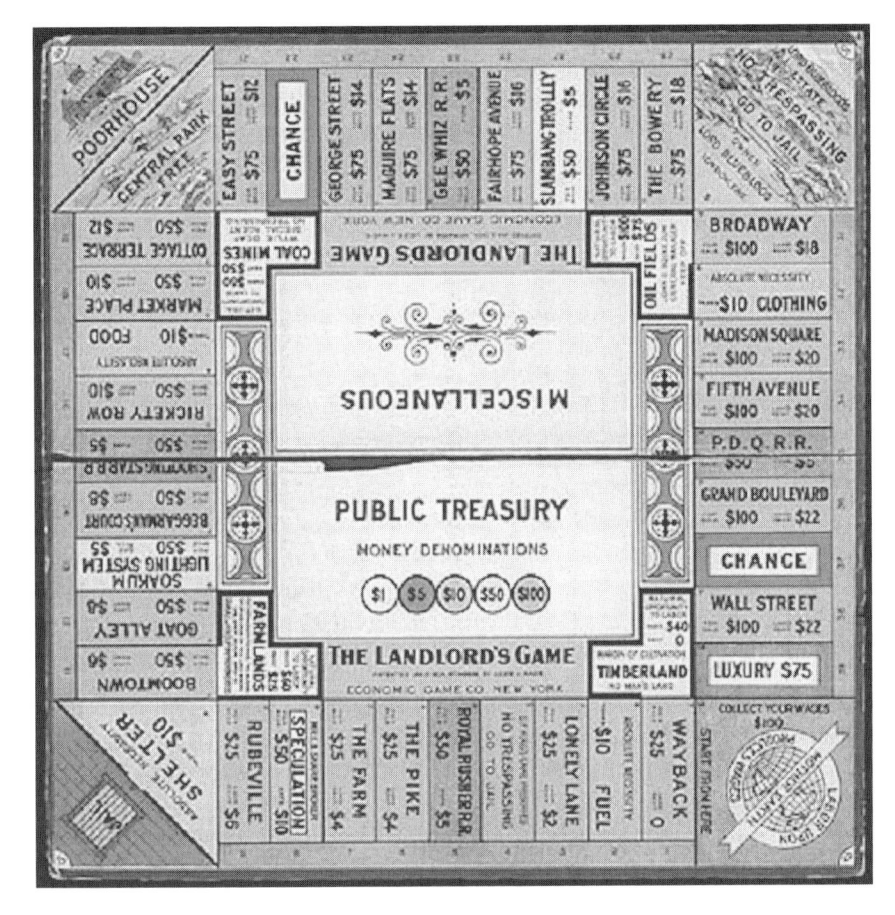

Figure 3.4 "Landlord's Game" board (1910)

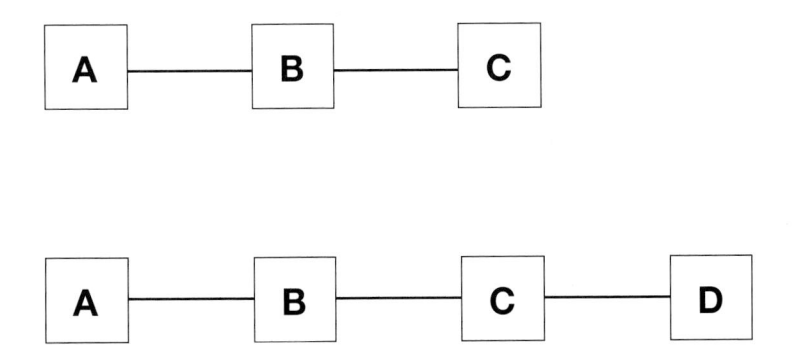

Figure 3.5 Example networks

one more link (for a total of 3), but the number of markets again increases substantially: in addition to the six markets already established, we now also have A-D, D-A, B-D, D-B, C-D and D-C. The number of markets doubled (we increased by 6 markets to a total of 12).

This phenomenon, dubbed the "Law of the Network" (and in a computer networking context, Metcalfe's Law, named for Robert Metcalfe, developer of the Ethernet networking standard) can be expressed as:

$$S = N(N-1)$$

where S = the size of the network (number of markets)
N = the number of nodes

(To illustrate: with 2 nodes: $S = 2*1 = 2$, with 3 nodes: $S = 3*2 = 6$, with 4 nodes: $S = 4*3 = 12$, etc.)

The value of S grows non-linearly as nodes are added to the network, until all nodes are connected. Clearly there is increasing value to the network as it gets larger. Since people are willing to pay more for goods of higher value, we would expect that people would pay more to belong to a larger network (live in a larger city). This is shown in Figure 3.6.

Measuring accessibility

The task of measuring accessibility is similar to that of measuring network size, but requires a slightly more sophisticated approach. Accessibility multiplies origins by destinations, but discounts that number by some function of the cost of connecting them. Walter Hansen's 1959 article is commonly referenced as the first application of accessibility within a context directly applicable to land use modeling. He presents a hypothetical model showing how differences in accessibility—arising from the construction of an express highway—could be used as the basis for a residential land use model. In this context and others, accessibility models showed how highways can help explain the characteristics of residential (or other) locations. Other applications, for example, analyzed automobile accessibility in Los Angeles as an indicator of quality of life.

A measure of accessibility to employment (for a given origin i) can be represented as follows:

$$A_i = \sum_j E_j f(C_{ij})$$

where: A_i = Accessibility to employment from zone i
E_j = Employment at destination j
$f(C_{ij})$ = function of the travel cost (time and money) between i and j. The higher the cost, the less the weight given to the employment location.

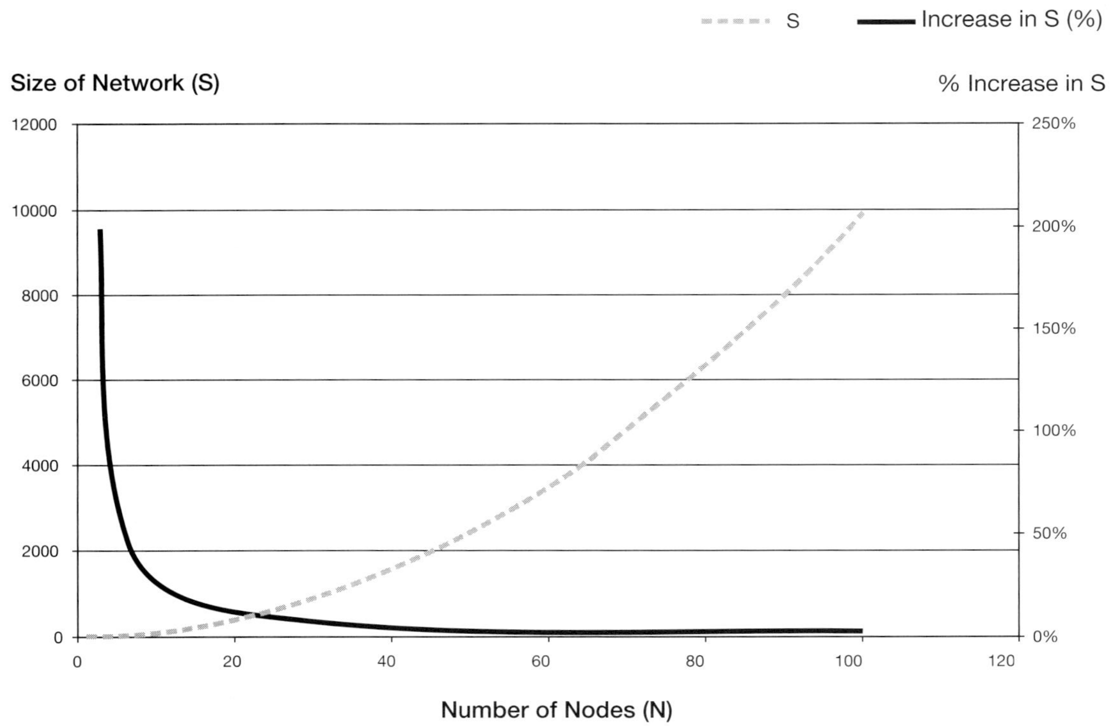

Size of Network (S)

– – – S ——— Increase in S

% Increase in S (%)

Number of Nodes (N)

Figure 3.6 Law of the network, showing increasing or decreasing returns

One can compute an overall accessibility measure comparable to the network size measure by multiplying the accessibility at each origin by a measure of the importance of each origin, and then summing across all origins, for instance:

$$A = \sum_i W_i A_i$$

where: A = overall accessibility
W_i = workers living at origin i

Accessibility (A) differs from Network Size (S) in that one multiplies each interaction by a function of the travel cost, such that distant interactions have less weight than nearby interactions. It is also common to replace the simple measure, number of nodes, with a slightly more sophisticated measure, number of jobs, to measure employment accessibility (or number of workers to measure labor force accessibility). This allows one to see how well the system connects workers with jobs. Box 3.2 illustrates accessibility calculations.

Connecting two nearby but previously unconnected destinations is more valuable than connecting two destinations that are similar in size and located farther from each other. Two cities can have the same number of workers and jobs but different accessibility if one has an efficient network and the other does not. However, larger cities, in general, have greater accessibility than smaller cities. This explains in part why land is more expensive in bigger cities: larger cities are more valuable. This also explains why, historically, downtown locations are more expensive than locations in the suburbs: downtown is more accessible. Furthermore, it explains why buildings downtown are taller than those in the suburbs: developers want to maximize profit by building more floor space on land with the highest value.[a] Granted, such factors are changing with the emergence of polycentric metropolitan areas and as some suburban settings (such as shopping malls) garner a high degree of accessibility.

Absolute vs. relative accessibility

The concept of accessibility also has several other aspects. First is absolute accessibility: the total measure of accessibility within a particular area. A transportation improvement increases overall accessibility—analogous to increasing the size of the pie. The second aspect is relative accessibility: the share of total accessibility associated with a particular place. A new transportation facility increases the relative accessibility of those points that can directly use the facility—analogous to increasing the percentage of the pie that a particular slice comprises. In the examples above, while society overall receives greater accessibility, the markets served by the improvement gain in both absolute and relative accessibility; this implies that other markets may lose relative, if not absolute position. New infrastructure benefits the area around the improvement but may make other areas worse off, at least in terms of relative

Box 3.2 Computing accessibility

Imagine a city with two traffic zones, Zone 1 and Zone 2. The land uses in the zones are distributed as show below:

Land use	Workers	Employment
Zone 1	100	200
Zone 2	200	100

The travel times between zones are shown as follows:

Travel time	Zone 1	Zone 2
Zone 1	5	10
Zone 2	10	6

All accessibility measures require an impedance function. This is often the most difficult (and arbitrary) measure to estimate. A simple and often-relied-on impedance function comes in the form of:

$$f(C_{ij}) = \frac{1}{C_{ij}^2}$$

The accessibility to Employment from Zone 1 is thus:

$$A_i = \sum_j E_j f(C_{ij}) = \frac{200}{5^2} + \frac{100}{10^2} = 9$$

Similarly, the accessibility to Employment from Zone 2 is:

$$A_i = \sum_j E_j f(C_{ij}) = \frac{200}{10^2} + \frac{100}{5^2} = 6$$

The overall accessibility is

$$A_i = \sum_i W_i A_i = 100(9) + 200(6) = 2100$$

A different accessibility pattern results if either the land use configuration or the travel times on the network change. All else equal, reducing the travel times or increasing the level of development raises accessibility.

position (and sometimes in absolute position). All infrastructure investments create winners and losers, especially when cities are competing with one another to "keep up with the Joneses."

Consistent with the standard model of urban economics, it has been observed that living in an area with relatively high accessibility to jobs is associated with shorter trips, as is working in an area of relatively high housing accessibility. [15–17] But in addition to the opportunities themselves, recognise that competitors absorb those opportunities. This relative location of houses and firms, measured using accessibility, is an important determinant of commuting duration.

Residential location decisions and accessibility's evolving nature

Accessibility comes home to roost when considered as a factor explaining how and where households decide to live, a decision that an estimated 16 percent of the US population makes every year. [18] The significance of the matter lies in the fact that housing units tend to be more complex than other consumer products. They are durable, location-tied goods providing a multitude of services and benefits to resident households. Housing not only represents the largest single financial investment made by most households, but in the context of this book, exerts a major influence on each household's daily interactions, travel, and activity patterns.

The bulk of the theory, analysis, and examples described so far have focused on home locations vis-à-vis major employment sites. Half a century ago, that is what mattered most and the study of land use and travel was much simpler than at present. The breadwinner in the home chose a workplace and the family determined home location largely based on the above theories. The home-to-work-and-back commute comprised the greatest share of daily travel. The models predicting this behavior worked well enough.

Analysts today widely recognize that travel is more complicated—in part because people's lives (and resulting travel) are more complex, and in part because analysts now explicitly recognize the complexity that was always there. There is wider variety in types and destinations of trips; the work commute definitely does not hold the significance it once did. Furthermore, the rise of the two-earner household and the increase in rates of vehicle ownership to one per licensed driver have expanded travel, and have both required and allowed households to "outsource" activities that used to be done at home, such as dining and child care.[b]

Despite the rise of out-of-home non-work travel, no single trip purpose (e.g., appointments, shopping, visiting) dominates like it once did. The work trip still receives more "hits" than any other single trip purpose (other than coming back home). But no longer does the breadwinner's workplace location dominate; instead, families may have two members in the labor force and therefore compromise between two work locations. Furthermore, the increasing variety of travel has prompted households to weigh not just access to work, but rather access to a broader array of activities. They may value access to neighborhood services, or prefer proximity to recreational amenities—and there is always the proverbial case where the quality of elementary schools trumps all other considerations, though for every family moving in to an existing house, one is moving out.

Another factor needs to be considered: mode of transportation. Accessibility depends on the network used. In an auto-dominated suburban landscape, regional accessibility by car tends to be higher than accessibility by transit, walking, or cycling. Smaller scales of geography may yield a different story. Different modes play different roles in providing attractive accessibility. For example, non-motorized modes serve shorter-distance trips and motorized modes tend to be more suitable for longer-distance travel. Performing a more extensive analysis requires one to assess accessibility to the various things people consider important (jobs, shops, others), controlling for things that affect the value of jobs (competing workers), by various modes. Some modes are more suitable for people with physical disabilities, low incomes, or who do not otherwise have convenient access to automobiles. Some modes are particularly important for certain industries.

Many different accessibility measures can be constructed. Conceptual frameworks (of the type presented in Figure 3.7) help planners understand why people choose the modes and locations they do; some modes and locations provide more accessibility in certain contexts than others.

The preceding sections highlight the ways in which individual location decisions are shaped by accessibility. However, by locating in a particular location, individuals also reshape the accessibility landscape for others. For example, employment opportunities are complements to employees and employees complement employment opportunities; but workers are competitors to other workers, and in terms of filling open jobs, firms are competitors with each other, bidding up the cost of labor.

Regional and local accessibility

Just as accessibility can be measured by destination and by mode, the level of geography is important to account for. Geographical area (i.e., drawing power) is a critical dimension in which measures of accessibility differ. The land use-transportation literature, rightfully, shows a keen interest in differentiating between what is commonly referred to as regional versus neighborhood accessibility.

Yehuda Gur [21] is widely credited as the first to recognize the necessity of distinguishing between scales of accessibility (and their travel implications). He contended that not merely the length of trips, but the rate of trip generation depend on at least two factors. The first factor was the ease and worth of travel to destinations far away: the easier the travel to various opportunities there, the more trips are going to be made to those far away destinations.

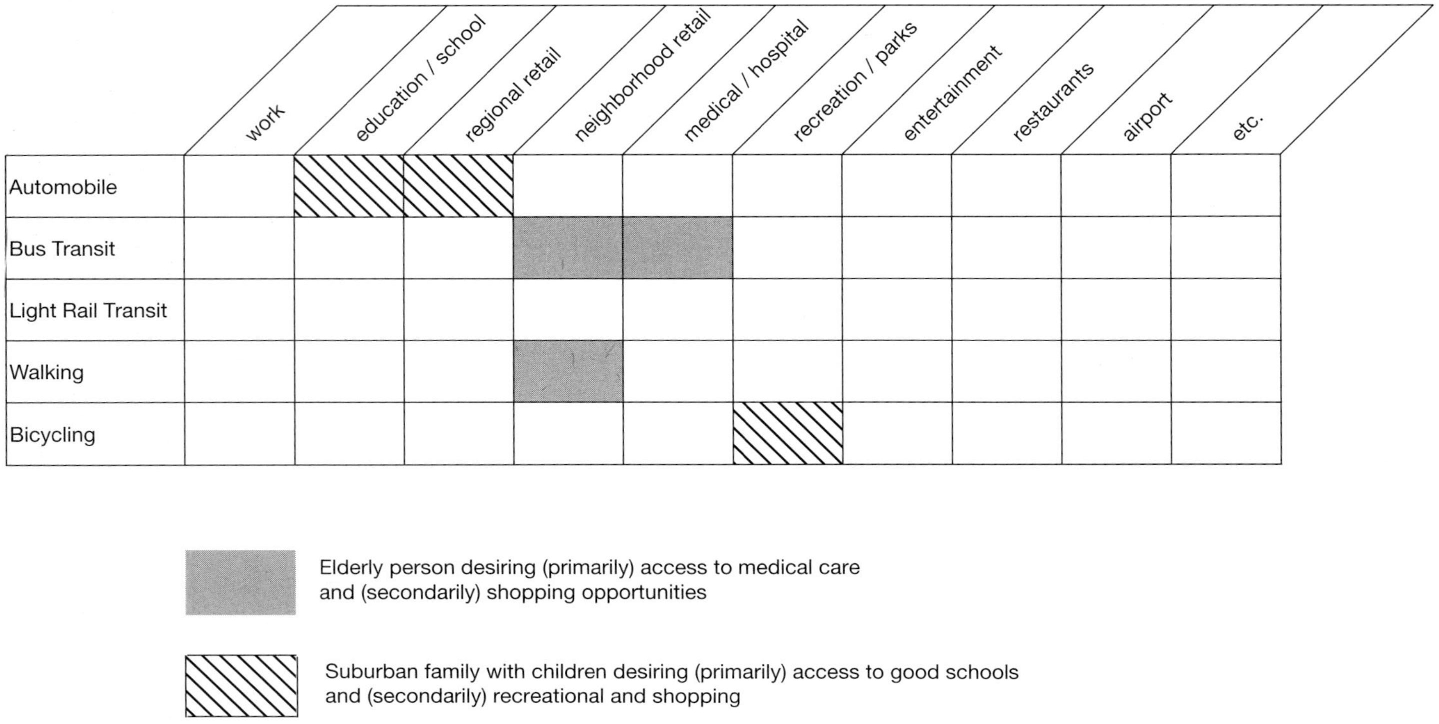

Figure 3.7 Accessibility matrix showing examples of modes and preferences

The second factor was the availability of opportunities close by: the more opportunities that are available close by, the higher the likelihood that an activity which may require a long trip is substituted for an activity close by. Gur showed that these factors have a significant effect on travel demand, and that an increase in the ease and worth of making a trip to distant destinations was positively related to trip generation.

Subsequent work [22–26] further reinforced the importance of drawing a separation between two different scales. Handy used the terms "regional accessibility" and "local accessibility," and measured each using different criteria. Regional accessibility was determined by the regional structure of a metropolitan area and incorporated variables such as location, type of activities, and size of activities that affect shopping behavior. Local accessibility, on the other hand, was primarily determined by nearby activity (where "nearby" is used to refer to the neighborhood unit, approximately one-half to one mile (800 to 1,600 m) in residential areas). Areas with higher local accessibility would be oriented to convenience goods, such as supermarkets and drug stores, and located in small centers.

Differentiating measures between the two levels of accessibility is a messy process. Many policy initiatives speak to increasing accessibility on both regional and local scales; and, while the two scales are intricately related, each calls for different policies. For example, regional land use-transportation policies may speak to issues of urban growth boundaries, increasing densification, and diversifying the geographical distribution of employment centers. It is not likely that such regional policies prescribe development regulations for specific neighborhoods. Neighborhood accessibility policy initiatives speak more to issues of mixing uses on a parcel or neighborhood scale, site design, and more directly, facilitating circulation patterns that enhance walking, bicycling and transit use.

In the end, any analyst needs to understand that the form of the region as a whole may have at least as great a role in influencing travel decisions as modal choice as neighborhood design, if not greater.

Aversion

While forces of high employment access, good schools, and proximity to parks are all pulling households to certain locations, there are opposing forces pushing them away from locations. For example, considering all other housing characteristics equal (e.g., size, quality, accessibility to certain goods), families get more for their money where property is cheaper. The fact that such areas are further from the centers of metropolitan areas creates an immediate push factor toward outlying regions. Over 80 percent of Americans see buying a home as a safe and smart investment. [27] One of the reasons Americans see homes as a good investment is the potential for homes to appreciate in value over time. History has shown that areas with the greatest appreciation are those developing services and other amenities—the suburbs and other outlying areas.c Of course, the buck does not stop there. Other factors commonly associated with city environs such as crime, older housing stock, and poor neighborhood quality contribute to push factors for most, certainly not all, households.

Unfortunately, aesthetics are not the only perceived benefit (or cost) one neighbor imposes on another. Massey and Denton argued that the fundamental cause of poverty among African Americans is segregation. [28] Despite considerable advances promoted by the US civil rights movement, progress on housing desegregation has been painfully slow. Race and location are a source of great inequities in modern society. The vast majority of large cities continue to divide geographically along racial bounds, establishing de facto racially segregated neighborhoods that have created and perpetuated an underclass by limiting the educational and employment opportunities for the residents of these neighborhoods. This happens because the wealthier or white residents (especially younger families with children) leave the area for the suburbs, decreasing the tax base, which hurts funding for education and in turn motivates everyone else who can afford to leave to do so, further decreasing the tax base and reducing funding for education. Few businesses choose to invest in an area with little money. Lack of opportunity encourages crime, which further discourages investment. The result is that only poor minorities remain behind. They have fewer opportunities for education or employment and are trapped in a vicious circle of poverty.

This phenomenon has had long lasting implications for where people choose to live. Of course, a number of factors come into play as they relate to preferences of different races. For example, a Detroit survey found that a majority of blacks preferred living in a neighborhood that was 50 percent black, whereas whites prefer a neighborhood *more than* 50 percent white. [29] Such preferences beg the urban analyst to question differences between racial preferences. In other words, how similar do one's neighbors have to be for that individual household to locate there?

Schelling's segregation model

To best demonstrate the powerful concept of segregation—perhaps the strongest push factor in land use-transportation—we rely on the work of the 2005 winner of the Nobel Prize in Economics, Thomas Schelling, whose segregation model [30] tells a revealing story.

Imagine a world with two types of agents (e.g., people or turtles): we can call them green and red turtles, or anything else, but the discussion is relevant to residential location if we think of them as black and white people. The important thing to note about the types is that, although they will tolerate neighbors of the opposite type, they do not want to be the only one of their type in a neighborhood. So, for instance, whites will live in a neighborhood that is 62.5 per cent black, but not more than 62.5 per cent black (if 62.5 per cent seems a strange number to pick, think of it as five of the eight neighboring homes in a square grid). The basic conclusions of the model do not depend too much on the particular percentage, as long as the general principle of isolation avoidance exists: whites do not want to be the only whites in the neighborhood, blacks do not want to be the only blacks in the neighborhood.

The model is illustrated in Figures 3.8(a) and (b). Figure 3.8(a) shows an initial distribution of "turtles" (light (green) and dark (red)). Here, on average, each turtle is surrounded by 50.4 percent of their own kind and 49.6 percent of the other, and each is willing to be surrounded by only 37 percent of their own kind and 63 percent others. However, in the initial set-up, which is randomized, 18.5 percent of the turtles are unhappy, because they are surrounded by fewer than 37 percent of their own kind. They want to move, and can move to a blank (unoccupied) square that makes them happy. By moving, they are changing the racial mix both in the place they left and in the place they are going to. This will make some turtles happier, and might make others unhappy. The unhappy can move again, and so on, until we have a pond (city) of happy turtles (people). Figure 3.8(b) illustrates that final equilibrium, which is considerably more organized and a lot less random than Figure 3.8(a).[d] It is also a lot more segregated.

Agents may be classified and categorized by any number of considerations, not just ancestry or race. Such factors might include age (retirement communities), wealth (exclusive or gated communities), drug use (a hippie commune), or clothing preferences (nudist colonies) and the so-called "happiness rules" are easily adjusted. The key point of the example, however, is that Schelling's

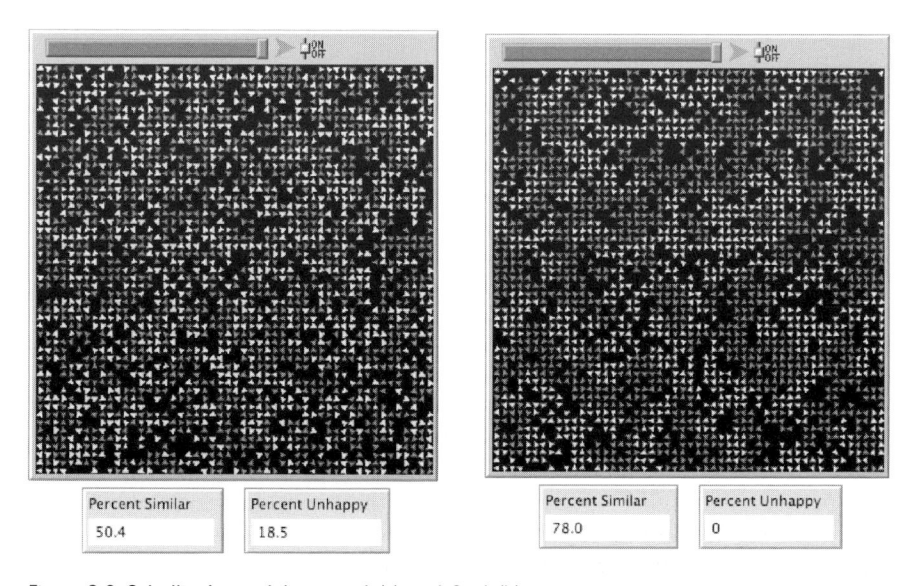

Figure 3.8 Schelling's model in initial (a) and final (b) states

model shows how the effects on residential patterns produced by small differences in household preferences provides some basic building blocks for our understanding of preferences, choices, and patterns—that is, a small degree of preference for one's neighbors to be of similar character could lead to total segregation. The outcome from a neighborhood planning standpoint is that once a cycle of separation-prejudice-discrimination-separation has begun, it has a self-sustaining momentum.

Race is certainly the predominant and most influential issue behind many of the "push" factors most people consider in deciding where to live. But it is not the only one. Neighborhood reputation, excessively high taxes or high traffic volumes often serve as primary deterrents for particular neighborhoods or even specific properties. In fact, each aversion dimension can be thought of as the reciprocal of an attraction dimension (though it is sometimes easier to understand certain dimensions as just a pull or a push factor).

Home purchases

People choose neighborhoods that reflect their preferences, lifestyle, self-image or other; through their choice they self-select the opportunities available to them (e.g., modes of travel that might be available) and constrain other choices (not having certain services available to them). This important idea of self-selection, discussed in Box 3.3, in turn affects the success of a variety of land use-transportation policies.

One way to best assess people's choices about where to live is to examine decisions from an estimated 70 percent of US residents who own their own home: how much they paid for their house. Real estate economists treat the house as a bundle of attributes, including location. It is not just a home. Rather, real estate economists consider 3,000 square feet broken into 3 bathrooms, 4 bedrooms, a 2-car garage, a fireplace, and air conditioning.

Then there are a host of location variables, for example, being five miles from downtown Minneapolis and four miles from downtown Saint Paul, a quarter mile from a small shopping center and two miles from a regional mall, down the street from a bus stop, in a municipality where taxes are declining, and in a school district where all children are above average (according to Garrison Keillor, host of the radio variety program A *Prairie Home Companion*, this last goal is easier to achieve in Minnesota). On top of these objectively assessed measures, there are more subjective or emotional attributes. For example, the US Federal National Mortgage Association (FNMA), commonly known as Fannie Mae, periodically surveys adult Americans about their housing and home ownership preferences. In a recent survey, four out of five stated that they would drive a longer distance to work if they could own, rather than rent, a home, and 73 percent responded that their ideal is a single-family detached house with a yard on all sides. Other survey responses consistently show that many consumers are unwilling to give up cul-de-sacs, large yards, and the privacy that comes with single-family detached homes set back from

Box 3.3 Self-selection

People who buy or rent homes are individuals who form and possess certain attitudes and preferences. For example, Kevin grew up in Glenview, Illinois. In terms of place and plexus, he was fortunate to live in a neighborhood where it was attractive to walk or cycle to nearly everything he needed, including Little League baseball fields, schools, grocery stores, friends' homes, trumpet lessons, and church. The affinity for walking access to such conveniences, to some extent, embodies who he is today; his preferences were formed, he liked them and they are now hard to change.

Kevin carries these tastes and values for more traditional neighborhoods with him in his search for where to live: he selects residential locations in part to match his travel preferences, including a marked preference for areas where he can drive less. He is not alone. Many residents move to a neighborhood where they can walk to the grocery store because this is an option that they prefer to have. From a perspective of land use and transportation, there are at least four reasons why planners much heed caution that relate to issues of self-selection.

1. To the extent that people who live in more traditional neighborhoods drive less, it draws into question the forces that prompt them to drive less. Is this behavior due to upbringing, the built environment, a combination of both, or something totally different? We do not know. Researchers since at least the 1970's [31] have been aiming to correlate one's neighborhood and their travel—a topic that has spurred a considerable amount of study as evidenced by the number of review articles and books [32–36]. But the bulk of available study has proven inadequate in some respects. This is because any differences that exist in travel between households with different neighborhood design should not be credited to the neighborhood design alone; the differences could be attributed to the broader preferences that triggered the choice to locate in a given neighborhood.

Thus, it is important to understand that documenting correlations between community design and travel does not necessarily support claims and claiming that community design can affect travel. Today, it is increasingly acknowledged that the majority of previous work on the subject has not adequately described the underlying or inherent preferences of individuals and households. An important point is that the effects of the two potential motivators—the neighborhood design versus personal preferences—need to be disentangled. This is no easy task and presents an emerging topic among researchers [37].

2. The above point is especially important because the relative magnitude of the independent effect of the neighborhood may become marginalized once preferences are accounted for. Put another way, efforts to alter the physical environment to induce unwilling auto-oriented households to drive less may be futile because their auto-using behavior may be a function of larger issues such as their overall preference for auto-oriented behavior. To twist the old adage, "you can take the family out of the suburbs but you can't take reliance on the Chevy Suburban out of the family."

3. In the US, the land of diverse tastes and values, these preferences run the gamut from high-rise urban dweller to the rural villager. They even include those who care less about their physical environment (e.g., their only concerns are school quality, school quality, and school quality). Is there reason to expect that auto-oriented households would locate in heavily transit-oriented neighborhoods in the first place? From Charles Tiebout's theory [38] that residents "vote with their feet," planners learned that individuals select communities to maximize their personal utility (which may or may not take into account considerations of accessibility). This case assumes the demand and supply of preferences and housing choice are in equilibrium.

4. If society is in equilibrium, it suggests a fourth point. It implies that the success of the "new urbanist" movement may be limited to the relatively small market of households who currently live in transit-oriented neighborhoods and/or those who will bring their non-auto-using preferences with them to newer neighborhoods. If there is a self-selection bias at work, policies designed to induce changes in household travel through the alteration of land uses may not have the expected or desired effect—or, their impact may be marginal. Too often, policy officials fail to recognize the role that basic attitudes and preferences play in influencing travel and residential location decisions. These are, perhaps, more important matters affecting people's travel.

Emerging research [39–41], however, suggests that some residents' dissatisfaction with their current neighborhoods stems from the fact that their neighborhood type does not match their transportation and land use preferences. This is not to suggest that Tiebout was wrong; it is just that some residents are unable to make a perfect match, thereby leading to some disequilibrium in the market. If this is the case, then there may be a demand that is "latent" (in the language of transportation studies). The magnitude of this latent demand would suggest how much success new community designs could expect to have in attracting residents. It may be that if residents had a fuller range of affordable choices, they would settle in a community of different character.

the street. [42] Residents may desire a sense of community and public gathering places, but only if the "essential" ingredients of space and privacy are provided.

These preferences, combined with relatively low mortgage interest rates, relatively lenient borrowing requirements, and a strong economy in the late 1990s, pushed home ownership to record levels. More than two-thirds of all Americans now own their own homes (including houses, townhouses and condominiums). If current trends and stated preferences are any indication of the future, this phenomenon suggests that the market for single-family detached homes will not only remain strong but continue to grow.

All of these factors—some easily measured, others difficult to measure—are used to estimate the price of a home according to the Hedonic model, from the same root as *Hedonism*, both from the Greek word *Hedonikos* meaning *pleasure*. Discerning the relative value of non-market goods using hedonic modeling techniques is a method with a long history. F. Taylor [43] used hedonic techniques to explain the price of cotton, and later applications by Kelvin Lancaster [44] and Sherman Rosen [45] standardized the method for consumer products such as houses. An extensive review of this literature [46] documents nearly 200 applications that have examined home purchases to estimate values of several home attributes including structural features (e.g., lot size, finished square footage, and number of bedrooms), internal and external features (e.g., fireplaces, air conditioning, garage spaces, and porches), natural environment features (e.g., scenic views), attributes of the neighborhood and location (e.g., crime, golf courses, and trees), public services (e.g., school and infrastructure quality), marketing, and financing.

The key point is that we obtain pleasure (or displeasure) from various attributes of our house. In modeling terms, this might be posited as:

> Price = f (House Size, Lot Size, Bedrooms, Bathrooms, Garage, Fireplace, Air-conditioning, Distance to Downtown, Distance to Shopping, Distance to Transit, Distance to School, Quality of School, etc.)

Households choose homes whose price and attributes provide the highest utility. Many of the attributes are internal to the property and the structure, but both price and attributes, which collectively comprise utility, also depend on the choices of others (both in what they do and what they are willing to pay). Thus, it is important to recognize that location attributes are the product of interacting agents, not simply something the homebuilder can supply. That Americans continue seeking the American Dream in terms of owning a single-family home (or upgrading to a bigger and better home) is not surprising. Such factors play out in two dimensions: those pulling home decisions to certain attributes (or locations) and those pushing home decisions away from other attributes (or locations). Box 3.4 details the conduct and interpretation of hedonic regression models. Hedonic models monetize people's aversion to mixing disparate populations to achieve social goals. Policies to promote racial and income integration raise a host of efficiency questions, some of which are discussed in Box 3.5.

Trends show that the housing market has produced a growing proportion of single-family units with steadily increasing floor area. In the half-century between 1950 and 2002, the average size of new houses increased 140 percent (see Figure 3.9).

Homebuying wrap up

When people, like the early pioneers of Levittown-like communities, search for homes, they want to be near some things and far from others. The choices of individuals interacting with others, each with different values and preferences, results in a varied, but still in many ways segregated, urban form. A single

Box 3.4 Hedonic regression analysis

To further demonstrate many of the concepts presented in this chapter, we predicted a hedonic model of home sale price in the Minneapolis-St. Paul metropolitan area. The Regional Multiple Listing Services of Minnesota, Inc. (RMLS) maintains home sale data from major real estate brokers throughout the state. This database includes all home sales in Anoka, Carver, Dakota, Hennepin, Ramsey, Scott, and Washington Counties in 2001, totaling 35,002 home sale purchases, including structural attributes of each home. The address of each home was mapped and married with GIS features for spatial analysis using ArcGIS.

In keeping with the prevailing literature, our model assumes a competitive market in which homebuyers are seeking a set of home attributes that can be tied to a location. Locations are defined by structural attributes (S) (including internal and external attributes), accessibility and spatial location attributes (L), and neighborhood characteristics (N). We build an equilibrium hedonic price function on these assumptions, where the market price of a home (P_h) depends on the quantities of its various attributes:

$$P_h = P(S, L, N)$$

Consistent with the bulk of house price analysis, we employed Ordinary Least Squares regression (OLS). We take the natural log of the dependent variable (home sale price) as well as several of the continuous independent variables to better replicate linear relationships (logged variables are indicated by an *ln* following the variable name). The results of our estimation are presented in Table 3.1 showing that our model successfully predicts 80 percent of the total variation in home sale price.

We break the variables into three groups and see that all structural and location variables are statistically significant and have the expected signs. In the first block of variables, we see by virtue of the positive coefficients how, on average, people are willing to pay more for homes with more bedrooms, bathrooms, square feet, fireplaces, garage stalls and that are on a larger lot. The negative coefficients suggest the price of the home is inversely related to age of the home (people prefer new properties).

In this model we included seven accessibility and other variables that are spatially dependent. For example, standardized test scores capitalize into home sale prices and are often mentioned as effective measure of perceived school quality. [47] This variable represents the sum of the average math and reading scores achieved by fifth-grade students taking the Minnesota Comprehensive Assessment. Scores associated with suburban homes are measured at the school district level, while Minneapolis and St. Paul scores are assigned to elementary school attendance areas. In each case, they are positively associated with home value. Spatially related demographic variables are derived from the 2000 United States Census. We include the percentage of people in the census tract who do not classify themselves as Caucasian and the average number of people in each household in the census tract. Census blocks with higher percentages of white people track with higher home price. Consistent with the bid-rent theory introduced earlier, we see that homes sell for more the closer they are to either of the CBDs. Proximity to a freeway has a negative effect on home value, which implies that the disamenity effects of freeways (e.g., noise, pollution) likely outweigh any accessibility benefits within particular neighborhoods. Similarly, homes that are on a busy street sell for less.

Are there other factors that predict home value? Certainly. However, data limitations preclude us from capturing every aspect that people consider. However, by virtue of predicting almost 80 percent of the variation (which is typical for hedonic regressions and also a reflection of the relatively large sample size), this application provides some convincing evidence.

Table 3.1 Hedonic regression results

Variable	Coefficient	Standard error	t-statistic
Structural attributes			
Number of bedrooms	0.033037	0.00157	21.05**
Number of bathrooms	0.079976	0.002018	39.63**
Homestead status	−0.027259	0.003481	−7.83**
Age of house (ln)	−0.092578	0.001759	−52.65**
Size of lot (square meters)	0.000003	0	21.68**
Finished square feet of floor space	0.000168	0.000002	82.14**
Number of fireplaces	0.068749	0.001768	38.89**
Number of garage stalls	0.075257	0.001268	59.37**
Accessibility and spatial attributes			
Distance to nearest major highway (meters)	0.000009	0.000001	10.35**
Distance to nearest central business district (ln)	−0.056065	0.006926	−8.09**
Home is on a busy street (y-n)	−0.033351	0.005096	−6.54**
Regional accessibility (gravity model)	−0.043551	0.008036	−6.23**
Standardized test score in school district	0.00016	0.00001	15.34**
Percent nonwhite in census tract	−0.004014	0.000183	−21.99**
Persons per household in census tract	0.038961	0.004481	8.7**
Constant	11.3148	0.079957	141.51**

Number of observations: 35,002

Adjusted R-squared: 0.80

Dependent variable: sale price of home (ln)

** Significant at $p < 0.01$

* Significant at $p < 0.05$

Box 3.5 Should affordable housing be new?

The above hedonic model shows that as houses age, they lose value. Although selected old homes such as those featured on the television program *This Old House* may appreciate, in general, all things considered, people prefer a newer house. Although "they don't build 'em like they used to," newer homes are thought to have fewer problems and require less remodeling; they generally have newer fixtures and appliances and are built to modern standards.

The US Department of Housing and Urban Development (HUD) suggests that housing is affordable when the occupant pays no more than 30 percent of their gross income for housing costs, including utilities. The belief is that families who pay more for housing will have fewer resources available for other necessities of life. There is debate about whether that is the appropriate definition, but the 30 percent threshold is widely used (e.g., the City of Minneapolis defines a unit as "affordable" if someone making less than half of the region's median income can rent a unit for less than 30 percent of their income). HUD has several programs (HOME: Title II of the Cranston-Gonzalez National Affordable Housing Act; SHOP: Section 11 of the Housing Opportunity Program Extension Act of 1996; the Homeownership Zone Initiative) that aim to build new affordable housing.

We ask the simple question: why should affordable housing be new? Affordable cars tend not to be new. They are used, and people new to the auto market (e.g., teenagers, those who have just stepped onto the economic ladder) often start with used cars, and as they become wealthier buy new ones. There are many reuse stores, selling things from clothing to books to music to athletic equipment. There is a large market in used homes, which are getting cheaper (relatively if not absolutely) every year. The desire to help people own homes (and there are good public policy reasons to support home ownership), does not require that people be given subsidies for new homes.

Imagine that instead of the government subsidizing new affordable housing in a particular subdivision, the subdivision be developed as market-rate housing (which presumably is more expensive). People would buy those homes, vacating existing homes (which are probably less expensive). Those homes would be sold to people vacating other homes, and so on. At the end of this housing exchange chain are relatively inexpensive houses that people who cannot afford market-rate new homes probably could afford. If subsidies are required, why not give them to lower-income persons? Such subsidies could be the form of vouchers or tax credits, so that the subsidies can be applied to the purchase of homes that are presently vacant, and widely distributed throughout the community, rather than concentrating a cluster of low-income individuals in a single new subdivision.

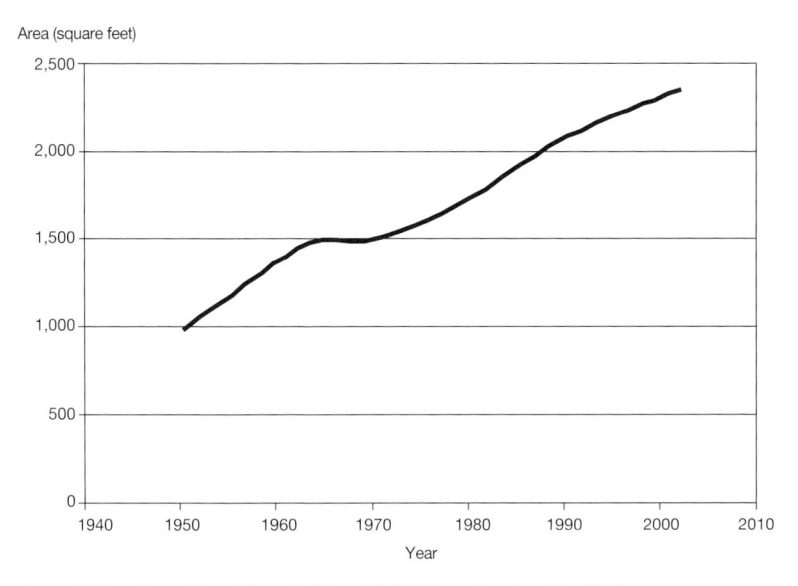

Source: National Association of Home Builders, This Old House Magazine (May, 2003)

Figure 3.9 Average size of new homes in the United States

chapter cannot do justice to five decades of literature related to residential location decisions. The choice of house and job are two location-based decisions that greatly affect the amount of travel undertaken. Choosing a home closer to work will reduce total work commuting, while choosing a job far from home will have the opposite effect. Commuting remains a critical issue in transportation as it is highly peaked (this concept will be discussed in later chapters), so the demand for transportation infrastructure (capacity) is driven by commuting choices. However, it may not be easily susceptible to policy, as people want the freedom to live and work where they choose, unfettered by regulation.

People choose their homes primarily based on the bundle of attributes of the house and its location. The hedonic model shows that these two classes of factors can largely explain the variation in the price of homes.

Realtors argue there are numerous attributes of the house to consider. And depending upon one's preferences, location relative to other places is also a complex matter. The more opportunities one can reach in less time, the more one is willing to pay. However, other factors, like not wanting to be near certain things (e.g., gravel pits, sewage treatment plants, and highway noise) or people (those of a different race, or income, or age, or clothing preference) push one away from certain places. These aversion factors explain the self-segregation that still takes place in American cities decades after legal segregation was prohibited.

66   Homebuying

In short, while this chapter provides evidence that transportation (and more specifically, accessibility) does indeed influence residential location preferences, its influence should not be oversold. In many cases, transportation policies may at best mitigate the negative effects of underlying demographic forces. Much of the rapid suburbanization of the period following World War II in the US is attributable to the rapid growth in household incomes, family formation, and other household characteristics giving rise to strong preferences for low-density housing. To attribute the bulk of suburbanization to transportation infrastructure changes (particularly federally funded highway investments) overstates their influence. Transportation is a necessary but not a sufficient factor for any development, and suburban residential subdivisions are no exception. Race is also key; a host of variables need to be accounted for. Similarly, the widely discussed recent in-migration of relatively small numbers of affluent, young households to central city areas may be more due to changing demography, particularly the increasing number of multi-worker households without children, than to any public action.

Policy makers should also recognize that a coordinated group of public policies involving accessibility, housing, and other actions will have far greater influence on residential preferences than any single-action alone. This decision process is itself shaped by a complex interplay of competitor and complementor agents (other individuals, governments, developers, businesses), who make investment and location decisions.

Notes

a   Although value is increasing, are returns increasing or decreasing? That is, is each additional unit of network size more or less valuable? That depends on whether people value the size of the network in an absolute sense, or whether they are more concerned with the relative change in the size of the network (which must diminish). The figure illustrates this point.
b   For example, households now dine out more often (the number of restaurants has increased from 491,000 in 1972 to 878,000 in 2004) [19], and child rearing now at day care more often (in 1947 only 12 percent of mothers with children under six were in the workforce, in 1997 it was 64 percent. [20]
c   This is compared to 39 percent of the respondents who said that an IRA or a 401K plan was a "safe investment with a lot of potential" and just 12 percent who felt that way about stocks.
d   Model run by authors, using model at http://ccl.northwestern.edu/netlogo/models/Segregation, http://ccl.northwestern.edu/netlogo/models/run?Segregation.637. 580.
   Note that this is a bounded world, so agents at edges and in corners have smaller numbers of potential neighbors. Note also that agents fail to count themselves when calculating the fraction of same type in their neighborhood.

References

[1]   *Time Magazine*, 'Up from the Potato Fields', July 3, 1950, LVI(1).
[2]   *New York Times*. 1952.

[3] Smith, A., *The Wealth of Nations*, New York: Modern Library, 1937.
[4] Gans, H., *The Levittowners: Life and Politics in a New Suburban Community*, New York: Columbia University Press, 1967.
[5] State Museum of Pennsylvania, *Levittown: Building the Suburban Dream*, 2003. Available at: www.fandm.edu/levittown/one/b.html (accessed January 26, 2006).
[6] Von Thünen, J.H., *The Isolated State*, vol. I, 1826, vol. II, 1850, and vol. III, 1867.
[7] Nelson, G.C., 'Introduction to the Special Issue on Spatial Analysis for Agricultural Economists', *Agricultural Economics*, 2002, vol. 27 (3): 197–200.
[8] Pivo, G., 'The Net of Mixed Beads: Suburban Office Development in Six Metropolitan Regions', *Journal of the American Planning Association*, 1990, vol. 56 (4): 457–469.
[9] Wachs, M. and Kumagi, T.G., 'Physical Accessibility as a Social Indicator', *Socio-Economic Planning Sciences*, 1973, vol. 7 (5): 437–456.
[10] Handy, S., 'Planning for Accessibility, in Theory and in Practice', in D.M. Levinson and K.J. Krizek (eds.), *Access to Destinations*, Elsevier, 2005, pp. 131–148.
[11] Handy, S. and Niemeier, D., 'Measuring Accessibility: An Exploration of Issues and Alternatives', *Environment and Planning A*, 1997, vol. 29 (7): 1175–1194.
[12] Hansen, W., 'How Accessibility Shapes Land Use', *Journal of the American Institue of Planners*, 1959, vol. 25 (2): 73–76.
[13] Krizek, K.J., 'Perspectives on Accessibility and Travel', in D.M. Levinson and K.J. Krizek (eds.), *Access to Destinations*, Elsevier, 2005, pp. 109–130.
[14] Wolfe, B., 'The Monopolization of Monopoly: The $500 Buyout', in *The San Francisco Bay Guardian*, 1976.
[15] Levinson, D.M., 'Accessibility and the Journey to Work', *Journal of Transport Geography*, 1998, vol. 6 (1): 11–21.
[16] Mills, E., *Studies in the Structure of the Urban Economy*, Baltimore, MD: Johns Hopkins University Press, 1972.
[17] Mills, E.S. and Hamilton, B.W., *Urban Economics*, 5th edition, New York: HarperCollins, 1994.
[18] Goetz, S.J., 'Migration and Local Labor Markets', in S. Loveridge (ed.), *The Web-Book of Regional Science*, Morgantown, WV: Regional Research Institute, 1999. Available at: www.rri.wvu.edu/regscweb.htm (accessed May 17, 2007).
[19] National Restaurant Association, *Industry at a Glance*, 2004. Available at: www.restaurant.org/research/ind_glance.cfm (accessed November 24, 2004).
[20] Ways and Means Committee of the United States House of Representatives, *Child Care*, 2000. Available at: www.policyalmanac.org/social_welfare/archive/child_care.shtml (accessed November 24, 2004).
[21] Gur, Y., *An Accessibility Sensitive Trip Generation Model*, Chicago, IL: The Chicago Area Transportation Study, 1971.
[22] Cervero, R. and Gorham, R., 'Commuting in Transit Versus Automobile Neighborhoods', *Journal of the American Planning Association*, 1995, vol. 61 (2): 210–225.
[23] Handy, S.L., 'Regional Versus Local Accessibility: Neotraditional Development and its Implications for Non-work Travel', *Built Environment*, 1992, vol. 18 (4): 253–267.
[24] Handy, S.L., 'Regional Versus Local Accessibility: Implications for Nonwork Travel', *Transportation Research Record*, 1993, vol. 1400: 58–66.

[25] Krizek, K.J., 'Residential Relocation and Changes in Urban Travel: Does Neighborhood-scale Urban Form Matter?' *Journal of the American Planning Association*, 2003, vol. 69 (3): 265–281.

[26] Krizek, K.J., 'Operationalizing Neighborhood Accessibility for Land Use-travel Behavior Research and Modeling', *Journal of Planning Education and Research*, 2003, vol. 22 (3): 270–287.

[27] Fannie Mae, *Understanding America's Homeownership Gaps: 2003 Fannie Mae National Housing Survey*, Washington, DC: Fannie Mae, 2004.

[28] Massey, D. and Denton, N., *American Apartheid*, Cambridge, MA: Harvard University Press, 1993.

[29] Farley, R., Schuman, H., Bianchi, S., Colasanto, D., and Hatchett, S., 'Chocolate City, Vanilla Suburbs: Will the Trend toward Racially Separate Communities Continue?' *Social Science Research*, 1978, vol. 7 (4): 319–344.

[30] Schelling, T.C., 'Dynamic Models of Segregation', *Journal of Mathematical Sociology*, 1971, vol. 1 (2): 143–186.

[31] Lansing, J.B., Marans, R.W., and Zehner, R.B., *Planned Residential Environments*, Ann Arbor, MI: Survey Research Center, Institute for Social Research. University of Michigan, 1970.

[32] Badoe, D.A. and Miller, E.J., 'Transportation-land-use Interaction: Empirical Findings in North America, and their Implications for Modeling', *Transportation Research Part D: Transport and Environment*, 2000, vol. 5 (4): 235–263.

[33] Boarnet, M.G. and Crane, R., *Travel by Design: The Influence of Urban Form on Travel*, New York: Oxford University Press, 2001.

[34] Ewing, R. and Cervero, R., 'Travel and the Built Environment: A Synthesis', *Transportation Research Record*, 2001, vol. 1780: 87–112.

[35] Kelly, E.D., 'The Transportation Land-Use Link', *Journal of Planning Literature*, 1994, vol. 9 (2): 128.

[36] NcNally, M.G. and Kulkarni, A., 'Assessment of Influence of Land Use-Transportation System on Travel Behavior', *Transportation Research Record*, 1997, vol. 1607: 105–115.

[37] Handy, S.L., Cao, X., and Mokhtarian, P.L., 'Self Selection and the Relationship between the Built Environment and Walking', *Journal of the American Planning Association*, 2006, vol. 72 (1): 55–74.

[38] Tiebout, C., 'A Pure Theory of Local Expenditures', *The Journal of Political Economy*, 1956, vol. 65 (5): 416–424.

[39] Inam, A., Levine, J., and Werbel, R., 'Production of Alternative Development in American Suburbs: Two Case Studies.' *Planning Practice and Research*, 2004, vol. 19 (2): 211–217.

[40] Schwanen, T. and Mokhtarian, P.L., 'The Extent and Determinants of Dissonance between Actual and Preferred Residential Neighborhood Type', *Environment and Planning B*, 2004, vol. 31 (5): 759–784.

[41] Schwanen, T. and Mokhtarian, P.L., 'What if You Live in the Wrong Neighborhood? The Impact of Residential Neighborhood Type Dissonance on Distance Traveled', *Transportation Research D*, 2005, vol. 10 (2): 127–151.

[42] National Association of Home Builders, *Consumer Preferences Survey*, National Association of Home Builders: Freddie Mac, 2004.

[43] Taylor, F., *Relation between Primary Market Prices and Qualities of Cotton*, U.S. Department of Agriculture, 1916.

[44] Lancaster, K.J., 'A New Approach to Consumer Theory', *The Journal of Political Economy*, 1966, vol. 74 (2): 132–157.
[45] Rosen, S., 'Hedonic Prices and Implicit Markets: Product Differentiation in Pure Competition', *The Journal of Political Economy*, 1974, vol. 82 (1): 34–55.
[46] Sirmans, G.S. and Macpherson, D.A., *The Composition of Hedonic Pricings Models: A Review of the Literature*, National Association of Realtors, National Center for Real Estate Research, 2003.
[47] Brasington, D.M., 'Which Measures of School Quality Does the Housing Market Value?' *Journal of Real Estate Research*, 1999, vol. 18 (3): 395–413.

Chapter 4

Jobseeking

> In every job that must be done, there is an element of fun. You find the fun and—SNAP—the job's a game.
>
> Mary Poppins

In April, 1994, the 454th commercial website opened its virtual doors. It aimed to link jobseekers with employers. The Monster Board, as it was then called, was envisioned by Jeff Taylor, president of recruiting firm Adion. Starting with 20 clients and 200 job openings, it grew quickly. The business was sold in 1995 to the world's largest recruitment and staffing organization, TMP Worldwide, and in 1999 was renamed Monster.com[a] and famously advertised on the Super Bowl with other dot-com stalwarts such as Pets.com. Today, Monster.com welcomes more than 12 million unique visitors per month (presumably most of them jobseekers), who search roughly one million job postings. Monster is one of a number of for-profit online sites (including CareerBuilder, owned by newspaper companies, and HotJobs. com, owned by Yahoo!), as well as more broad-based communities such as Craigslist, that have changed how many people find formal information about employment.[b] Such strategies are underscored by Careerbuilder.com, now the largest online job site in the US and also aggressively advertised during the Super Bowls in 2005, 2006, and 2007.

Like many on-line resources, these websites claim to provide a superior way of doing old things. Job search has been around since formal employment, and the classifieds have been around perhaps since the dawn of the newspaper.

But Monster and Careerbuilder also do new things. The online ads often contain an "Apply Now" link. Jobseekers can also advertise themselves by posting a résumé, to be bid on by employers. This almost certainly changes how some people find employment. Previously, it was accepted practice for jobseekers to engage in local social networks, carefully nursing personal relationships and attending meetings of professional associations. Biologists refer to such a plan for carefully managing a few relationships as a "female" strategy. Now, recent entrants to the job market increasingly rely on a "male" strategy: spreading their seed widely, applying to hundreds of jobs simultaneously, and seeing who bites. These two jobseeking strategies (dubbed male and female) are largely unrelated to the sex of the jobseeker, and many jobseekers certainly adopt a strategy associated with the opposite gender, or utilize both "male" and "female" strategies, but we posit that these strategies have significant other impacts. Does the depersonalization of jobseeking and the nationalization of job-markets affect community and how we connect with others? It might seem that as people see less value in personal networking, less networking will take place, and all of the social and community structures associated with personal networks will be similarly diminished. We return to this question later in the chapter.

How did you find your first job? Was it through classified ads, family contacts, friends, a referral service, or from industry contacts? What factors were central to you deciding to take this job? Did they include proximity to your existing home, the salary, the opportunity to advance your career or some other factors?

Like residential location decisions, multiple factors rise in importance and may even vary for a single individual depending on matters such as how long one has been in a particular career, stage in a career life cycle, lifestyle, and the presence of children. Our intent is to better explain people's employment and its relationship to land use-transportation, particularly honing in on the role geography plays in such decisions. To the chagrin of many in the fields of traditional urban economics and neoclassical labor market theory, location is but one of many factors *and* information about jobs is not equally available to everyone.

People find jobs using one or both of two processes: a formal job search (reflected in advertisements and human resources departments), and an informal approach based on contacts. However, these two decision processes are themselves shaped by the decisions of other agents (individuals, governments, developers, businesses), who make investment and location decisions. Gravity models tend not to explain why an individual chooses a particular job. Social network models have identified the long-neglected informal jobseeking process, but cannot explain formal jobseeking processes, and do not comprehensively account for physical constraints such as travel time. It is against this backdrop that we now turn to the task of describing and understanding not only how individuals find jobs, but also how they decide between competing employment offers.

Anchors away

Residential location (examined in the previous chapter) and place of employment are the two principal anchors for working-age adults, and the shapers of cities for those who depend on current employment for their livelihoods: working adults and their dependents. Retirees and the independently wealthy have less to anchor them to place, though they may anchor themselves intentionally to be near (or far) from family or friends. Home and work locations are relatively stable in duration (i.e., lasting several years), comprise the vast majority of trip origins or destinations, and represent land use policies on which many avenues of policy intervention are based.

If one were to recall (or idealize) life in the 1950s—or in a 1950s television situation comedy—one would find these matters to be considerably simpler. The primary breadwinner located an employment opportunity, signed a contract with the employer, and subsequently, a residential location decision followed. Corporate loyalty was highly respected and career longevity was not an issue.

Times have seemingly changed; everything is in flux. Or is it? The median job duration for adult males was largely stable from 1963 through the early 1990s [1]. Individuals in Montgomery County, Maryland, change employers, on average, every six years. [2] Younger individuals hold the same job for a shorter period, older individuals for a longer span. Household and individual decisions of whether to work, where to work, and what hours to work are becoming increasingly difficult to understand, much less model.

It turns out that typical travel behavior exceeds the minimum required to hold a job and live in a suitable house. The concept of excess driving was perhaps first introduced in the mid-1980s under the rubric of "wasteful commuting," as a part of research questioning the degree to which people economize their commuting costs. Several studies [3, 4] demonstrated a healthy number of metropolitan areas with excess commuting: cases where actual average commutes in cities are much longer than predicted by standard urban models (sometimes on the order of eight times greater!). Research shows this excess commuting is not caused by mismatches between the location of jobs for specific occupational groups and the location of houses suitable for members of those groups (what is termed the non-wasteful part of the commute). This finding challenges the very foundation of urban land economics: the premise that cities are shaped by people economizing on commuting. [5] Those results suggest that commuting distance and time are *not* very sensitive to variations in urban structure, and are far in excess (the wasteful component) of what can be explained by jobs-housing imbalances.

The "wasteful" part of the commute arises from other considerations than minimizing commuting resources, allowing Giuliano and others to conclude that the behavioral assumption of commuting cost minimization in the standard urban planning models is inadequate to explain commuting. Thus, large-scale changes in urban structure designed to promote jobs-housing balance (see more in Chapter 8) would have only *small* effects on commuting. This work has resurfaced as researchers are identifying instances in which travel may be pursued for fun, recreation, or health (see more in Chapter 6). But while the location of a job may not be the sole determinant of household location, one cannot conclude the opposite, that job location has no effect. The relationships between origins and destinations over networks do matter.

Gravity-based explanations

Chapter 3 implied that matters were considerably simpler in previous decades. This was especially true when it came to where to live and where to work. The conventional wisdom is that job tenure was long (on the order of 20 to 30 years) and people stayed in their homes (which tended to be close to their jobs) for the same amount of time. But how did people find these jobs in the first place? People were only willing to travel a certain distance (some more than others). These behaviors were modeled using traditional gravity models.

The key feature of gravity models is that distance (or travel time) matters. Employment search is constrained by how far jobseekers are willing to travel on the days of the week that they work. Constraints vary by individual and by profession. In seeking new employment opportunities, most people are constrained to the metropolitan area (or town) in which they live (there is always the exception of the investment banker who may be transferred or may switch metropolitan areas depending on where the next great opportunity arises). But even within metropolitan areas, most people have a fairly rigid timeframe for tolerable work-related travel (again, the exception being the long-distance commuters, often found in larger metropolitan areas, who are willing to take on multiple hour one-way commutes).[c]

Using the logic of gravity models, we can model the likelihood that someone who lives in place i will work in place j. One of the more straightforward applications is called the trip distribution or destination choice model. The interaction between places is inversely proportional to travel cost (e.g., time, distance, money), and may use the same function $f(C_{ij})$ that was introduced in Chapter 3 when considering accessibility. See Boxes 4.1 and 4.2 for a more technical description of the evolution of gravity models and their application.

Choosing a house depends in part on how close you are to jobs. Which job you take depends on how close that job is to where you live. The cost of travel (C_{ij}) depends on how many people are traveling, which depends on how many people live at i and are going to j. Disentangling this decision process is not simple but in the interest of parsimony, we need to make some assumptions.

Figure 4.1 presents a visualization of the model. There are three balls representing towns of different sizes (A the largest, C the smallest), and three links connecting them; the width of these links represents the amount of traffic generated between the balls, while the length of the links is proportional to the generalized cost of travel. Assume that the size of the balls is proportional to both the number of jobs and the number of workers, a topic we will

Box 4.1 History of gravity models

Isaac Newton asserted, in his masterpiece *Philosophiae Naturalis Principia Mathematica* (1687), that "Every object in the Universe attracts every other object with a force directed along the line of centers for the two objects that is proportional to the product of their masses and inversely proportional to the square of the separation between the two objects."*

Mathematically, this principle can be expressed as:

$$F = \frac{G m_1 m_2}{r_{12}^2}$$

where:
F = gravitational force between two objects
m_1, m_2 = mass of object 1, 2
r_{12} = distance between objects 1 and 2
G = universal constant of gravitation

In the late nineteenth century, Ernest Ravenstein [6–8] developed a similar idea in the context of the social sciences. Ravenstein proposed seven laws governing migration, which held that:

1. Most migrants only proceed a short distance, and toward centers of absorption.
2. As migrants move toward absorption centers, they leave "gaps" that are filled up by migrants from more remote districts, creating migration flows that reach to "the most remote corner of the kingdom."
3. The process of dispersion is inverse to that of absorption.
4. Each main current of migration produces a compensating counter-current.
5. Migrants proceeding long distances generally go by preference to one of the great centers of commerce or industry.
6. The natives of towns are less migratory than those of the rural parts of the country.
7. Females are more migratory than males. [9]

Ravenstein's theories were subsequently extended by several other researchers. Isard [10] notes that H.C. Carey's *Principles of Social Science* (1858–1859) observed a gravitational force in social phenomena that was in direct ratio to mass and inverse to distance. Lill [11] applied the gravity model to railway traffic. In a 1931 book, William J. Reilly developed a "Law of Retail Gravitation" [12] suggesting that:

$$B = \frac{r_{12}}{1 + \sqrt{\frac{P_1}{P_2}}}$$

where:
    B = the distance from the city center to a "breaking point" or "boundary point" between two market areas
    P_1, P_2 = population of areas 1 and 2

Reilly posited that individuals at point B were indifferent to shopping in Area 1 or Area 2. One might object that this formulation ignores many factors, such as incomes, quality and type of stores, etc. Although these objections are valid, Reilly's theory was intended simply to estimate the size of market areas, and as an average it is probably a reasonable first approximation.

Shopping is but one of many interactions that involve different places. John Q. Stewart developed the notion of demographic force (F) between places, and this demographic force equation forms the basis of the gravity model used in many transportation planning models [13]:

$$F = \frac{P_i P_j}{r_{ij}^2}$$

* In addition to inventing Calculus and major contributions to physics, Isaac Newton was important as a member of Parliament from 1689–1690 and Master of the Mint in Queen Anne's reign, for which he was knighted. We are most interested here not in what he did, but how his work was adapted in the analysis of place and plexus.

consider in more depth when we talk about jobs-housing balance in Chapter 8. Though A is closer to C, because B is larger, there is more traffic between A and B (i.e., more people who live in A work in B). The shorter the distance between two objects, and the greater their mass, the greater the gravitational pull between them.

Figure 4.2 illustrates the idea that observed journey-to-work time distributions result from combining willingness to travel and the attraction of greater opportunity. Employment opportunities increase roughly with the square of the distance (or time if speed is uniform) traveled from a point, assuming jobs are distributed uniformly and the region goes on forever. Although not strictly true, this is consistent with the idea that the number of jobs available will be greater in a ten-kilometer radius than a one-kilometer radius in any metropolitan area. In a spatial analogy, the area of a circle (of job opportunities) increases with the square of the radius (trip distance). However, as travel time increases, commuters are less willing to travel—the classic friction factor of gravity models. Interaction declines as the cost of interaction (e.g., distance, travel time, dollars) increases.

This gravity model suggests several things. First, as city size increases, mean commuting time increases (we have a left-truncated distribution, so as the right branch extends outward, the average must increase). However, the increase

Box 4.2 Performing gravity model calculations

Ever since Alan Voorhees first applied the gravity model to address problems of urban transportation planning, [14] engineers, economists, and planners alike have had a love-hate relationship with it. Nothwithstanding early and sharp criticisms, [15] it has withstood the test of time and will likely to continue to do so in the future.

Economists critical of the gravity model approach suggest that it is an ad hoc tactic and largely bereft of supporting theory. But think about it another way—what is $f(C_{ij})$ but a demand curve? As cost (time) increases, willingness to pursue (purchase) a trip between two points (a market) declines. Behaviorist critics argue that the relationships described by gravity models are aggregate relationships, which can match population totals but cannot predict what an individual will do. All is true. It is important to note that although gravity models can predict on average the number of trips that are five minutes or ten minutes long, or the number of trips between two large areas, they exhibit relatively poor performance in matching the number of trips between small areas.

Transportation modelers have, since the 1950s, used traffic zones to simplify the analysis of large regions. All trips generated in a zone are imagined to originate at the centroid (a point near the center) of that zone. In modeling, more zones yield more accurate results, but at the price of requiring more input data and more computational time.

For example, Traffic Zone E, located in the eastern portion of a mid-size metropolitan area that is divided into 1,000 zones of 1,000 persons each, may send 0, 1, 2, or 3 people to Traffic Zone W in the western suburbs. A gravity model will predict a continuous number (e.g., 0.47) of trips from E to W. But even that (very precise but not very accurate) number won't reveal which person in E goes to W.

The reason for the inaccuracy is easier to see when you think about the process from the modeler's point of view. When building predictive trip distribution models as part of a regional transportation forecasting process (four-step models), modelers are typically trying to estimate a trip table (see Table 4.2). In our fictional mid-size metropolis with 1,000 zones, there are 1,000 × 1,000 zonal interactions. The mathematically observant will note that 1,000 × 1,000 = 1,000,000 zonal interactions. So we would need to know not only the number of jobs and the number of workers in each zone, but also the generalized costs (travel time) for all 1,000,000 cells in a matrix of zonal interactions. Fortunately, performing a million calculations is now easy for even desktop personal computers, but when these models were first developed in the 1950s, they were extremely demanding, cutting-edge applications requiring the use of large mainframe computers. While computing hardware has improved significantly, the models—at least those widely used in practice—have not.

Lest you think there are merely a million calculations, note that this type of calculation must be repeated for all different types of trip purpose (home to work, home to shop, home to school, home to other, work to shop ...), for every mode (drive alone, carpool, take transit, walk ...). Other calculations are required to determine the free-flow and congested travel times on all routes connecting every origin and destination.

Mathematically, gravity models often take the following form:

$$T_{ij} = K_i K_j \frac{T_i T_j}{f(C_{ij})}$$

$$\sum_j T_{ij} = T_i$$

$$\sum_i T_{ij} = T_j$$

$$K_i = \frac{1}{\sum_j K_j T_j f(C_{ij})}$$

$$K_j = \frac{1}{\sum_i K_i T_i f(C_{ij})}$$

where

T_{ij}	=	trips between origin i and destination j
T_i	=	trips originating at i (for example, workers)
T_j	=	trips destined for j (for example, jobs)
f	=	distance decay factor, as in the accessibility model
C_{ij}	=	generalized travel cost between i and j
K_i, K_j	=	balancing factors solved iteratively

The balancing factors are required because this model is of a type referred to in modeling jargon as a "doubly constrained" model—it is designed to guarantee that the total number of trips from the origin zone (and to the destination zone) is equal to the total number of trips for that zone forecast at the trip generation/frequency stage.

Table 4.1 Illustrative trip table

Origin\Destination	1	2	3	Z
1	T_{11}	T_{12}	T_{13}	T_{1Z}
2	T_{21}			
3	T_{31}			
Z	T_{Z1}			T_{ZZ}

Where: T_{ij} = Trips from origin "i" to destination "j".

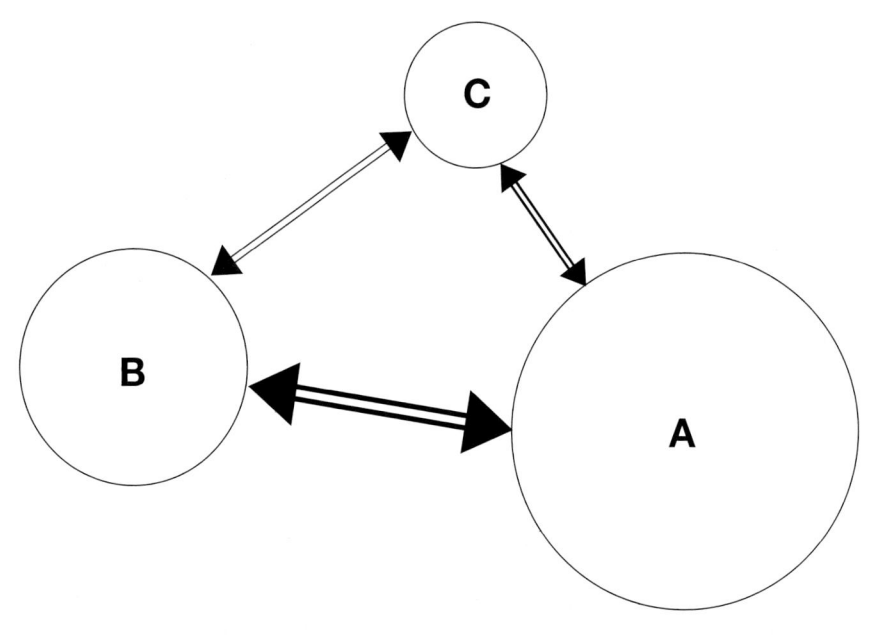

Figure 4.1 Illustration of a gravity model

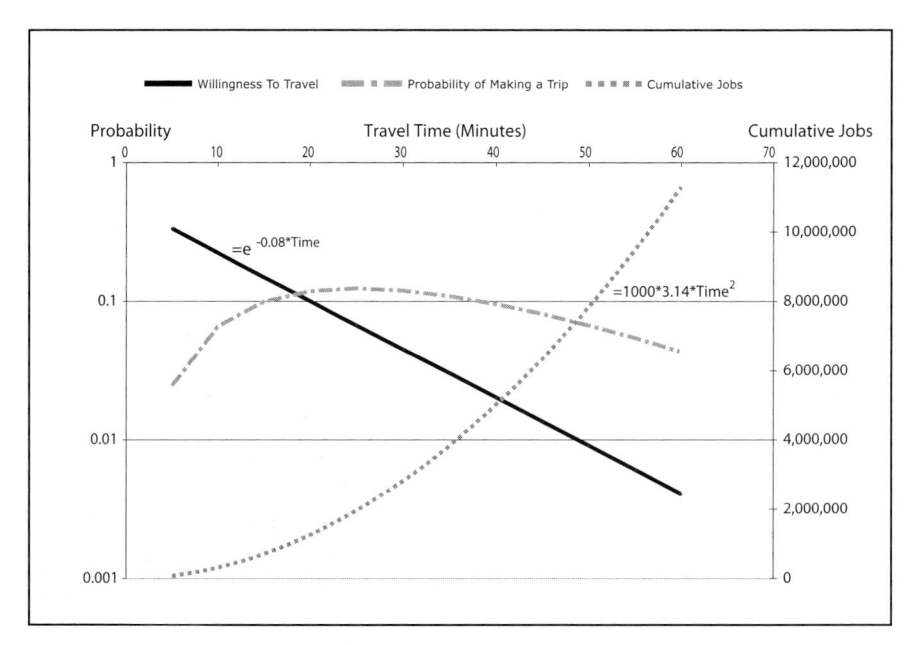

Figure 4.2 Gravity model of relationship between willingness to travel and available opportunities

is non-linear, so as cities get larger, additions have a smaller and smaller effect on travel time. This is illustrated in Figure 4.3. The structure of gravity models implies diminishing marginal returns to job opportunities at the edge, since each additional job is less and less likely to be taken and thus less likely to increase travel time.

Second, these models are largely independent of density—except to the extent that density changes network speed. A uniform increase in density increases the opportunities within each time band (e.g., five minutes, ten minutes) proportionately, and thus does not change the distribution of travel times. Third, if preferences shift, mean travel time will change inward or outward. Fourth, if congestion rises, more opportunities will be farther away in terms of travel time, and fewer nearby—implying that average commuting time will rise.[d] Box 4.3 illustrates an aggregate model of average travel time.

Social networks

Theories of behavior based on gravity models assume that geography plays a prominent role in predicting who interacts with whom and how frequently. Clearly there are macro-structures (e.g., travel time is an important constraint, whose size indicates the market you are willing to search given a residence) that play important roles in such factors. But does access along a transportation network determine such decisions? No. There are other networks, most prominently social networks, that come into play in finding a job.

For example, in 1973 a sociologist from Stanford University, Mark Granovetter, published what later became a classic paper, "The Strength of Weak Ties," [16] which analyzed how people found jobs, observing that most people (roughly 56 percent) found their job through a personal contact. He further documents this in his 1974 book *Getting a Job*, [17] describing how people often learn about job prospects through personal relationships. Such networks serve as information channels to the benefit of potential employees and employers, not only because they are conduits of information, but also because the information that face-to-face relationships provide is richer in content than what is available through impersonal mechanisms. The most surprising results were that only 16 percent of people found jobs through a contact they saw "often" (close friend or family) and 84 percent got their job through a contact they saw "occasionally" or "rarely."

Granovetter suggests that a key reason for these results is that the people you know well are likely to also know about the same opportunities. It is the people who you interact with infrequently—who operate in different social and professional circles—who are likely to know about different opportunities, and hence be the most valuable contacts for finding a job. This research highlighted the importance of weak ties in linking people within larger social networks. Hanson and Pratt [18] further emphasized the role of informal networks in job-finding, especially among working-class women, giving some

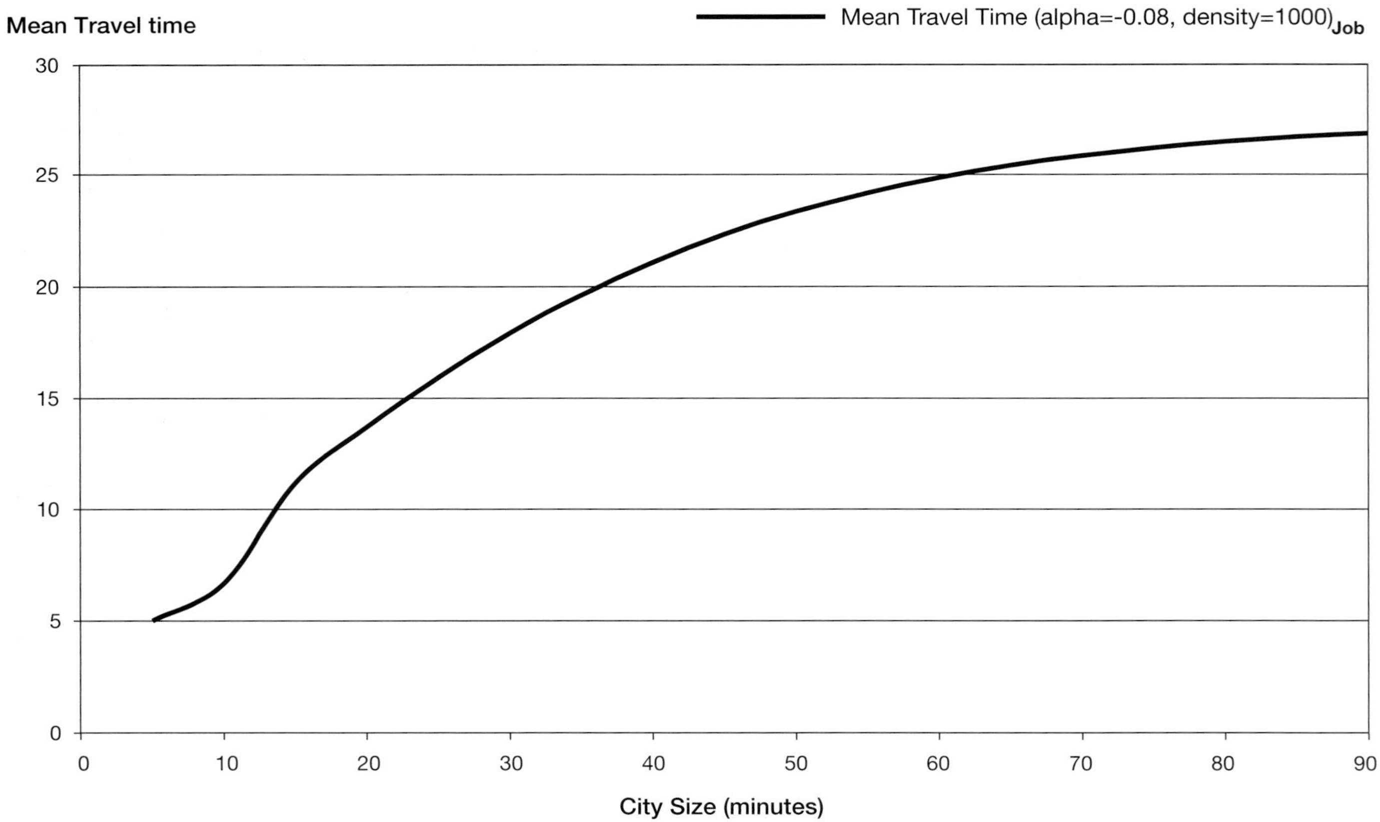

Figure 4.3 Relationship of travel time and city size

Box 4.3 Modeling aggregate home to work travel times in the aggregate

So, what affects the travel time between home and work? There are a number of approaches for estimating such. One would be to apply regression analysis to predict mean metropolitan commuting time from the 2000 Census.[a] The Census Bureau defines Metropolitan Statistical Areas (MSAs), Primary Metropolitan Statistical Areas (PMSAs), and Consolidated Metropolitan Statistical Areas (CMSAs). There are 65 metropolitan areas in the US; 39 are MSAs and 26 are PMSAs (we chose not to use any CMSAs because the congestion data from the Texas Transportation Institute correspond only to MSAs and PMSAs).

The model is given by:

$$T = f(C, P, D, A, I)$$

where:

T = mean metropolitan journey-to-work time,
C = congestion index from TTI Urban Mobility Indicators,[b]
P = population,
D = population density,[c]
A = area, and
I = median household income.

Table 4.4 shows the results of several regressions: linear, log-linear, and log-log. The models differ only slightly, the linear model providing the best fit, with an adjusted R-Square of 0.71. All variables are significant at the 0.05 level or better except Population and Area. Despite what some may consider a gross aggregation—that is, looking only at metropolitan areas—the relatively simple model has significant explanatory power. The R-Square value indicates that only about 30 percent of the variability in mean metropolitan commuting time remains to be explained by excluded factors.

The constant term is the greatest contributor to commuting time and is the most significant of the independent variables. These observations point to the existence of a large underlying determinant of commuting time that is largely independent of the metropolitan characteristics and congestion in the sample. Redmond and Mokhtarian posit a positive utility to commutes, suggesting why commutes are higher than are minimally necessary to locate everyone relative to their workplace. [19] This constant term may suggest the minimum temporal separation between home and work, related both to location constraints and to positive utility. In rural areas, and where congestion is non-existent, limited job opportunities may affect the constant term.

The congestion index has both a substantial and significant effect on commuting time. The next largest contributor is income, though this is not extremely significant. This observation supports, and is supported by, other research suggesting that wealth translates to more time in the car. Population size and population density are both positive contributors to mean travel time to work. Again, this finding is supported by other research that finds higher metropolitan density does not necessarily reduce travel time for commuters. Last, area is the smallest and least significant predictor of mean commute time.

[a] The travel time distributions for each metropolitan area suggest a Poisson distribution. The Poisson model is common in transportation, and using it to estimate travel time offers two advantages. The first is that it is appropriate for categorical data, which is what the Census Bureau provides, as discussed in the second note The second advantage is that estimation is straightforward because just one parameter describes the Poisson distribution. One dependent variable is convenient and greatly simplifies regression. The Poisson distribution maximum likelihood parameter is estimated for every metropolitan area. Figure 7 depicts typical commute times and Poisson estimations for the first two cities in the sample. Table 4.2 shows the range of estimated parameters for all metropolitan areas in this study. All Chi-Square goodness of fit statistics are significant at the 0.01 level, so we do not reject the Poisson model.

[b] The Texas Transportation Institute's 2001 Urban Mobility Report covers 68 US cities with populations above 100,000. They base their analysis chiefly on the Federal Highway Administration's Highway Performance Monitoring System. We use their Roadway Congestion Index (RCI), which provides a general measure of vehicle travel relative to roadway capacity on major roadways.

The Census Bureau provides journey-to-work data for 330 MSAs and PMSAs with populations over 100,000. However, three metropolitan areas from the Urban Mobility Report are not included in the journey-to-work Census statistics. These three are Laredo, TX, Louisville, KY, and Norfolk-Virginia Beach, VA-NC. The data set therefore contains 65 observations.

TTI also defines a Travel Rate Index (TRI) which "shows the additional time required to complete a trip during congested times versus other times of the day." For our purposes, the TRI is too sensitive to the relationship between population or number of trips and the available roadway capacity. A small city and a large city may have similar TRIs if the relative availability of roadway capacity is similar. The Congestion Index is a better measure of general mobility for the purpose of this investigation. In addition, the journey-to-work generally takes place around a peak travel period, and the RCI does represent the travel conditions during these times. Nonetheless, a future study may find interesting results while incorporating the TRI. Road network expanse, or roadway density, is available from the Texas Transportation Institute (TTI) but is not included because TTI has shown it to be a very poor predictor of congestion and travel time.

[c] One might think that $D = P/A$. However the variable Density is the population divided by land area only (the same way the Census Bureau reports it); the Area includes water area, so it's a unique measure of expanse. Thus for some cities, it makes little difference, but for lake cities, dividing P by A gives a value much less than their reported D. The reason we did it this way is because two cities could have the same population and the same density, but the metropolis with a lake or bay in the middle of it (e.g., Seattle or San Francisco) will have a bigger area and therefore longer commutes. Observation of the descriptive statistics (Table 4.2) suggest heteroscedasticity, so corrected robust estimates using the Huber-White estimator of variance are presented.

Table 4.2 Summary statistics and regression results to predict mean commuting time

Variable	Mean	Std. Dev.	Minimum	Maximum	Linear	Log-Linear	Log-Log
Mean Commute Time (minutes)	25.7	3.7	19.1	41.0			
Constant					14.40 (10.10)	2.78 (48.65)	1.17 (1.80)
Congestion (C)	1.00	0.17	0.69	1.52	6.089 (2.74)	0.250 (2.84)	0.17 (1.66)
Population (P)	2,088,800	1,922,507	291,288	9,519,338	4.64E-07 (1.62)	1.79E-08 (4.69)	0.06 (1.05)
Population Density (persons/mi² land) (D)	730.6	1,046.8	39.7	8,158.7	1.54E-03 (5.87)	4.51E-05 (4.69)	0.05 (0.95)
Area (mi²)(land and water) (A)	4,490.4	5,760.6	751.4	39,719.1	9.13E-05 (1.73)	3.38E-06 (1.80)	0.02 (0.39)
Median Income ($) (I)	42,714	8,313	23,992	76,752	(1.90)	6.38E-05 (2.21)	2.91E-06 0.07 (1.28)
Adj. R-Square					0.715	0.674	0.63

Note: (number in parenthesis indicates t-statistic using Huber-White Robust Estimator). The above regression performed with Peter Rafferty.

grounds for the classification of networking as a "female strategy," as suggested in the introduction to this chapter.

Granovetter's theory has much to commend it. Still, is the finding really so surprising? One might ask how often you see people who you see often? How often do you see people who you see rarely? The terms are vague.

Perhaps we can illustrate with an example. First, how many people do you see often? This number must be less than the number you see occasionally or rarely, since you know far fewer people than you don't know, and you know well far fewer people than you know incidentally. If I have 100 encounters a week with people for 10 minutes each, only four of whom I see often, say averaging four times a week, I am spending 160 minutes a week with people I see often, and 840 minutes a week with people I see occasionally or rarely (less than once a week). In other words, I am probably getting 16 percent of my information from people I see often and 84 percent from others. These numbers were made up to illustrate a point. The observations supporting the theory that many jobs come through weak ties make sense, but it is difficult to establish to what extent this effect is due to different information being exchanged across weak and strong links, and to what extent it simply reflects the fact that we receive a greater volume of information from our large number of encounters with people who we do not consider close friends or family members.

The preceding argument is intended to illuminate the important role of social networks and, to a certain extent, to highlight the complex but indirect role geography plays.

In 1967, psychologist Stanley Milgram sent letters to 60 individuals in Wichita, Kansas. He asked them to forward the letters to a complete stranger, the wife of a divinity student living in Cambridge, Massachusetts, a small college town outside Boston. Rather than simply re-mailing the letters, the study participants were to pass the letters by hand to personal acquaintances who, they hoped, would have a better chance of reaching the stranger. Milgram published his findings in 1967 [20] stating that one of letters reached the target in just four days. However, what was not reported widely is that of the 60 letters, only three (or 5 percent) actually reached their target, and legend has it that two of the three chains went through the same people.

Later studies suggest that most of the social networks that were identified had a hub-and-spoke structure, so a few hubs would connect most people. Chains that were completed had, on average, six intermediaries, giving rise to the "six degrees of separation" claim: that all people are separated from everyone else by only six other people; that is, they know someone, who knows someone, who knows someone, who knows someone, who knows someone, who knows someone, who knows the person of interest. The formation and power of social networks is described by Watts. [21] A surprising illustration of less than six degrees of separation is shown in Box 4.4.

Social networks—the relationships between individuals, their kith and kin—can be mapped similarly to transportation networks. Think of yourself as a

Jobseeking 85

Box 4.4 Social networks, Bush, and Hinckley

The introduction to Chapter 2 discussed the March 30, 1981 attempt to assassinate then-President Ronald Reagan. In one of the stranger coincidences (or conspiracies, if your mind works that way) associated with this story, it turns out that Scott Hinckley, brother of would-be assassin John Hinckley Jr., was to have dined with Neil Bush and his wife on March 31, 1981. Neil Bush is the son of then-Vice President George H.W. Bush, and brother of future President George W. Bush. Scott Hinckley was Vice President of his father's firm Vanderbilt Energy, while Neil Bush was a "land man" for Amoco Petroleum (now a part of BP), and had been campaign manager for his brother's unsuccessful 1978 Congressional campaign. At the time both Neil and George W. Bush lived in Lubbock, Texas, as did John Hinckley Jr., who was attending Texas Tech; however, there is no evidence they were ever in contact. In short, George H.W. Bush, who would have received a great promotion had the assassination succeeded, was three links away (and perhaps fewer) from the assassin, as shown in Figure 4.4.

node. Draw everyone you deal with (say weekly) as nodes around you and draw links between yourself and these other nodes. Connect those nodes and you should have a star shape. Then, consider everyone those individuals deal with on a weekly basis, and draw them as nodes. The process quickly becomes messy, with lots of overlapping lines; in Figure 4.5, we focused on the relationships with co-workers, for example. The co-worker shares contacts with you (you have a common friend and both know your boss), but he also has a wife

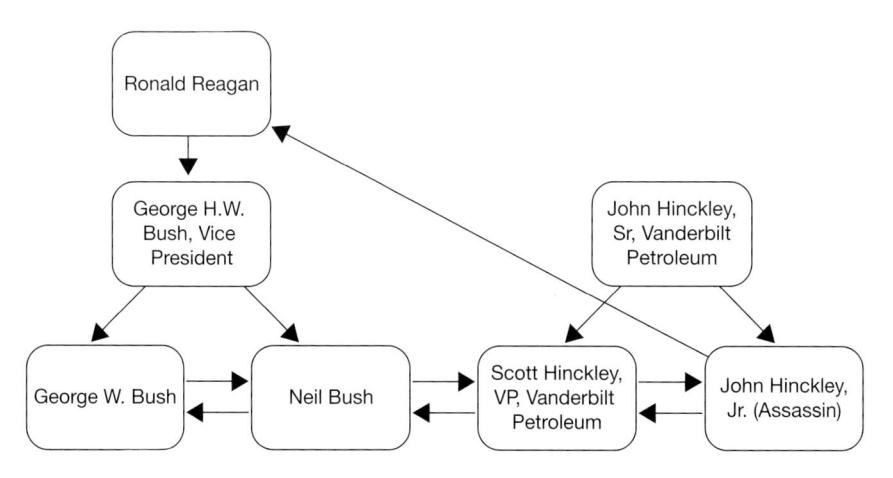

Figure 4.4 Social network connecting the Reagan, Bush, and Hinckley families

and colleagues with whom he deals regularly but with whom you have little or no contact. Maybe it is those individuals who provide the informal knowledge about job openings (or good restaurants, or potential dates) that constitute weak ties. You meet them occasionally (at parties at your co-worker's house), but never know them well enough to invite to your own house.

These social networks are created and maintained through communication, but many forms of communication require proximity (e.g., attending a party). Thus, transportation and geography are links that tie individuals to their social networks and activities. Accordingly, a lack of transportation contributes to isolation, and constrains access to social networks.

Social capital is earned by participating in social networks. Robert Putnam's *Bowling Alone* [22] contends that social capital is declining. He measures capital in a number of ways, one of which is involvement with formal organizations (such as clubs or lodges). He sets out to ask four questions:

1. What has been happening to civic engagement and social connectedness over the past three decades?
2. Why has this happened?
3. So what? What are the consequences of a decline in social capital?
4. What can we do about it?

He concludes, with some sadness, that traditional social clubs (e.g., Lions, Rotary, Elks) are declining for a variety of reasons, including the increasing

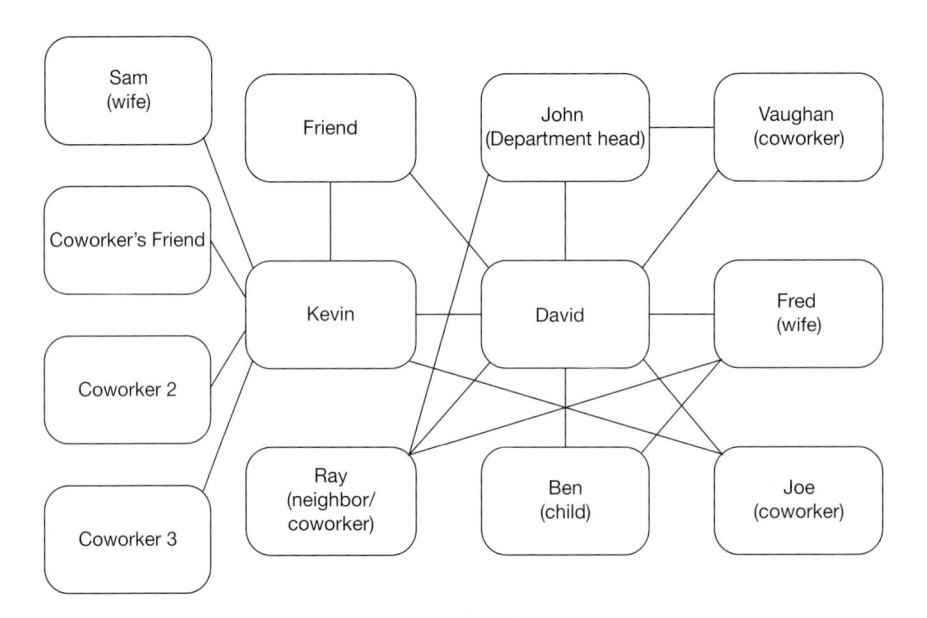

Figure 4.5 A simple example of an individual's social network

complexity of life and an ever-greater focus on work; these factors, in turn, are causing a deterioration in our civic lives.

But if the nature of social networks is changing and becoming less formal (or perhaps more work-centered), Putnam's measures may be misleading. His ranking of the social capital in US states (Table 4.3) is dominated by northern-tier states. People in North Dakota, which has among the lowest population densities in the country and has been suffering depopulation, have the highest level of social capital according to his measures. What does this mean? Is old formal social capital an indicator of lack of opportunities for informal social interaction? When it is cold, it is hard to meet your neighbors except when shoveling snow or drinking draughts from the bar at the Odd Fellows Lodge. Is it a cultural phenomenon (are descendants of northern Europeans inherently joiners)? Is it a measure of decline, or failure to advance?

Social networks may be formalized or informal, and conclusions are not so easy to draw. The fluidity of job markets in places such as California's Silicon Valley, [23] where we suspect few join the Moose Lodge, clearly indicates that social networks are plentiful in informal areas. Although we cannot comment too much on quality of life, casual observations suggest that more people would rather live in California than North Dakota.

On the auction block?

Consider for a moment similarities between considering what job to take and that action pursued in an auction. Internet auction sites, such as eBay (founded in 1995), are becoming a familiar way to transact business. A seller posts a good, along with a description of its attributes; buyers bid in competition with other buyers, so each buyer bids a higher price. Ultimately, time runs out and the highest bid is accepted. But not only is there competition among

Table 4.3 Social capital scores, by state

State	Score
North Dakota	1.712
South Dakota	1.693
Vermont	1.424
Minnesota	1.325
Montana	1.296
Nebraska	1.157
Iowa	0.988
New Hampshire	0.779
Wyoming	0.6710
Washington	0.6511
Wisconsin	0.5912
Oregon	0.57

(Source: Putnam 2000)

buyers for a good, there are multiple similar (if not identical) goods competing for attention. This competition among sellers can be thought of as increasing opportunities, while other buyers bid up the price, and thereby decrease the chances to buy at a preferred or reservation price. We can look at it the other way as well: competition among buyers increases opportunities for sellers, but other sellers decrease their chances. However, from a buyer's perspective, the presence of other buyers does create the market in the first place, providing many opportunities that were not available before the advent of eBay. Thus, buyers are mostly complementary to other buyers (and sellers to other sellers) in that they create the network necessary for the existence of the marketplace. This competitive behavior within the market is a classic push situation; the complemenarity is a classic network/agglomeration effect.

While there is no auctioneer speaking faster than one can understand, formal job searches have been traditionally conducted as auctions. A jobseeker posts her résumé to a number of firms who have advertised positions and attends several interviews. A firm interviews several candidates. The firm makes an offer; the jobseeker evaluates offers and goes with the best one (usually, but not always, the most lucrative). If the jobseeker receives no (adequate) offers, the process is repeated. If the firm receives no (adequate) applications, or its offer is rejected, the job remains open and the process repeats. A subset of game theory, known as auction theory, [24] may be used to describe such interactions.

All of the market equilibration produces some interesting outcomes. People have different lengths of commute depending on their job type and salary. Certainly, jobs requiring a higher level of skill are scarcer, but they also pay more. So why do people who earn more money travel more? Box 4.5 asks this question.

Jobseeking wrap up

The job may or may not be a game, as Mary Poppins, P.L. Travers' magical English nanny, declares in the opening epigraph. But *finding* a job surely is a game (and perhaps more than one at that). In one game there are two actors, potential employee and potential employer, who have strategies. The employer can offer or not offer, the jobseeker can accept or reject an offer. Of course it is more complex than that; the jobseeker can counter-offer, and the employer can accept/reject/counter-offer again. There is also an aspect of game-playing in whether to offer an interview, and whether to take the interview if offered. There is a further game centered on positioning for later recruitment. Individuals not in the job market still want to keep future options open. Not only is there a game structure, there is a complex pay-off function. Workers are interested in salary, benefits, location relative to current housing, relocation costs and packages if available or necessary, and so on. Employers care about salary as well as how much can employees do, and how well can they do it. We argue that social plexus is as important as transportation plexus in understanding why people work where they do.

Box 4.5 Do the rich travel more?

Most travel behavior analyses of the subject provide evidence that suggests the rich travel more than the poor. This evidence includes both private vehicle trips and total trips (though the poor ride transit more than the rich). In Minnesota's Twin Cities, trip rates per household increase from 6.0 trips per day for households with less than $15,000 to 16.1 for households with incomes between $75,000 and $100,000. [25] It should be noted that this result does not control for household size, and larger households probably have more workers and more income, as well as more travel. A more rigorous analysis [26] shows that individuals with household incomes of $100,000 have an annual vehicle kilometers traveled (VKT) on the order of 40,000 while those with a $20,000 household income have an annual VKT of 21,000, which is explained by associating VKT more closely with number of workers and number of vehicles. The rich make longer trips than the poor. Although the rich are more likely to have jobs than the poor (as jobs often determine income), it is true that rich workers travel more than poor workers. Whether this effect is great or small depends on how you look at it. Barnes and Davis [27] suggest the income effect on total travel time of workers is weak: 0.2 minutes for $10,000 in additional income. This effect also applies to non-commuting travel. Rajamani et al. [28] show that income is positively associated with driving alone for income groups for non-work trips. The evidence is certainly not without some cautions, in particular, the effects, although statistically significant, are not always huge.

Why should a positive correlation exist between income and travel (total time, number of trips, etc.)? The rich have a higher value of time, which would imply they would pay extra to save time. The rich can certainly afford to live closer in than they currently do; after all, poor people live close to the city center. This would reduce their daily work commute times. But the wealthy live in the suburbs more often. The poor have less money, which implies they would travel farther for a better job and spend more time comparison shopping. Yet none of these factors seems sufficient to change the facts noted in the preceding paragraph.

A number of reasons have been proposed to explain these observations.

First, we could look at mobility. The wealthy are more likely to own vehicles, and to own more vehicles per person than the poor, increasing overall mobility and thus lowering the time cost of travel. Murakami and Young, citing evidence from the 1995 NPTS and other sources, note that low income groups (defined in Table 4.4) own fewer vehicles (0.7 per adult on average, compared with a US average of about one vehicle per adult), in part because discretionary income is spent on food and shelter. [29] The poor family's car is also older (and presumably less reliable).

Second, we could look at land use. The hedonic model shows that newer homes (and homes that have the characteristics of newer homes) are more expensive, and thus are more likely to be owned by wealthier individuals. People will pay more for new construction (and the rich can pay more than the poor); new construction tends to be in undeveloped land, which tends to be in the suburbs.

Third, we could consider differing preferences. Perhaps the rich (relatively) prefer space to time more than lower income groups. If we believe the gravity model, the evidence implies that wealthier individuals have a higher tolerance for longer trips, suggesting a smaller α in their friction factor: $f(C_{ij}) = (e^{-\alpha t})$. So, perhaps wealthy individuals enjoy their travel more than poor individuals, because of nicer vehicles and the like, giving some support to the "positive utility of travel" hypothesis of Redmond and Mohktarian.

Finally, we might be able to explain increased non-work travel as a result of motive and opportunity. By opportunity, we mean that the rich have additional resources to substitute out-of-home goods and services for in-home. By motive, consider scarcity of time. Wealthier people on average work longer hours than less wealthy people.

Table 4.4 Definition of "low-income" households, 1995 NPTS

Number of persons (regardless of age)	Household income
1–2 persons	Under $10,000
3–4 persons	Under $20,000
5+ persons	Under $25,000

Source: Murakami and Young (1997)

A variety of forces constrain individuals in the jobs they seek or take. Arriving at a consistent taxonomy is no easy task. Of course the reasons people *do not* take jobs (should they be available) are just as diverse as the reasons they do take jobs. Following the framework of constraints described in Chapter 2, such factors include (but are not limited to) the following:

- Responsibility: Dual worker households: Some 70 percent of all households contain multiple workers, [30] begging the question of how multiple individuals balance multiple demands (from either a residential or employment location perspective). For example, an exceptional job held by one family member may sway a residential location towards a community or outlying area that may make the commute to certain employment for the second adult intolerable.
- Money: Lack of free parking: In Downs' typology of the American vision, one tenet is choosing to work in attractively landscaped environs with adequate parking. This implies that not having parking both available and free is a deterrent.

- Geography: Spatial mismatch: Transportation-disadvantaged individuals do not have convenient access to certain employment opportunities (due to the lack of a personal vehicle) deterring them from taking certain jobs. Spatial mismatch theory, proposed by Kain (1968) [31] and reviewed in Ihlanfeldt and Sjoquist (1998), [32] addresses this issue. We discuss it in Chapter 8.

Work is found formally through classic information networks (newspaper classifieds, signs in the window, online job boards), and informally through social networks (friends of friends telling you about job opportunities). The finding of work through these "ephemeral" networks results in travel between home and work that occurs on the physical transportation network. Work trips are not the majority of trips, and time spent at work is less than one-fourth of all time even for workers, yet work travel still dominates urban networks. This is because it is peaked. Morning and afternoon rush hours drive demand for capacity. Were the peaks less peaked, less pavement would be in place. However, unlike a red carpet, it is difficult to only roll out pavement when it is needed, and pull it back after the crowd has gone. Therefore, facilities are sized to accommodate the peak driven by work (though not to accommodate it without queuing—see Chapters 10 through 13). One way of reducing peak work-based auto demand is to encourage more people to use transit (see Chapter 5); another way is to reschedule the demand (see Chapter 6); and a possible third approach is to shorten trips by moving workplaces and workers, or sellers and shoppers, closer together (Chapters 7 through 9).

Notes

a   After 1999, Monster.com dropped the .com suffix.
b   Ever wonder where reality TV shows find their participants? Craigslist is one place. Under the category tv/film/video/radio jobs are listed topics such as "New TV Show Looking for Troubled Couples!," "WE, Women's Entertainment, is proud to announce Daddy's Girl, a new one-hour reality television series." "Fear Factor is coming to town" and "A&E Series: Were you in a CULT, GANG or NEO-NAZI GROUP?" Participating in reality shows is a booming occupation.
c   One such example is Mariposa, CA resident Dave Givens who makes a 186-mile drive—each way—five days a week to his job in San Jose. The electrical engineer has been doing that commute since 1989, spending seven hours every day getting to and from work at Cisco Systems Inc. (see:http://milwaukee. bizjournals.com/sanjose/stories/2006/04/10/daily41.html, accessed April 19, 2007). Givens is the "ultimate road warrior," according to Midas Inc. and drove home with its first-place prize in the nationwide search for "America's Longest Commute."
d   One might think that a large city with great density means greater accessibility to jobs within a given travel time. Therefore, by controlling for congestion, and assuming comparable transportation infrastructure, higher density implies a lower average journey-to-work time. Consistent with this idea, another model, called the intervening opportunities model, would make a prediction different from the gravity model about the effect of uniform density increases (higher density would suggest shorter trips in an intervening opportunities model, after controlling for population

and congestion). However, if commute time preferences are inelastic, people may take advantage of the density and accessibility to trade-off travel time for a better job or house and to maintain their commute time.

References

[1] Farber, H.S., *Are Lifetime Jobs Disappearing? Job Duration in the United States: 1973–1993*, Princeton, NJ: Princeton University Press, 1995.
[2] Levinson, D., 'Job and Housing Tenure and the Journey to Work', *Annals of Regional Science*, 1997, vol. 31 (4): 451–471.
[3] Hamilton, B., 'Wasteful Commuting', *Journal of Political Economy*, 1982, vol. 90 (5): 1035–1053.
[4] Giuliano, G. and Small, K.A., 'Is the Journey to Work Explained By Urban Structure?' *Urban Studies*, 1993, vol. 30 (9): 1485–1500.
[5] Cervero, R. and Landis, J., 'The Transportation-Land Use Connection Still Matters', *Access*, 1995, vol. 7 (Fall): 2–10.
[6] Ravenstein, E.G., 'The Birthplace of the People and the Laws of Migration', *The Geographical Magazine*, 1876, vol. 3: 173–177, 201–206, 229–233.
[7] Ravenstein, E.G., 'The Laws of Migration', *Journal of the Statistical Society of London*, 1885, vol. 48 (2): 167–235.
[8] Ravenstein, E.G., 'The Laws of Migration', *Journal of the Royal Statistical Society*, 1889, vol. 52 (2): 241–305.
[9] Corbett, J., *Ernest George Ravenstein: The Laws of Migration, 1885*, Center for Spatially Integrated Social Science, 2005. Available at: www.csiss.org/classics/content/90 (accessed October 23, 2007).
[10] Isard, W., *Methods of Regional Analysis*, Cambridge, MA: MIT Press, 1960.
[11] Lill, E., 'Das Reisegesetz und seine Anwendung auf den Eisenbahnverkehr', in S. Erlander and N.F. Stewart (eds.), *The Gravity Model in Transportation Analysis: Theory and Practice*, Utrecht: VSP, 1989.
[12] Reilly, W.J., *The Law of Retail Gravitation*, New York: The Knickerbocker Press, 1931.
[13] Stewart, J.Q., 'The Development of Social Physics', *American Journal of Physics*, 1950, vol. 18 (5): 239–253.
[14] Voorhees, A.M., 'A General Theory of Traffic Movement', in *1955 Proceedings, Institute of Traffic Engineers*, New Haven, CT: 1956.
[15] Heggie, I.G., 'Are Gravity and Interactance Models a Valid Technique for Planning Regional Transport Facilities?' *Operational Research Quarterly*, 1969, vol. 20 (1): 93–110.
[16] Granovetter, M., 'The Strength of Weak Ties', *American Journal of Sociology*, 1973, vol. 78 (6): 1360–1380.
[17] Granovetter, M., *Getting a Job*, Cambridge, MA: Harvard University Press, 1974.
[18] Hanson, S. and Pratt, G., *Gender Work and Space*, New York: Routledge, 1995.
[19] Redmond, L.S. and Mokhtarian, P.L., 'The Positive Utility of the Commute: Modeling Ideal Commute Time and Relative Desired Commute Amount', *Transportation*, 2001, vol. 28 (2): 179–205.
[20] Milgram, S., 'The Small World Problem', *Psychology Today*, 1967, vol. 1 (1): 61–67.
[21] Watts, D., *Six Degrees: The Science of a Connected Age*, New York: W.W. Norton, 2003.

[22] Putnam, R.D., *Bowling Alone: The Collapse and Revival of American Community*, New York: Simon & Schuster, 2000.
[23] Saxenian, A., *Regional Advantage: Culture and Competition in Silicon Valley and Route 128*, Cambridge, MA: Harvard University Press, 1994.
[24] Klemperer, P., *Why Every Economist Should Learn Some Auction Theory*, Invited Lecture to the 8th World Congress of the Econometric Society, Seattle, 2001: Cambridge University Press, 2003.
[25] Metropolitan Council, *2001 Twin Cities Transportation System Audit*. Available at: www.metrocouncil.org/planning/transportation/Audit2001/ExecSumm.pdf (accessed July 2007).
[26] Kweon, Y.J. and Kockelman, K., *Non-parametric Regression Estimation of Household VMT*, Presented at 2004 Transportation Research Board annual meeting, Washington, DC.
[27] Barnes, G. and Davis, G., *Land Use and Travel Choices in the Twin Cities, 1958–1990*, Report No. 6 in the Series: Transportation and Regional Growth Study, Minneapolis, MN: Center for Transportation Studies, 2001.
[28] Rajamani, J., Bhat, C.R., Handy, S.L., Knaap, G., and Song, Y., 'Assessing the Impact of Urban Form Measures in Nonwork Trip Mode Choice after Controlling for Demographic and Level-of-service Effects', *Transportation Research Record*, 2003, vol. 1831: 158–165.
[29] Murakami, E. and Young, J., *Daily Travel by Persons with Low Income*, Paper for NPTS Symposium, Betheda, MD, October 29–31, 1997. Originally presented with six month NPTS dataset at African American Mobility Symposium, Tampa, FL, April 3–May 2, 1997.
[30] United States Department of Transportation, 'People, Energy, and the Environment', in *The Changing Face of Transportation*, BTS00-007, 2000. Available at: www.bts.gov/publications/the_changing_face_of_transportation/chapter_05.html (accessed May 15, 2007).
[31] Kain, J.F., 'Housing Segregation, Negro Employment, and Metropolitan Decentralization', *Quarterly Journal of Economics*, 1968, vol. 82 (2): 175–197.
[32] Ihlanfeldt, K.R. and Sjoquist, D.L., 'The Spatial Mismatch Hypothesis: A Review of Recent Studies and Their Implications for Welfare Reform', *Housing Policy Debate*, 1998, vol. 9 (4): 849–892.

Chapter 5

Traveling

"Take the bus ... I'll be glad you did."

The Onion,
November 29, 2000

In March 2004, the Amalgamated Transit Union workers of the Twin Cities Metro Transit bus system voted to strike, protesting their employer's proposal to increase the cost of insurance premiums for workers, cut health care for retirees, and freeze wages. Despite receiving hundreds of millions of dollars in funding for the new Hiawatha light rail line, scheduled to open later in 2004, the transit agency had been reducing service and increasing fares as the basis for financing the system was changed from property taxes to general revenue. The strike lasted six weeks; the settlement resulted in slightly higher salaries for unionized employees, the new money coming from savings generated by not operating a transit system during the strike (unlike most businesses, transit systems operate at a loss, so not operating saves money).

Twenty-one months later, the Transport Workers Union (TWU) in New York City voted to strike. They rejected a new contract in which the Metropolitan Transit Authority (MTA) proposed to increase either the retirement age for future employees or the amount they contribute to finance their pensions. It was the first transit strike in New York City since an 11-day 1980 strike and was illegal according to the New York State Civil Service Law, more commonly referred to as the Taylor Law. The city filed a court injunction over the illegal nature of the strike and with union leaders facing the threat of jail, it lasted only 60 hours.

The difference in the impacts between the two strikes was striking. When the Metro Transit strike took place in the Twin Cities, there was little noticeable increase in congestion. A relatively minor side effect of the six-week strike was to delay the opening of the Hiawatha light rail transit line by three months. Other than transit-dependent travelers, however, few people appeared to notice.

New York City's transit strike, on the other hand, forced millions of people to find alternative means of travel, or to forego traveling altogether. The strike, just days prior to Christmas and Chanukah, cost the city a reported one billion dollars, with shops, theatres and restaurants deprived of annual holiday business. The city government implemented an aggressive contingency plan in advance of the impending strike, mandating that all cars below Ninety-Sixth Street in Manhattan contain four passengers, banning all vehicles along both Fifth and Madison Avenues, permitting taxis to pick up multiple fares, and opening several bridges for pedestrian and bicycle traffic only. Millions of people were forced to find alternative modes—many of whom walked staggering distances in freezing temperatures. Despite the contingency plan, many of the city's major thoroughfares were clogged.

Both strikes provided opportunities for transit advocates to justify the existence of transit. Although bus and subways services are justified on many grounds, the most prominent is the threat of paralyzing congestion. In one case (New York), this held true; in the other (Twin Cities), there was little effect. Those most affected by the strike, in both cases, were "captive" riders—those with little choice but to take transit. Newspaper accounts from both cities featured anecdotal accounts of captive riders' plights [1–4]. In our own classes, some students had to leave early, or arrive late, or even miss class because they relied on the goodwill of others to get them around town. Residents had built a lifestyle assuming the reliability of public transit, and for six weeks—or two days—it was kicked out from under them. Understanding professors and employers might cut them some slack, others might have less sympathy and see them fail or fired. (Box 5.1 illustrates how to count the number of users in the Twin Cities.)

Transit users build transit use into their lifestyles. Their lives are organized around decisions such as where to live and work, whether or not to own a car, and how to construct a daily schedule of activities that permits them to run errands and engage in other activities. Their lifestyle also relies on a variety of assumptions: that their house, job, car, and bus will be there tomorrow, just as they were today. When any of those assumptions is violated, the best laid plans go awry. With the onset of the two transit strikes, captive transit travelers had to find alternatives, such as walking, bicycling, getting a ride from a friend or family member, taking a taxi, or trying to find a place in a carpool or vanpool. All of these solutions are more time-consuming, inconvenient or expensive than taking transit; if they were not, transit riders would be using them every day. Consuming time, money, and the goodwill of friends and family has a cost, and that cost involves doing less of other things, spending less time with family, shopping less frequently for food, or doing favors for friends and family so as to not dip too deeply into one's social capital.

Continual transit use, like automobile use or walking, tends to be informed by habitual behavior. A habit is a behavioral tendency to repeat responses [5]

Box 5.1 Understanding (and questioning) mode split values

Before and during the Twin Cities (2004) bus strike, many newspapers and politicians contended that, "40 percent of downtown Minneapolis workers commute by bus"—a number that those close to the transportation industry thought was a bit high. Nevertheless, this figure was cited frequently by governments, newspapers, transit advocates, and other groups weighing in on the bus strike, particularly those who wanted to stress the importance of bus transit. Although 40 percent mode share became accepted as popular dogma, the only source provided was the Metropolitan Council—the regional planning agency and sister agency to the transit provider. It was unclear what was being measured, what data sources were used, or what calculations were employed. We sought to better understand how this value was derived; in order to do so, it was necessary to resolve several issues that had not been rigidly formulated.

First, it was necessary to specify the measurements. Many different kinds of trips enter downtown every day, including trips for purposes of work, shopping, appointments, and leisure. The sum of all these is the "total trips" entering downtown. Second, we had to formulate a firm definition of "downtown." The core of the downtown area is used much more intensively than are the fringes. Transit service at what many consider to be the core of downtown (specified as the intersection of Seventh Street and Nicollet Avenue) is far superior to the service available in "edge" areas surrounded by surface parking. Third, there is great temporal variation in mode split. Inbound peak period riders rely on bus service more than evening traffic.

So the question is: What is the best measure of mode split? Should we use peak period trips? What about the peak hour? Or should we look at a 24-hour span? Should we consider the entire downtown, or focus on the core? Should we consider all trips, or only those related to employment? Answers to these questions determine what is reported as "the downtown Minneapolis mode split."

According to the 2000 census, downtown Minneapolis is home to an estimated 136,000 jobs (about 8 percent of a regional total 1.75 million). If, as is often claimed, 40 percent of downtown workers take the bus to work, 54,400 people should be riding into downtown every day, just for work. The census question "how do you get to work?" for downtown Minneapolis workers provides a 25 percent transit use rate (or 34,000 people) for the entire downtown, 24 hours a day, work-trips only. The following table documents more situations, some perhaps leading to the popularly misinterpreted "40 percent." When we hear or read such claims, we must wonder where and how they are derived, by whom, and for what purpose. We must also keep the claims in context—even 40 percent of 8 percent is just 3 percent of all regional workers.

The above information was adapted from an exercise completed by Charles Carlson for CE5212 (University of Minnesota) in the fall of 2004.

Table 5.1 Sources of information for commute mode share to downtown Minneapolis

Source	Transit mode share (%)	Scope
Census results (2000)	25	All downtown, all day, work trips only
Cordon Count- Minneapolis plan (1995)	34	All trips, peak period (Survey teams at 100+ entrance points counting entering downtown)
Employer survey (SRF Consulting, 2000 Downtown Transportation Study)	40	Work trips, peak hour
Regional travel survey (2001)	36–41	All downtown, peak period, work trips (5% sample of regional households)
Regional travel survey (2001)	43–44	All downtown, peak hour, work trips
Minneapolis downtown transportation plan	24–58	Depending on location, peak period
Metropolitan Council, TBI	26.5	Entire day (avg. inbound/outbound)
Metropolitan Council, TBI	39	Peak period (avg. inbound/outbound periods)
Metropolitan Council, TBI	44	Peak Hour (avg.)

Table 5.2 Regional travel survey estimates of transit mode share of inbound travelers broken out by time of day

Geography	A.M. peak hour	A.M. peak period	P.M. peak hour	P.M. peak period	Off peak	Entire day
Core	55.8	52.2	68.2	17.0	29.7	39.6
Outer core	29.2	31.6	39.1	28.3	19.1	24.8
Frame	26.3	20.2	8.5	13.8	6.8	10.8
All	43.4	40.6	31.0	19.7	18.8	26.8

Note: Small sample size of afternoon peak period inbound may make estimates unreliable.

and the discipline of psychology defines habit as, "learned sequences of acts that have become automatic responses to specific cues, and are functional in obtaining certain goals or end-states." [6] Decisions about where to live and work, although informed by habitual behavior, are made a dozen or so times over a lifetime. What mode of transportation to take—transit, automobile, or walking—tends to be habitual and informed by upbringing, personal preferences, and individual tolerance for being inconvenienced.

But, you may ask, isn't the decision of whether to drive to work one that individuals make every day, or several times a day? Why are mode choices not classified as short-term decisions? Although such decisions are made daily, it is more useful to consider them as instances of long-term behavior repeated on a daily basis. That is, modal decisions are important in their own regard, and operate on a time scale between that of where to live (every half-decade or more) and which pair of shoes to wear (daily).

We know, for instance, that in the US, 86.6 percent of all trips are made in a personal vehicle (single- or multiple-occupant), [7] a number that sends chills up the spines of many planners. The goal of many land use-transportation planning initiatives is to decrease the percentage of trips by private automobiles and increase the percentage of trips by transit, walking, or bicycling. The question of why nearly nine out of every ten trips in the US are made by private automobile has been answered succinctly by many researchers: no other transportation mode, on average, rivals the freedom, flexibility, and overall cost of the automobile. But there are additional reasons underlying the relative attractiveness of the auto in comparison to other modes of transportation.

Consider the quotation opening this chapter, a headline from the satirical newspaper, *The Onion*. While flippant in nature, this statement has theoretical underpinnings deserving a more thought-out discussion. First, it suggests a dichotomy between transit users and auto users. But more importantly, it demonstrates a psychological distinction between individual and group motivations. What is good for society may not be good for an individual. Again, individual motivations are influenced by *pull* effects enabled by the size (and availability) of the transportation network, and by *push* effects affected by what everyone else is doing. These effects produce patterns of modal decision-making. Our aim in this chapter is to describe theoretical arguments, define prevailing concepts, and outline issues affecting why people make the mode choice decisions they do. We do so by introducing two broad, but distinct, concepts. The first addresses the nature in which transportation services come available by mode and how such services subsequently affect choices related to transportation mode. The second applies concepts of game theory to demonstrate how mode choices play out.

Vicious and virtuous circles

As introduced in Chapter 2, complementors and competitors are central to understanding individual behavior when it comes to how, when, and where people travel. Just as there are advantages to locating where other people locate, there are advantages to choosing the same mode as others. At first, this may seem a puzzling statement. If it is congested and I drive, won't I be stuck in traffic? If it is crowded on the bus, won't I have to stand? Network effects play out in multiple ways across different temporal dimensions and modes. Assuming networks are fixed, there is a negative feedback relationship in the short term between supply and demand; additional demand makes the network more congested. Congested networks run more slowly. Lower speeds decrease demand. Thus, the interaction of congestion and demand is a self-limiting negative feedback process. Congestion applies not only to road capacity but also to transit service; with more people riding the bus, each stop takes longer as more individuals board and alight. This makes taking the bus a slower way to travel, which diminishes demand until equilibrium is reached. So, in a scenario of increasing demand, congestion is a short-term response and network expansion is the long–term response.

Complementors stimulating plexus

Conversely, consider what would happen if no one were driving. Would there be any roads at all? If no one were using public transit, wouldn't you still be waiting at the stop for a nonexistent bus? The reason there are roads and transit services is because others, like you, demand these services. Consider a potential transit trip. With one bus an hour, on average you have to wait 30 minutes for the next bus. If you know the schedule, you can perhaps wait indoors, but you must still wait. That bus can serve 50 seated passengers. If there are 100 potential passengers, the bus company can send out two buses. A savvy bus company will spread them out, so that one comes on the hour, the other at half-past the hour. On average, a passenger would only wait 15 minutes.

Suppose there were 150 potential passengers; in that case, three buses would be deployed, running every 20 minutes, leading to an average wait time of ten minutes. When bus frequency increases on a given route, users benefit from reduced waiting times, an increasing returns property of networks dubbed the "Mohring Effect," after University of Minnesota economist Herbert Mohring, who first described the phenomenon. [8] Each additional bus reduces your wait time as shown in Figure 5.1. The positive feedback loop is shown in Figure 5.2, where additional ridership increases revenue, increased revenue improves the rate or frequency of service, and more service induces additional riders.

The more frequent the service, all else equal, the more likely people are to ride transit. The more people who ride transit, the more buses will be provided. In other words, there is a long-term positive feedback relationship between transit supply and transit demand (or between the supply of any transportation infrastructure facility—highways, bicycle paths, sidewalks—and the demand for that facility). Additional users are complementors. The same principle applies in a spatial context, where instead of more frequent service, there would be denser networks. Imagine once more that there were few automobile drivers; there might only be one transcontinental highway connecting New York City to San Francisco through Chicago. Your travel time from other points (e.g., Baltimore to St. Louis) would be higher, as you would need to go from Baltimore to New York to Chicago to St. Louis, adding at least six hours to your trip. With more drivers, another freeway—directly connecting Baltimore and St. Louis—might be built.

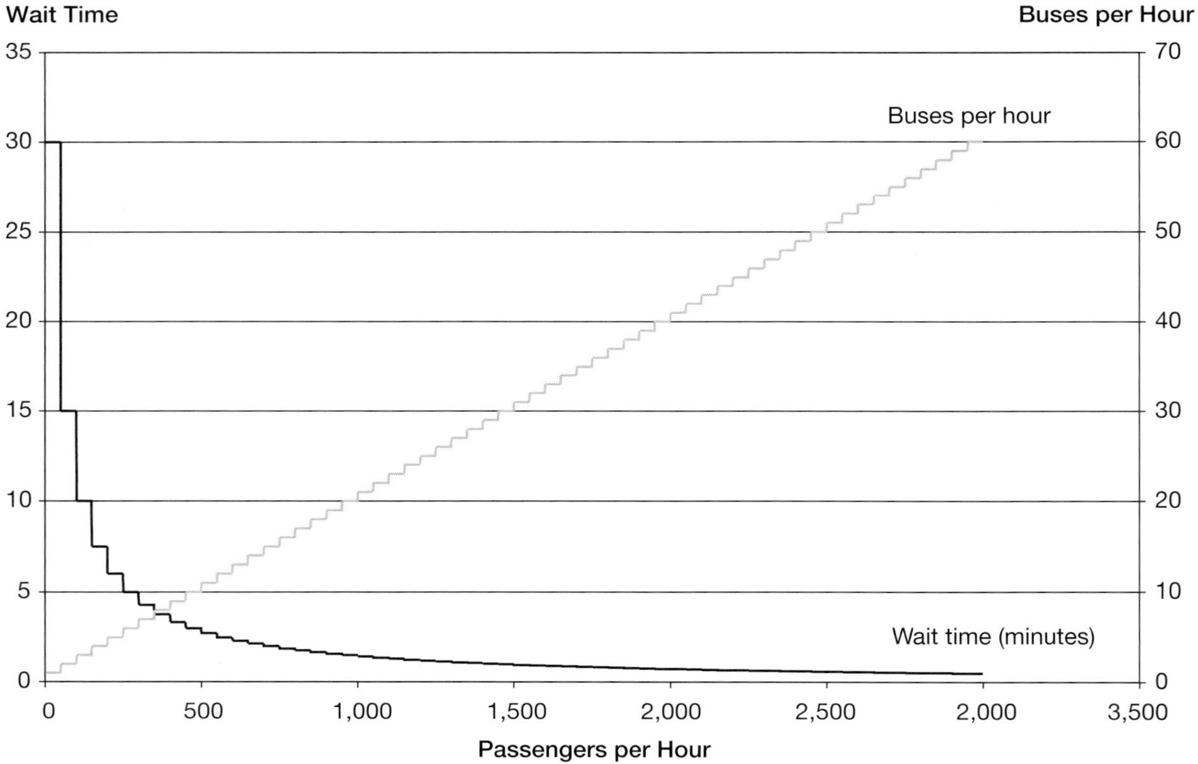

Figure 5.1 The Mohring Effect

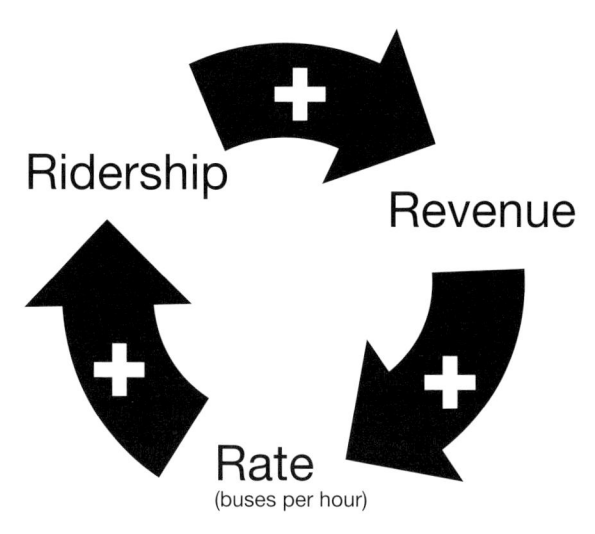

Figure 5.2 A positive feedback loop operating on bus service

Ridership → Revenue → Rate (buses per hour) → Ridership (all positive)

Such network effects are a structural property of transportation systems. The same principle aplies for different transportation modes. Additional supply creates additional demand and additional demand, in turn, creates additional supply. Positive feedback loops lead to "virtuous circles" (where "positive" characterizes an enlarging, not dampening, effect). Small [9] has used this term to describe changes in vehicle delay and transit use in London, which resulted in a 17 percent reduction and a 16 percent increase, respectively, in response to the introduction of congestion pricing. Positive feedback loops, on the other hand, can also lead to "vicious circles." Such positive and negative relationships, however, tend to be nonlinear and subject to diminishing marginal returns, meaning that the first increase in supply likely leads to a greater increase in demand than does the second increase in supply.

Similar processes can also work in reverse. Disinvestment in a given category of infrastructure (e.g., transit, roadways, sidewalks) increasingly turns travelers away, which leads to greater disinvestment. Any external event that lowers demand can also trigger such a vicious circle. Mogridge [10] discusses how highway investment can encourage transit passengers to drive, which will make transit travel less attractive (either by increasing fares or reducing service), which will attract more transit passengers to highways, which may make congestion worse than it would have been if there were no highway improvement at all. The following sections discuss this inter-modal competition in more depth.

Although similar processes are at work for roadways and transit systems, they work differently. A new bus should be relatively easy to deploy. So when a bus gets full, an additional bus (or a longer bus with more doors) can be used. On the other hand, roads are more difficult to deploy. When a road gets full, a planning process to add another lane or another road might begin and take several years to result in more blacktop. In general, for transit, the long-term positive feedback process should be more important, while for highways, the short-term congestion/negative feedback process appears to dominate.

Complementors and competitors affecting destination choice

Competition also affects the choice of destinations. People desire restaurants, for example, that others find desirable and that provide good food and service. If others find such a place too desirable, however, people are forced to wait or they are charged a premium (if the restaurant tries to exploits its newfound popularity). This reminds us of the Yogi Berra maxim: "No one goes there anymore … it's too crowded." Popular places, like popular routes, suffer; fellow urbanites compete for the same limited roadway space, thereby triggering seemingly unbearable congestion.

Complementors and competitors affecting route choice

The choice of a route or path (which is simply a collection of roads or transit lines connecting two points) between an origin and a destination depends upon the nature of the network and upon the choices of other travelers. The most important factor in the choice of route is the travel time on the route compared with travel times on alternative routes. All else being equal, transportation models assume that people choose the route with the lowest travel time. There are other factors besides travel time (among them number of stops, complexity of the route, availability of information, reliability of the route, aesthetics, and familiarity), but for regular trips, time is generally the leading factor.

When a potential route takes longer between an origin and destination than the shortest route, Wardrop's Principle of User Equilibrium [11] states that "the journey times in all routes actually used are equal and less than those which would be experienced by a single vehicle on any unused route." Bottom line: a potential route that takes longer than existing routes will not be used. It is important to note that this means users are minimizing their own time; they are not minimizing society's overall travel time—referred to as the "System Optimal" time. In other words, users behave selfishly.

Several factors affect the travel time on a route. When there is no other traffic, the time is referred to as "free-flow," which means that travel time depends only on the length of the route and the free-flow speeds on the links comprising the route. However, when there is traffic, the travel time is higher, as we have to add the additional delay caused by other travelers. This queuing process is discussed in more detail in Chapter 12. In short, as traffic levels approach the capacity of the road (which is really a shorthand for the willingness of drivers to travel closely together at high speeds), the travel time substantially increases. When users compete with each other for use of scarce road space, they are, in effect, bidding up the cost for everyone.

But matters of congestion come back full circle; the availability of scarce road space, in the first place, depends on there being enough other customers or travelers to warrant building it. If there were no travelers, no one would build a local road, much less an interstate highway system. Furthermore, society could not afford to pay for those roads with the limited gas tax revenue that would be generated. Users are complementing each other in the construction of facilities. So, the next time you are stuck in congestion and cursing all the other drivers, thank them instead—for without them, you would not have the opportunity to use the road at all.

We chart these two processes in Figure 5.3. The construction costs per user drop with additional users, but congestion costs per user rises. In this example, construction costs are $1,000,000 per year (representing the total expenditure as an annualized cost), while congestion costs follow the classic Bureau of Public Roads equation:

$$T = 0.15 \, (V/C)^4$$

where T is travel time, V is vehicles per hour, and C is capacity, assumed to be 2,000 vehicles per hour.

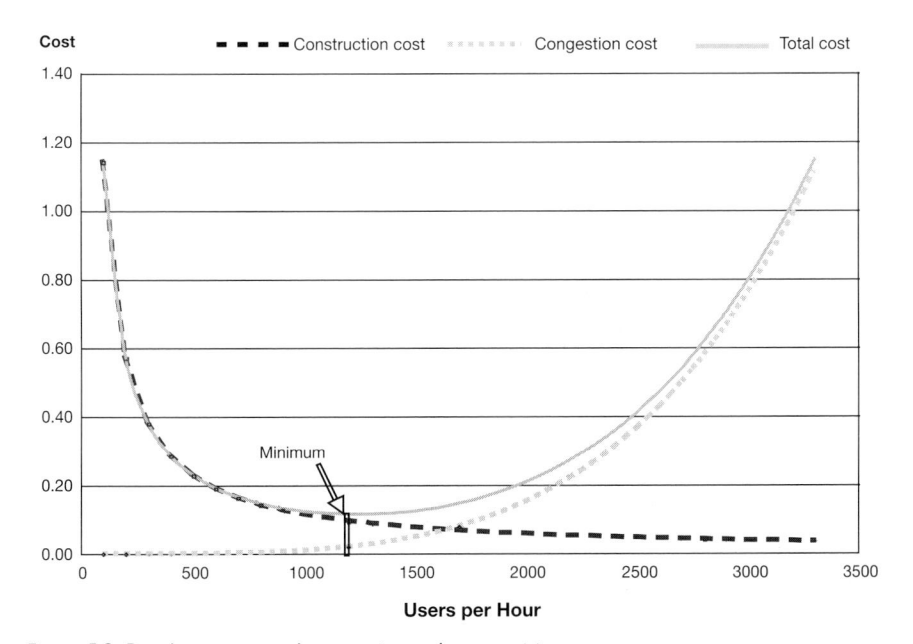

Figure 5.3 Road costs: complementarity and competition

Game theory applied to mode choice

The model of vicious and virtuous circles described above play out largely in the context of providing new or additional transportation services. Given an existing transportation system, what factors determine how people decide to travel and what mode to use? What affects the perceived utility of alternative modes? We next turn to tenets of game theory, specifically the "prisoner's dilemma" (a staple of television police dramas) and "arms race" scenarios, to help understand human decision-making.

Prisoner's dilemma

Imagine that two suspects are arrested. To make this a transportation example, let us say the crime is grand theft auto and the suspects are named Thelma and Louise.[a] They are both brought in for questioning, but interrogated individually. They therefore do not have the opportunity to discuss strategies between themselves. During their interrogations, the prisoners are presented with an offer. The first to confess and testify against her partner will be released with a pardon, while the partner who is convicted will be sentenced to six years in jail. If they both confess (cooperate with authorities, not each other), they will both go to jail for three years. If they both deny the crime, they can only be convicted on a lesser charge (resisting arrest), with a one-year sentence. In this game, collectively the players are better off if neither confesses (i.e., both deny). However, being separated, they have an incentive to confess since they do not know what the other may do. If Thelma denies and Louise confesses, Thelma will pay with six years in prison. Similarly for Louise, if she denies and Thelma confesses. We can represent the game in tabular notation as below:

		Louise	
		Confess	Deny
Thelma	Confess	[3, 3]*	[0, 6]
	Deny	[6, 0]	[1, 1]

where [x,y] indicates jail term for [Thelma, Louise] and payoff is years in jail (and each player's objective is to minimize time in jail). * indicates Nash Equilibrium in a one-time game.

What happens? Economists assume that if this scenario is played out only once and that Thelma and Louise will have no further relations, both players will choose to confess. Why? Imagine Thelma is thinking about denying. If Thelma denies, Louise will get one year in jail if she denies, and zero years in jail if she confesses and testifies against Thelma. Zero is better than one, so Louise will confess. Imagine Thelma is thinking about confessing. Then Louise will get six years in jail if she denies, and three years if she also confesses. Three is better than six, so Louise will confess again. Given that Louise confesses, Thelma is better off confessing (three years in prison is better than six). This solution is called a Nash Equilibrium (named for mathematician John Forbes Nash, recipient of the Nobel Prize in Economics and the subject of the book and film *A Beautiful Mind*). [12] The theory of Nash Equilibrium states that Thelma can do no better given what Louise is doing, and vice versa. In addition, an inferior outcome results (3, 3). In technical terms, the strategy that emerges is Pareto inefficient—one party could be made better without the other being made worse (here, both parties could reduce their jail time). Both Thelma and Louise would be collectively better off if they both denied (one year in prison each is better than three years each).

One key to understanding the outcome of the prisoner's dilemma is that it is a one-time game. If there were a sequence of games or repeated games (repeated an infinite or indefinite number of times), the players would learn what the other would do. The rule of thumb that may emerge is dubbed the *tit-for-tat* strategy. [13] Players will optimize their payoff if they play whatever their opponent played in the previous round. The reasoning for this is as follows: if the number of repeats is known, there will be no difference in the character of the equilibrium than with a one-time game. If a game is repeated exactly *n* times, the last time is like a one-time game since "there is no tomorrow." The round before that will also be like a one-time game since if players do not cooperate on the last round, they have no incentive to cooperate on next-to-last round. This reasoning continues to the first round.

However, if the number of repeats is unknown, players will cooperate (deny) in the hope that cooperation will induce cooperation in the future; this line of reasoning requires that there will always be a possibility of future play. As long as both players care about the future payoffs, the threat of non-cooperation in the future may be enough to convince parties to play the *Pareto efficient* strategy—the one in which neither party can improve their outcome without making the other worse off.

Prisoner's dilemma applied to land use-transportation policy

Consider, for example, the situation faced by landlords in a declining neighborhood who must decide whether to improve their rental property or invest their money elsewhere. If a few landlords improve their properties and all the others do not, the neighborhood will continue to decline, rendering the investment futile and financially inadvisable. On the other hand, if the same few landlords choose not to improve their properties while the others improve theirs, the general improvement of the neighborhood will allow all landlords—including those who have invested no money in improvements—to raise rents. As a result, it is in each individual's self-interest to make no improvements; however, if they all refuse to do so, the neighborhood will decline further, making things worse for everyone. An identical inevitable logic leads the competitive market to overutilize "common pool" resources that have limited supply and free access.

Applying game theory to matters of land use-transportation policy, imagine the lives of Thelma and Louise before their crime spree. They live on the same block and work for the same downtown firm that only has one free parking space. Knowing that transit is less convenient than driving alone, and has a higher out-of-pocket cost ($4 round-trip vs. $5/day for parking paid only every other day (—the day they cannot beat the other to the free space), the workers are in a dilemma (not quite the classic prisoner's dilemma, but a dilemma nonetheless). Louise would prefer that Thelma take transit every day so that she can drive (and use the parking space without conflict). Thelma feels the same way toward Louise. Given two options of driving alone or taking transit, they are collectively best off if one drives and the other takes transit. However, the one that takes transit loses out.

We can now tweak the example a bit. If Thelma and Louise realize they are both taking transit, they could also carpool. This would be faster than transit, have a lower out-of-pocket cost, but still require some coordination (waiting for the other, ensuring they both are ready to leave at the same time). Suppose this cost is valued at $1 per person per day. In this case, the problem is a true prisoner's dilemma. Together, they are both better off carpooling than if one took transit and the other drove, or even if both drove alone. However, if one chooses to take transit or carpool, the other will be better off driving alone and getting the parking space to herself.

This variation on the prisoner's dilemma illustrates a conflict between individual and group rationality. In a group whose members pursue rational self-interest, everyone may end up worse off than a group whose members act contrary to rational self-interest. The scarcity in parking is analogous to scarcity in road space, and the price of parking is analogous to delay.

| | | Thelma | | |
		Drive Alone	Transit	Carpool
Louise	Drive Alone	[0.5 P, 0.5 P]	[0, F]	
	Transit	[F, 0]	[F, F]	
	Carpool			[C, C]

Supposing the values of each were:

    Parking $(P) = 5$
    Transit Fare $(F) = 4$
    Carpool penalty $(C) = 1$

... would yield the following:

| | | Thelma | |
		Drive Alone	Transit
Louise	Drive Alone	[2.5, 2.5]*	[0, 4]
	Transit	[4, 0]	[4, 4]

or:

		Thelma	
		Drive Alone	Transit/Carpool
Louise	Drive Alone	[2.5, 2.5]*	[0, 4]
	Transit/Carpool	[4, 0]	[1, 1]

* Indicates Nash Equilibrium

Although not nearly as explicit, most individuals implicitly make calculations not unlike those described above as part of their daily travel decisions. Such contexts, however, are usually not *one-on-one* but rather involve a single individual attempting to out-guess many other travelers. Similar games are played out for the prizes of limited parking space, roadway capacity, or seating on a bus.

As a final example, let us turn to a specific issue at the heart of countless planning initiatives—using private automobiles or public transit (in this case, buses). If we assume that the bus system operates on the same street system as private automobiles—and must make stops to pick up and discharge passengers—its travel time will always be greater (assuming no high-occupancy vehicle lanes, queue jumpers, bus rapid transit, or dedicated busway facilities) than the automobile for a given automobile mode share (in this case, bus mode share equals 100 percent minus automobile mode share). The system average travel time (T_S) can then be represented by:

T_S = $(M_A * T_A) + (M_B * T_B)$
M_A = Automobile Mode Share
M_B = Bus Mode Share = (1—Auto Mode Share)
T_A = Automobile Travel Time
T_B = Bus Travel Time

Assuming there are congestion effects (noticeable, say, after the automobile mode share hits 75 percent) causing delay, and that the higher the bus demand, the more bus service, and the lower the bus wait. Figure 5.4 demonstrates that although automobile travel is always faster than bus travel, it would be better for everyone (in terms of faster travel) if fewer people drove.[b] Whereas it is individually rational for each person to drive, it would be collectively better for automobile drivers as well as bus passengers if, beyond some point (in this case 75 percent automobile mode share) everyone rode transit. The travel time for automobile travelers at 80 percent automobile mode share is higher than the system travel time (or bus time) at 0 or 10 percent automobile mode share.[c] In terms of game theory, this is a variant of the prisoner's dilemma in which competing actions may be pursued based on one's global or personal best interest.

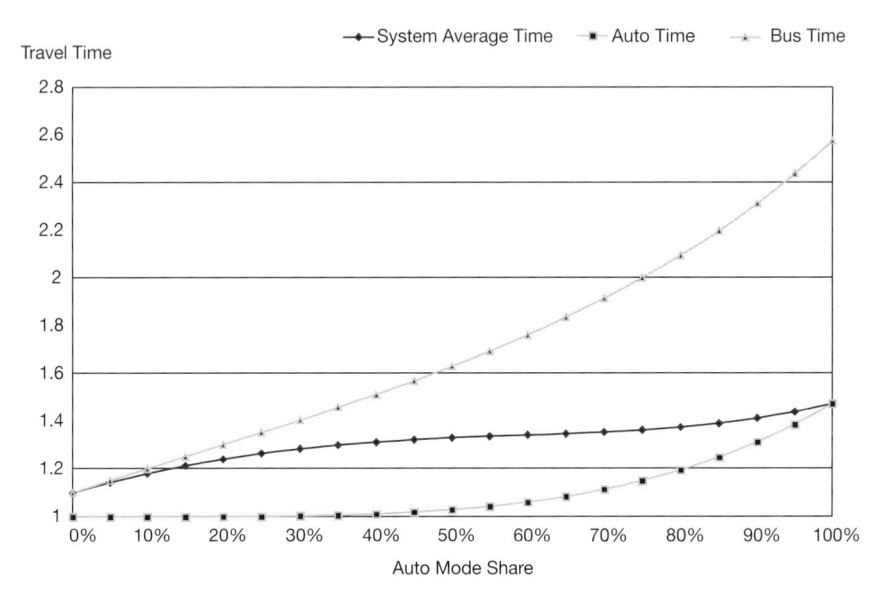

Figure 5.4 Modal competition model

By combining individual demand, the idea of modal competition, and network effects, yields results shown in Figure 5.5, ridership vs. wait time.[d] There are two stable equilibria in this case, one at very low demand (high wait time yields zero ridership, which returns high wait time), and one at very high demand (low wait time yields high ridership, which returns low wait times). There is a third point where the initial wait time and resulting wait time curves cross (wait time of 9 gives 141 riders), but this is not especially stable. Slightly less bus service than the ridership deserves will reduce demand and drive down service. Slightly more bus service will increase demand and drive up service. Points to the left of this crossing point can be seen as a vicious circle, points to the right as a virtuous circle.[e]

The cases described so far involve coordination with perfect information. Much information in land use-transportation related decision-making, however, is tacit. Box 5.2 considers how successful people are at coordinating with only tacit information.

Arms race

While the prisoner's dilemma shows one application of game theory to land use and transportation decision-making, other rubrics exist. An alternative way of uncovering the relative attractiveness people assign to different modes is not unlike that of an arms race. The term 'arms race' was originally coined by physicist and psychologist Lewis Fry Richardson (an ardent pacifist) to

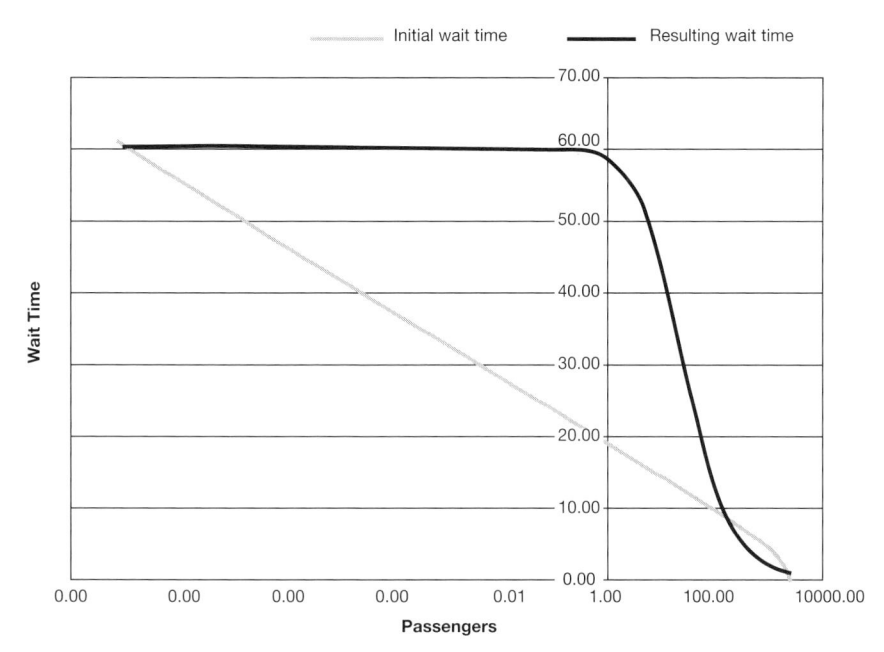

Figure 5.5 Relationship of bus ridership to wait time

describe the competition between two or more nations for military supremacy. In Richardson's model, nations compete to produce superior military technology, either by enhancing the quality of their technology or by increasing the quantity of weapons and materiel produced. Richardson explained an arms race as an interaction between two states. Having a large available arsenal makes a given nation more likely to engage in conflicts; specifically, it constitutes a prelude to war. Richardson aimed to examine the stability of an arms race between two nations in order to predict whether a large conflict could be precipitated by a small event. Because weapons are costly, their expense creates fatigue that decreases future purchases. This in turn has traumatic effects for the degree to which resources are applied to achieve other goals of the nation, state or community.

More generally, "arms race" is used generically to describe any competition in which the sense of a common goal is obscured by the relative goals of staying ahead of other competitors. Last we checked, we were unaware of any arms races in planning contexts that could be directly tied to war. However, a variety of struggles related to planning also affect the decisions people make. For example, competition and grievances between cities and states cause them to acquire arms (e.g., successes in economic development) that are used against one another.

Some examples drawn directly from the discipline of urban planning may be instructive here. We, the authors, are currently writing from Minneapolis

Box 5.2 Grand Central Station

You must meet someone in New York, but you don't know where (or when) you are supposed to meet, and there is no way to communicate with the other person ahead of time. Where would you go?

Game theorist (and recipient of the 2005 Nobel Prize in Economics) Thomas Schelling posed a version of this question to Yale law students in 1958. A majority of students said they would go to the information booth in Grand Central Station, one of New York's two main passenger rail stations (incidentally, Grand Central is the station serving trains to and from New Haven, Connecticut, where Yale is located) and almost all the students said they would meet at noon. James Surowiecki, a columnist for *The New Yorker*, wrote that "if you dropped two law students at either end of the biggest city in the world and told them to find each other, there was a very good chance they would end up having lunch together." [14] Schelling posited that people's expectations (of what other people think) centers on important landmarks or "focal points." There is community knowledge, which is tacit, about what these places are. Notably for this book, Grand Central Station is an interface of transportation networks and the land use system, as is any transportation system terminal or station. The advantage from a way-finding perspective is that a lot of people, a plurality if not a majority, flow through this relatively small gateway (especially in 1958, before widespread adoption of the jet airplane, in crowded New York City). Schelling also notes the results are culturally biased, and someone without the same grounding and experiences might draw a different conclusion. As someone not from greater metropolitan New York (including New Haven), we might have said the top of the Empire State Building, but perhaps we are influenced by the 1957 film *An Affair to Remember*.

We have informally repeated this experiment with Minnesota-based planning and engineering students (mostly first-year graduates and senior undergraduates who were unaware of Schelling's experiment or Surowiecki's book) for both New York and Minneapolis (and a few other cities) and found no overwhelming consensus. Though Grand Central Station received about 10 percent of the students' votes, the largest number went to Times Square (33 percent). For Minneapolis, the plurality (40 percent) of students identified locations in downtown, on the Nicollet Mall or in the public areas of the IDS tower (located on the Nicollet Mall), although those are big places. The Mall of America (again a bit vague on the location, probably the center, which houses an indoor amusement park)—which, notably, is not even in Minneapolis—came in second with 20 percent. (The problem of finding appropriate Minneapolis landmarks recently reappeared in a new version of Monopoly that Hasbro is vetting by allowing people to vote on landmarks in 22 US cities; for Minneapolis, the Mall of America came in first.) In addition to being the largest mall in the United States, the Mall of America is also the southern terminus of the metropolitan light rail line. The airport (also not in Minneapolis proper), came in third. For St. Paul, the State Capitol—a notable landmark on the skyline, though difficult to reach except

> by car—came out first (45 percent); for Chicago, the Sears Tower (5 percent); and for Los Angeles, the airport (LAX), with (29 percent). In terms of meeting time, noon received a plurality of votes (41 percent), but other times were widely spread throughout the day. It is highly likely that two University of Minnesota students dropped off at either end of a large city today would pass like ships in the night, and unlike the Yale law students of 1958, would have only a small chance of having lunch.
>
> Thomas C. Schelling, *The Strategy of Conflict,* Cambridge, MA: Harvard University Press, 1960. pp. 54–67.
>
> Discussed in *The Wisdom of Crowds: Why the Many Are Smarter Than the Few and How Collective Wisdom Shapes Business, Economies, Societies and Nations* by James Surowiecki.

and St. Paul. These two adjacent cities comprise the Twin Cities of Minnesota, but one would be remiss not to notice a minor non-military arms race between them. Although it is foolish for one city to completely duplicate the services and amenities of the other (the metropolitan area is not large enough to support such redundancy), each city is driven to respond to the latest accomplishment or coup of its neighbor: Minneapolis wins a franchise from the National Basketball Association, St. Paul announces a National Hockey League franchise; Minneapolis boasts a symphony orchestra, St. Paul promotes its chamber orchestra; St. Paul is home to *A Prairie Home Companion* (a nationally syndicated radio variety program), Minneapolis draws top name musicals; St. Paul cherishes an annual winter carnival, Minneapolis launches an annual cross-country ski race; Minneapolis has a chapter of the American Automobile Association (AAA), St. Paul has its own chapter. In many respects, these are local instances of the constant economic competition between municipalities. For example, the Boeing Company's decision to relocate their corporate headquarters from Seattle pitted Dallas, Denver, and Chicago against each other in a battle for the new headquarters and its large pool of high-paying jobs. Alternatively, ski resort operators compete for business by developing bigger and better resorts: American Ski Resorts, owner of Stowe (Vermont) Whistler, (British Columbia), and The Canyons (Utah) competes with Vail Resorts Management Company, owner of Vail, Beaver Creek, Breckenridge, and Keystone (Colorado) for the right to claim the largest skiable area, greatest number of lifts, and other service benchmarks.

But the connection of an arms race to travel has direct applications to transportation. One of the more obvious ones is the recent (implicit) competition between the so-called "sport utility vehicles" or SUVs (or in the United Kingdom, "Chelsea Tractors"). For a decade or more, vehicle manufacturers (and, subsequently, consumers) have been competing against one another for larger vehicles that are higher and higher off the ground. If I am a driver, I would like to maximize the information available to me when driving. I do this by seeing ahead of the driver immediately in front of me, but I can only do this if I am taller than that vehicle and can see over it. But there are other, more behavioral arms races going on. "I want to travel faster or better than others, who are interfering with my prefered travel pattern."

The arms race model is not limited to car vs. transit; car vs. bicycle is another example. Recently, Shanghai announced that it would prohibit bicycles on major arterials, as bicycles cause excess delay for automobiles. [15] The war between bicycles and cars is an example of an arms race. Once, China was renowned for its high bicycle mode share, a phenomenon largely attributed to communism and poverty. When bicycles ruled the road, conditions for bicycles were quite good. However, once wealth gave China some private automobiles, the cars pushed the bicycles farther and farther to the side of the road (Figure 5.6). In a fight between a two-ton car and a bicycle weighing 20 kilograms, the car usually wins.

Amsterdam provides a counter-example; there, law and culture give bicycles pre-eminence. There are several reasons for this. The existence of an extensive network of bicycle paths is an important factor, but bicycles often dominate even on city streets. Another reason is a generally pleasant climate, at least compared with many parts of North America. Before World War II, about

Figure 5.6 Bicyclists in Shanghai (photograph courtesy of David Loutzenheiser)

three-fourths of all trips in Amsterdam were by bicycle. During the war, bicycles—especially good bicycles—disappeared. After the war, affluence helped the automobile gain market share. A pro-cycling protest movement in the 1970s helped redirect policy; speed limits were lowered, investments in bicycle facilities were increased, and the law was changed so that automobile drivers are presumed to be at fault in the event of a collision with a cyclist. Today, about 28 percent of trips in the city are by bicycle (and a large number are by the extensive streetcar system) and there is about one bicycle per person.

The public image of modes makes a difference. In China, the bicycle is perceived to be the mode of the poor, whereas in Amsterdam, cycling does not bear that social stigma. Transit faces similar problems; Box 5.3 investigates some alternatives.

Traveling wrap up

During World War II, the United States government enacted numerous measures to conserve scarce resources like fuel and rubber, including rationing. But, in addition, the government attempted to persuade those on the "home front" of the military and moral necessity of behaving in socially beneficial ways. The outcome included posters such as Figure 5.7 reminding citizens that "When you ride ALONE you ride with Hitler!." The promotional campaign suggests, perhaps more directly than some are comfortable with, a link between travel mode choice decisions and competition (in the case of the poster, the competition is between the Axis powers and the Allied forces). Our perspective in this book highlights mode share and traveling as a system involving competition.

Competition between modes, under certain circumstances (without subsidies for positive feedback industries, and without penalties for negative externalities), may result in socially sub-optimal results. The degree to which the results are deemed sub-optimal, and subsidies justified, depends upon the public's belief that government can actually understand the dynamics of the system under question and figure out where to direct subsidies (the "pork-barrel" problem). Not all subsidies are warranted; they are often blamed for leading to inefficient conditions and many are incorrectly justified based on the logic outlined above.

Encouraging citizens to join in the fight against either congestion or sprawl principally relies on a combination of moral suasion and education. Such strategies are in contrast to the ideas presented in the previous two chapters. However, some optimists hope that preferences can be changed, be it through moral suasion, education, marketing, harping, or a combination thereof—land use-transportation planners have at their disposal a host of strategies for influencing public opinion and behavior. Evidence about real transit systems suggests that marketing alone is not the answer, though it could certainly be improved. In the Minneapolis-St. Paul metropolitan area, Metro Transit serves only a fraction of users, though with a large share in a few markets like downtown Minneapolis. Because the overall fraction is so small, the 2004

Box 5.3 Rebranding transit

When buses replaced streetcars in the 1950s, buses were seen as new technology: flexible, comfortable, and air conditioned, while the old streetcars (which had been capital-starved for a number of years) were old and dowdy. In the 1980s, this perception was reversed; light rail transit became the new high-tech mode, and buses were uncomfortable, smelly, and subject to mechanical breakdowns. Rail was "sexy," buses weren't. We now see bus stops in many parts of the US where one cannot even learn the route number of the bus that stops there, much less consult a map of its route or a schedule. It is little wonder fewer and fewer people use buses. Imagine, for example, an airport without flight numbers or destinations posted, in which you have to go outside, look at the front of the plane, then go back home and call the airline to find out a flight's destination. Buses and planes are very similar from a transportation service perspective (after all, a major airplane manufacturer calls itself Airbus). Yet despite spending $200,000 on a bus, transit service providers fail to spend $1.75 per stop to post a laminated schedule, much less provide real-time information about when the next bus will arrive.

A few downtowns have systems like London's "Countdown" or San Francisco's "NextBus" to inform patrons when the next bus is due to arrive. There is limited evidence, however, suggesting such strategies actually increase ridership (though they certainly don't decrease it) as opposed to increasing customer satisfaction. In concert with other technologies, information, especially real-time information, provides riders with the confidence that bus transit is a real and reliable alternative (while real-time information is probably superior, confidence in the system can be enhanced with even minimal information such as the schedules of routes and maps).

Just as the image of excessive numbers of toll booths leading to great delays has set back the cause of toll roads, the cause of transit has been set back by bus systems operated as if riders do not have a choice by managers who don't take care of even minimal details. To promote transit, it seems to be conventional wisdom that cities must build entirely new systems, at great cost, to route around the bureaucratic inertia set in place with existing bus services. Surely there is a better way.

Perhaps rail is preferred to bus because it is newer and cleaner—perhaps because it is harder to get lost when there are maps and schedules. Alternatively, it might be because there are a limited number of clearly named rail routes (the "Green Line" vs. Route #651-X-N) that stop at known places.

The last approach is one that has been employed in a variety of places including the Boulder, Colorado area where their core transit system consists of seven routes creatively labeled HOP, SKIP, JUMP, BOUND, DASH, BOLT and STAMPEDE. All have a unique identity and amenities shaped with community input and direction. In 1990, the area's transit ridership was about 5,000 riders daily for all local and regional routes in and out of Boulder. In 2002, the reported daily average ridership stood at about 26,000—an increase of over 500 percent.

Figure 5.7 Poster encouraging car sharing, Second World War

Twin Cities strike was imperceptible in traffic count data. New York's transit system is much more significant, not because of better marketing, but because of better markets.

The bottom line is that positive feedback systems such as transit supply and demand have two stable states: low and high. Low may be above 0 percent and high below 100 percent, but interim states tend not to be stable without large subsidies to prop them up. The long-sought "balanced" transportation system will require enormous subsidies to keep from tipping over. Under certain circumstances, it might be collectively rational for society to invest in transit and to give the proper incentives for people to use it, as a high transit mode share would lower collective travel times. This result depends on the specific characteristics of the system, in particular the bus travel times at high and low bus usage, and automobile congestion levels.

Identifying sources of investment for new technologies (the cycles of transportation capital) and describing the structural mechanisms for resources going to new technologies is only half the story—an analysis limited to the supply side, taking demand as a given. Why would people demand personal automobiles instead of streetcars? A model based on the theory that people's primary concerns are their time and their money explains much (though not everything). All else equal, people seek the fastest, cheapest mode. Introducing a slower and more expensive mode is unlikely to get one very far as an entrepreneurial investor or a mercantilist metropolis.

There are several take-away points to the narrative presented above. The first is that you can model me as an individual as if I only care about my own travel time (or cost). Unless you provide a mechanism for me to consider the effects of my behavior on others, I am unlikely to voluntarily do so. My time is faster if others take transit. Perhaps society's overall travel time would be lower if most people took transit (which would eliminate street congestion). The modal competition model makes that argument.

The second point is that it is difficult to develop a mechanism to achieve a collective good while still allowing individual freedom. There are mechanisms to ensure individuals consider their effects on others that are addressed in later chapters. One such example is where travel time in a corridor by private modes exceeds that of public modes (grade separated transit), additional demand switches to transit, ensuring equilibrium in travel times between the two modes. [16] However, this observation—like so much in transportation policy—is disputed. In particular, the problem is that the equilibrium that is established is point-to-point. Transit often has advantages where the points are the access and egress points of the rail line. However people seldom live on top of one station and work on the other; instead, they travel from their front door to the door of their destination. Transit can seldom beat the automobile in terms of door-to-door travel time (although in some cases transit is certainly the more expedient option).

Basic economic theory tells us that the actual consumption of goods and services (be they roads, real estate, or reefer) is determined by the equilibrium point between the demand and supply curves. Without transportation service, few (if any) goods would be consumed, regardless of how much demand existed. On the other hand, increasing the supply of transportation services (by increasing network density, reducing headways, and/or lowering fares) serves to lower the perceived cost of travel; this moves the demand and supply equilibrium point and results in increased travel (overall). To the chagrin of many planners, travelers are being individually rational by selecting the automobile when they do, and they are also being rational those times when they select transit. Whether it is better for society to support more highways or more transit, and whether one mode should be subsidized or not, is also an important question. Further, if buses and trains are packed full and service supply is insufficient to accommodate demand, increased service supply will lead to increased consumption of transit trips by accommodating demand that had been previously suppressed due to the inadequacy of supply.

Notes

a The two hypothetical suspects, Thelma and Louise, are based on the movie of the same name in which, throughout the film, they perform a series of crimes that they find easier and easier to commit.

b The Modal Competition Model relating auto travel time, bus travel time, as a function of mode shares was illustrated with specific equations relating time to mode share. The Alternate Theory makes a somewhat different assumption about bus times:

$$T_B = T_A + D + 0.1$$

where: D = Schedule Delay

$$D = \frac{0.5}{M_B}$$

This model is designed to account for schedule delay (a little) more rigorously. Here the schedule delay (the wait between the buses) is proportional to the half the time between buses. When there are 60 buses per hour, there is essentially no wait (say 30 seconds). When there is 1 bus per hour, the wait is about 30 minutes. It assumes buses per hour are linearly proportional to demand – at 100 percent bus mode share there are 60 buses, at 50 percent bus mode share there are 30 buses.

c The details for the modal competition model assumed the following:

$$T_A = 1 + 0.15 \left(\frac{M_A}{0.75}\right)^4$$

The auto travel time increases with auto mode share (particularly when auto mode share exceed 75 percent) to account for congestion. This form is basically the classical Bureau of Public Roads (aka BPR) equation, which has been used to estimate travel time on segments in transportation planning applications

$$T_B = T_A + M_A + 0.1$$

The Bus time equals the auto time (they are running on the same roads) plus the auto mode share (to account for less frequent bus service when auto use increases) plus a fixed amount, to account for bus stops and starts. The exact form of this equation is unimportant to illustrate the general point.

d This model has two built in relationships and several parameters.

$$v_B = 2 - 0.5\, Cost - 0.5\, Wait_{Initial}$$

$$P_B = \frac{e^{vB}}{1 + e^{vB}}$$

$$Passengers = 3000 P_B$$

$$Wait_{Resulting} = 0.5*60* \left(\frac{1}{0.5 + 0.02 Passengers}\right)$$

Where $Cost = 1$

e The astute reader might ask why we can't over-invest at a small level of transit demand and drive up service, and thereby increase demand? Suppose initial waiting time is 15 minutes in the above example that results in seven riders per hour, which returns enough service to justify 45 minute waits. If we lowered the time to 14 minutes, we increase ridership to 12, which justifies 40 minute waits, definitely an improvement, but not enough to get more than 12 riders. Transit systems still require subsidy. On the other hand, if ridership exceeds 140 riders, any improvement in service will improve wait time sufficiently to gain even more riders.

References

[1] *City Pages*, 'Bring Back the Buses', vol. 25 (1218), April 7, 2004.
[2] McGeehan, P., 'The New Trail to Work: Longer and More Inventive, Complicated and Crowded', *The New York Times*, December 22, 2005: B6.
[3] Piña, P. and Sullivan, J., 'Bus Strike is on', *St. Paul Pioneer Press*, March 4, 2004: A1.
[4] Williams, B., 'Bus Strike is More than Inconvenience for Some', Minnesota Public Radio, March 11, 2004.
[5] Ouellette, J.A. and Wood, W., 'Habit and Intention in Everyday Life: The Multiple Processes by Which Past Behavior Predicts Future Behavior', *Psychological Bulletin*, 1998, vol. 124 (1): 54–74.
[6] Verplanken, B. and Aarts, H., 'Habit, Attitude, and Planned Behavior: Is Habit an Empty Construct or an Interesting Case of Goal-directed Automaticity?' *European Review of Social Psychology*, 1999, vol. 10: 101–134.
[7] Bureau of Transportation Statistics, *Highlights of the 2001 National Household Travel Survey*, Washington, DC: United States Department of Transportation, 2004.
[8] Mohring, H., 'Optimization and Scale Economies in Urban Bus Transportation', *American Economic Review*, 1972, vol. 62 (4): 591–604.
[9] Small, K., 'Unnoticed Lessons from London: Road Pricing and Public Transit', *Access*, 2005, vol. 26 (Spring): 10–15.
[10] Mogridge, M.J.H., Holden, D.J., Bird, J., and Teris, G.C., 'The Downs/Thomson Paradox and the Transportation Planning Process', *International Journal of Transport Economics*, 1987, vol. 14 (3): 283–311.
[11] Wardrop, J.G., 'Some Theoretical Aspects of Road Traffic Research', *Proceedings of the Institute of Civil Engineers Part II*, 1952, vol. 1: 325–378.
[12] Nasar, S., *A Beautiful Mind: A Biography of John Forbes Nash, Jr.*: New York: Simon & Schuster, 1998.
[13] Axelrod, R., *The Evolution of Cooperation*, New York: Basic Books, 1984.
[14] Surowiecki, J., *The Wisdom of Crowds*, New York: Random House, 2004.
[15] Luard, T., *Shanghai Ends Reign of the Bicycle*, 2003. Available at: http://news.bbc.co.uk/2/hi/asia-pacific/3303655.stm (accessed December 9, 2003).
[16] Lewis, D. and Williams, F.L., *Policy and Planning as Public Choice: Mass Transit in the United States*, Aldershot: Ashgate, 1999.

Chapter 6

Scheduling

"If I work in the house, why am I always driving the car?"
    Bumper sticker as seen on Chrysler Minivan

When Robert Levine, a professor of social psychology, began a sabbatical in Brazil, he was in for a relatively rude awakening. Levine was stunned at how students typically showed up late for his class, causing him to initially think they were rude or not interested in his lecture. However, he also found they failed to leave at the scheduled time, but remained in their seats to ask questions and discuss the topics in the lecture until he would plead hunger, thirst or a call of nature. Such occurrences were in stark contrast to his home institution in California where he was accustomed to listening for the rustling of students' papers and notebooks to signal the impending end of class—provoked by a pain that, in his words, "usually becomes unbearable at two minutes to the hour for undergraduates and at about five minutes to the hour for graduate students." Levine, who studies what he refers to as the "tempo" of cultures, cities or regions, contends that "time talks with an accent." [1]

In his work, Levine has attempted to uncover how conceptions of time differ in different countries and cities. He found that the pace of life and time consciousness are more intense in the Northern industrialized countries (as shown, for example, by the presence of clocks, watches, and calendars) and less intense in the Southern countries where "event time" is more commonly used. He followed his cross-cultural observations with a systematic comparison of 36 cities across the United States. In one example, he measured the average walking speed of randomly selected pedestrians over a distance of 60 feet (18 m) in downtown environments. He timed how long it took bank tellers to make change, he counted the number of people who wore wristwatches and, in the spirit of personal sacrifice for the greater good of science, spoke with postal clerks in various cities, asking them to explain the differences among regular, certified and insured mail, tape-recording their responses for later analysis. His research group then played back the tapes and calculated 'articulation rates' by dividing the number of uttered syllables by the total time of the response.

Combining such data into four different factors, he uncovered that the most time-conscious cities in the US were Boston, Buffalo (New York), New York City, Salt Lake City and Columbus (Ohio); the bottom five were Memphis, San Jose (California), Shreveport (Louisiana), Sacramento (California), and Los Angeles. Based on such findings, one can conclude that the Northeast is a more time-conscious region than the sunny South or laid-back California.

At last count, everyone technically has 24 hours (1,440 minutes) in their day. But the result of Levine's research suggests that there are geographic variations in how people use their minutes; his findings are also transferable to punctuality, and in some respects, to how quickly they travel. Through his work, we learn where people are the most generous with their time and where they talk the fastest, as well as gaining deeper insights into cultures of South America, Japan, and other areas through understanding inhabitants' perceptions of scheduling and time use. This chapter deals with the ways in which people schedule their time spent in activities, and how travel behaviorists describe time spent in transportation. We begin by discussing characteristics of time spent in travel versus time spent in activities. We then introduce the reader to a useful vocabulary for studying time use, and to strategies used by analysts to break down and understand time spent in travel.

Travel time budgets

A principal point of interest for transportationists is how people break down their daily 1,440 minutes. We already examined this question in Chapter 2, recounting the average time Americans spend in 12 different types of activities. The problem with the accounting scheme presented earlier is that although the activities sum to 24 hours, no travel time is denoted; the travel is combined with the activity. As transportationists, however, we are interested in the minutes getting from one activity to another.

Transportation folklore suggests that ever since the ancient Babylonians, and for today's industrialized and non-industrialized populations, allocations of time for travel and for other activities have been relatively constant. This folklore has been introduced to recent times by researchers who coined the term "travel time budget," suggesting that people have an amount of time that they are willing (or may even prefer) to spend on travel. [2] The idea has merit and *in the aggregate* the theory appears to hold water. Average people from all walks of life generally appear to travel between 1.1 to 1.3 hours per day. This idea is most associated with transportation researcher Yacov Zahavi (1926–1983), who published his surveys of travel throughout the world in the 1970s and 1980s.

Consider the matter in somewhat simplified terms: there is time devoted to different types of activities (e.g., at work or at the opera) and time spent in travel between different activities (e.g., actually traveling to work or traveling to the opera). We show data in Figure 6.1 from two different metropolitan areas and for different populations (males and females, workers and non-workers). We break time in activity into five different categories.

By the metrics presented in Figure 6.1, one can see that matters have not drastically changed over the past 30 or so years (and things are relatively similar across metropolitan areas). Results for certain groups (e.g., people in the workforce versus those who are not), tend to be similar in terms of the cumulative time spent in different types of activities. Any trend we can discern is towards more travel, not less, and more work, not less (among those who work).

Several important points emerge from this analysis. First, for most people, the overwhelming majority of time in every day is spent at home. If the roughly 500 minutes or so per day sleeping are removed, then work activities takes on increased prominence (for workers). Second, work imposes a major constraint on the time available to pursue other activities because it leaves considerably less waking time to engage in shopping and other activities, which are then restricted to the afternoon peak as workers return home. Third, there is a not a clear winner in terms of other activities. For both workers and non-workers, the "other" category often rivals the "travel" category. Fourth, non-workers have more flexibility (non-workers include people who did not travel to their primary place of work on the survey day, who have no job, or who may simply have had the day off); despite this flexibility, however, they still travel 85 to 100 minutes per day, even without a work commute.

Consulting relatively exhaustive reviews on the subject, however, we are pressed to think a bit deeper about whether a travel time budget really exists. One needs to be aware that almost all travel models aim to predict variations in travel distance or time as a function of socio-demographic conditions, occupations, residential location or even the type of car a subject drives. Quite simply, these models assume that variations exist and reviewing their results suggests that "the claim of the definitive existence of *constant* travel time and money budgets in time and space is not supported." [3] Whether or not a travel time budget exists in the aggregate—but may break down in the disaggregate—it is worth considering the idea that any moderate reduction in travel time will result in increased time in activity.[a] If there is a fixed daily time budget (1,440 minutes) that can be allocated between travel and activities, Figure 6.2 shows what might happen if there were a capacity expansion that reduced delay (travel time), before any other change in behavior (such as making additional trips, or different trips) took place. Box 6.1 considers the implications of multi-tasking.

Time use and travel data

Considering that out-of-home activities are responsible for the bulk of traffic, we focus on the causes of travel. How do analysts gather detailed data on

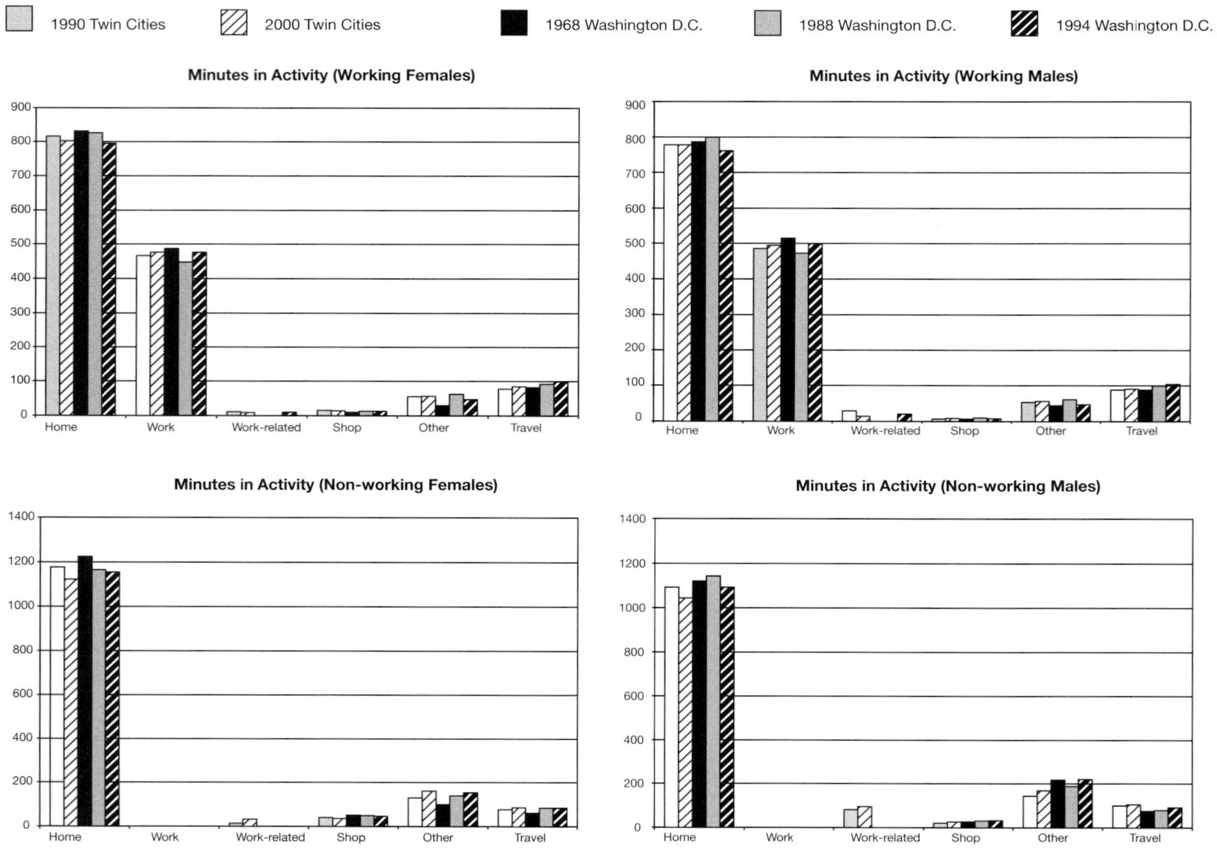

Figure 6.1 Mean minutes spent in various activities, Minneapolis-St. Paul and Washington, DC

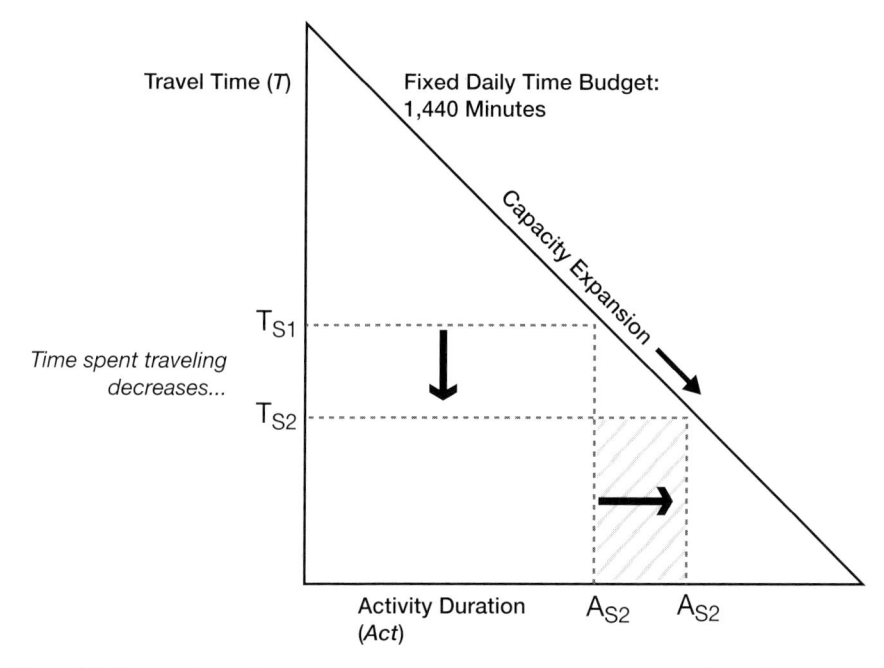

Figure 6.2 Comparison of time spent in various activities versus required travel time

travel and activities to begin with? The most common approach is to rely on information from travel surveys, also known as travel diaries or home interview surveys (see Box 6.2 for background). These surveys come in a variety of forms and have evolved over the years. Usually, researchers ask a sample of individuals to complete a travel diary form (Figure 6.3), in which respondents record the time, mode, purpose, and destination of travel to different locations. Often, the agency administering the survey will complement mailings with telephone calls to verify responses and collect additional information.

For purposes of analysis, data from the surveys is organized in specific ways. The information is typically stored in a series of relational database tables normalized to share a key field, which allows the information in different tables to be linked. A typical travel survey database might have four tables: Household, Person, Vehicle, and Trip. Each entry in the Household table will be keyed with a household identification number, as well as a variety of variables that describe each household (e.g., location, income, household size, number of adults, number of children, house type). Entries in the Person table identify the household to which each person belongs (the Household identification number), and contain a number of variables about the individual (e.g., age, employment status, gender). Entries in the Vehicle table contain an identifier to match them to entries in the Household table (vehicles are used by households), and for each motor vehicle the household owns lists the year,

Box 6.1 More than 24 hours in a day?

We used to be more confident that the daily time people devoted to activities (what they did) and travel (how they got there) summed to 1,440 minutes. Is it possible that there are now more than 24 hours (1,440 minutes) in a day? Two related and recent phenomena cloud the issue, making it increasingly possible for people to perform multiple tasks simultaneously and thereby log more than 24 cumulative hours of activity and travel.

The first phenomenon is that our society is increasingly well versed in multi-tasking (aided in part by technology, no doubt). For example, people increasingly watch television while eating dinner, socialize via email while working at the office, and use cellular telephones and text messaging to trade stocks while vacationing on the ski slopes.

The second is that technology allows—and encourages—us to work while traveling. People all over the country carry on business via cell phones while commuting; many transit agencies (Seattle, Washington is one example) offer wireless technology on transit vehicles, allowing people access to their work files or to check emails while commuting. These phenomena are leading researchers to question the value of typical data collection methods. Do our ways of collecting data accurately represent behavioral patterns and decision-making processes?

Box 6.2 Travel surveys

The 1927 Cleveland Regional Area Traffic Study was the first metropolitan planning attempt sponsored by the federal government of the United States, but the lack of comprehensive survey methods and standards at that time precluded the systematic collection of information such as travel time, origin and destination, and traffic counts. The first systematic US travel surveys were carried out in urban areas after the Federal-Aid Highway Act of 1944, which permitted the spending of federal funds on urban highways. [4] A new home-interview survey method was developed in which households were asked about the number of trips, purpose, mode choice, origin and destination of the trips conducted on a daily basis. In 1944, the United States Bureau of Public Roads published the *Manual of Procedures for Home Interview Traffic Studies*. [5] This new procedure was first implemented in several small to mid-size areas.[a] Highway engineers and urban planners made use of the new data collected after the passage of the 1944 Highway Act by extending federally sponsored planning to include data collected from travel surveys, in addition to traffic counts, highway capacity studies, pavement condition studies and cost-benefit analyses. Soon the method was diffused to metropolitan areas across the US and then to other developed countries, and more recently to developing countries.

[a] The first surveys with the new procedure were in: Lincoln, Nebraska; Little Rock, Arkansas; Kansas City, Missouri; Memphis, Tennessee; New Orleans, Louisiana; Savannah, Georgia; and Tulsa, Oklahoma

01/17/01 traveldiary
NATIONAL HOUSEHOLD TRAVEL SURVEY
TRAVEL DIARY

Respondent Label with Name
Travel day and date

At the beginning of my travel day (4:00 a.m.) I was:
☐ Home ☐ Some other place

WHERE did you go? (Name of place)	What **TIME** did you start and end each trip?		**WHY** did you go there?	**HOW** did you travel?	How **FAR** was it? (blocks or miles)
	Started at:	Arrived at:			
EXAMPLE: *West Park Theater*	*2:00 p.m.*	*2:55 p.m.*	*To see a movie*	*walk, bus, walk*	*6 miles*
1.					
2.					
3.					
4.					
5.					
6.					
7.					
8.					
9.					
10.					

Please remember, a trip is whenever you go from one address to another.
Have this travel diary by the phone when the interviewer calls. You do not need to mail the diary back to us. Thanks.

OMB Nos.: 2139-0008; 2125-0545
Expiration date: xx/xx/xx

Figure 6.3 Example travel diary form

make, and model of vehicle, and perhaps a reference to the primary driver in the Person table. The Trip table lists every trip recorded, and for each trip has keys to link to the Vehicle, Person, and Household tables (described in greater detail in Box 6.3).

Trip records, by their very nature, are relatively detailed. However, many short trips go unaccounted for. Walking from one store to another while the car remains parked may technically be a trip, but is unlikely to be recorded in a diary for a variety of reasons. First, people tend not to adequately remember (or even consider) their walking trips because these trips are so short or take so little time. Second, instructions included with travel surveys or diaries typically encourage respondents to record trips as defined by differing addresses. Although multiple trips within a mall or commercial center may technically visit different addresses, they are often considered to be part of the same location. Third, some surveys ask respondents not to even consider trips less than a prescribed distance, such as a quarter-mile (400 m). The good news is that advancing technologies are making it easier to gather detailed data. Newer techniques, in which Global Positioning System (GPS) tracking devices are carried by individuals over the course of a day or week, are able to better record information—so long as travel occurs outside of buildings, where GPS signals are available.

Box 6.3 Language of travel behavior

The task of transportationists is to make sense of the morass of people's travel; to understand it for purposes of predicting, modeling, or prescribing effective policies. Doing so requires both a vocabulary and a protocol to codify such behavior.

The most basic unit of travel is the point-to-point trip by an individual traveler. Information about trips is typically recorded in a 24-hour diary format in the Trips table described in the above text. The data reveals the location of the destination to which an individual traveled, the purpose of the travel to that destination, the time and mode of travel, and whether the traveler was accompanied by any other people. Because activity data are not always available in travel diaries (we only have trip information), activities must be inferred from trip purposes. That is, we can reconstruct their time before the next recorded trip. The drawback, of course, is that in-home activities are undifferentiated (e.g., watching television is the same as eating a meal). As shown in the last column in Table 6.1, we learn that the travel time was 75 minutes, spread out over five trips, while the time spent at activities was 1,365 minutes.

Trip-related data is commonly analyzed in terms of trip frequency, distance, and mode split (i.e., the portion of all trips taken by a specific mode of travel). Cumulative travel statistics are gleaned by summing trip information to acquire total travel time or total travel distance. These surveys, however, regard each segment of travel as an independent observation when in reality travel is a sequence of trips that are linked together when one leaves home. An analytical technique for taming the complexity of travel involves organizing travel into multi-stop trips, commonly known as tours or trip chains. Tours are the sequence of trips that combine all travel every time one leaves home. Simple tours would involve one stop (i.e., home to work to home) whereas complex tours would involve more than one stop.

Individual trip purposes depend on the level of detail of the travel survey and typically include such destinations as work, school, shopping centers, appointments, others' homes, religious locations, and others. Using the information from the last column of Table 6.1, it is possible to sum the activity duration for each location. Given the vast array of types of activity, however, parsimony has its role. Therefore, many efforts follow the lead of Israeli geographer Shalom Reichman [6] and group activities into three major classes of travel-related activities. These activities represent subsistence activities (to which members of the households supply their work and business services; travel associated with this activity is most commonly commuting); maintenance activities (consisting of the purchase and consumption of goods or personal services needed by the individual or household); and leisure or discretionary activities (comprising multiple voluntary activities performed during free time, not allocated to work or maintenance activities).

Table 6.1 Example records in a trip file

Person ID	Trip ID	Origin zone	Dest. zone	Origin purpose	Dest. purpose	Mode	Travel time	Start time	End time	Time spent (min)
1	1	301	415	Home	Dropoff	Car	15	8:30	8:45	30
1	2	415	225	Dropoff	Work	Car	15	9:15	9:30	360
1	3	225	607	Work	Shop	Car	15	15:30	15:45	105
1	4	607	609	Shop	Pickup	Car	10	17:30	17:40	20
1	5	609	301	Pickup	Home	Car	20	18:00	18:20	850
					Totals		75			1365

Hagerstrand's space-time prism

One powerful method for describing a person's travel is to map out a space-time prism. First popularized by Torsten Hagerstrand, [7] a space-time prism represents travel duration, activity duration and distance from home in three dimensions (similar to the two-dimensional Diamond of Action introduced in Chapter 2). The prism maps the duration and complexity of travel away from home by enclosing the locations a person can reach considering various time and distance constraints. The principal advantage of space-time prisms is that they are able to represent the total travel and its relation to activities and individual trips.

As shown in Figure 6.4, the shape of the prism represents the set of opportunities (chances) one has over time and space. The boundaries of the prism are determined in large part by the temporal and geographical constraints the individual imposes. The vertical axis represents time, the X and Y axes represent space over which travel can take place. If one remains at a location, time advances but the location stays the same, represented by a dashed vertical line. The diagonal solid lines signify travel, which takes place over both time and space. Everyone's time-space prism differs, but everyone has one.

A problem is that space time-prisms differ considerably in terms of the destinations to which people travel. The average household in the United States currently makes almost ten person-trips per day. [8] When many itineraries are depicted, these assorted trips begin to resemble games of "pick up sticks," with lines in all directions. Because of this, travel behaviorists tend to analyze aggregate information across populations.

Travel behavior: trips or tours

To the extent that travel is derived from the demand for activities, the prism proves a good starting point. Most travel is routine; people travel to familiar destinations, traversing similar routes. For many trips, there tends to be little variation among the physical places they frequent daily (or at least semi-daily)

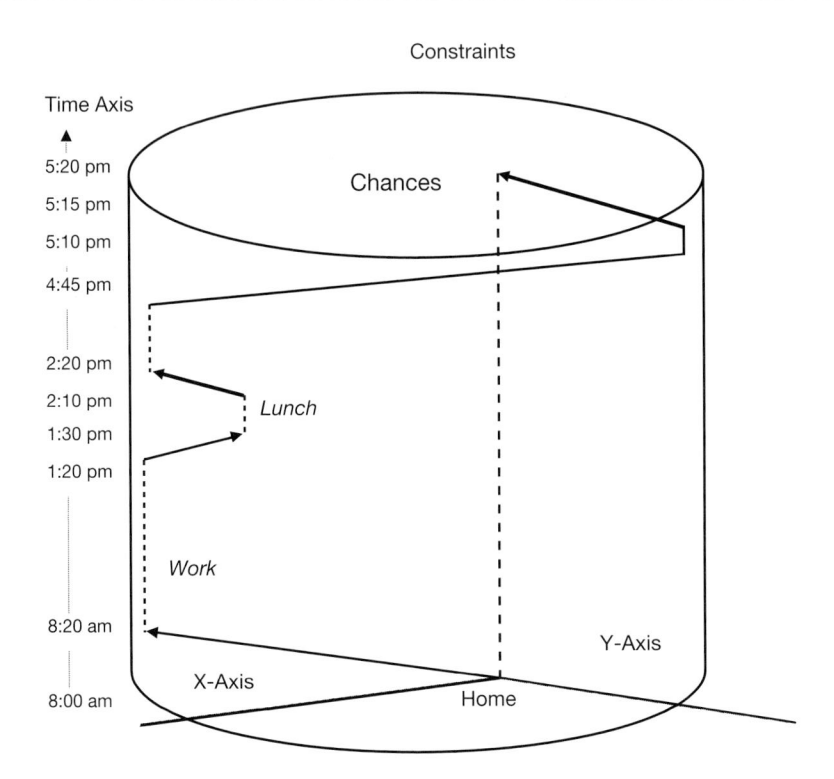

Figure 6.4 A space-time prism

such as workplaces, day care centers, recreational areas, or grocery stores. Wider variation exists in the places people go to once a month (e.g., the hardware store, department store, etc.).

The United States Department of Transportation uses the National Household Travel Survey, administered on average each half-decade, to keep tabs on these matters. Figure 6.5 shows the average trip distance for different purposes for five different iterations of the survey. Although trip distances have been slowly climbing over the years, there are two purposes that stand out from a distance standpoint: work and leisure. Conversely, shopping-related travel enjoys the shortest distances.

But how frequent are these trips? Figure 6.6 provides a more detailed breakdown by purpose. We learn that travel for purposes related to family or personal business and social/recreational travel comprise over 70 percent of all trips. Other types of trips include going to work, to the doctor, to dinner, to visit friends or to the movies. Aside from work, these latter trips are less frequent and tend to be more temporally and geographically dispersed throughout a region.

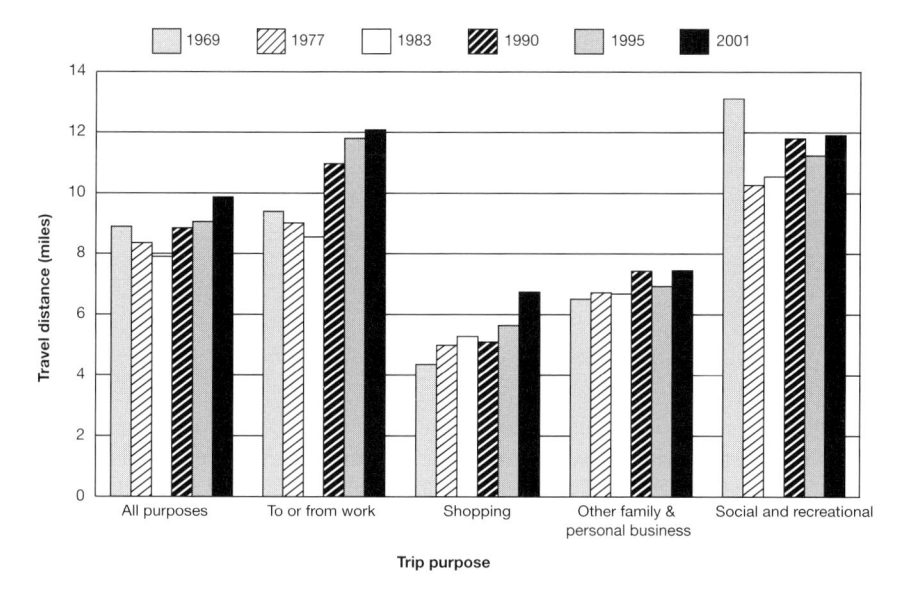

Figure 6.5 Average trip distances, by trip purpose.

Source: National Personal and Household Travel Surveys of respective years, US Department of Transportation

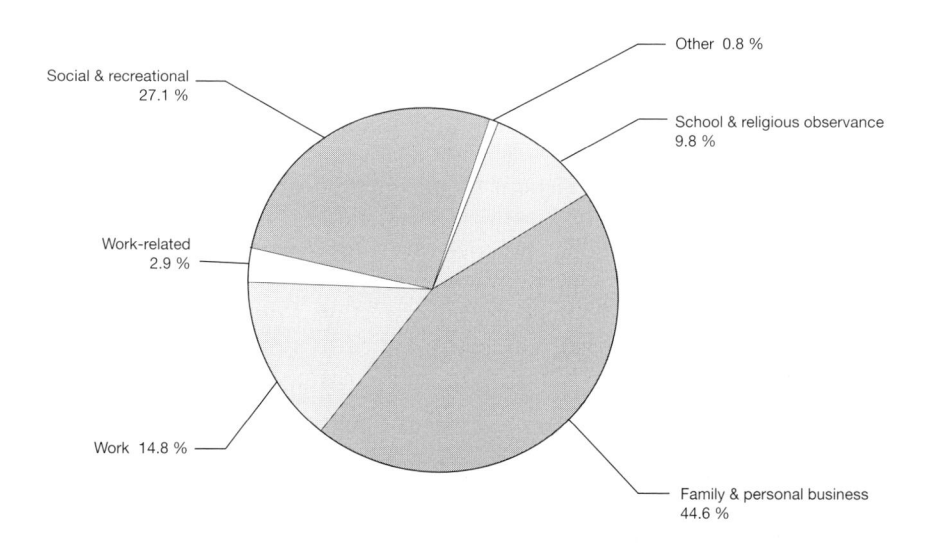

Source: 2001 National Household Travel Survey, Daily Trip File, US Department of Transportation

Figure 6.6 Frequency of travel, by trip purpose

Figure 6.6 catches many off guard because of the relatively low rates of the work trip, not even exceeding 15 percent of all trip-making. If work trips are clearly not the most frequent type of trip, why do they receive the bulk of attention from a transportation perspective? The answer is in large part due to complementors; people need to work and interact with others when they work (similarly, people shop when stores are open, when complementors shop). One need only witness the dramatic onset of trips (all purposes) between the hours of 7A.M. and 7P.M., when people complete most business transactions (Figure 6.7a).

For the employed population, most of their travel involves heading to work between 7.00 and 9.00A.M. Monday through Friday and returning home between 4.00 and 6.00P.M. (thus the dramatic spikes in Figure 6.7b). Although work travel is far from the largest portion of travel, together with travel related to school and religious observance, it is clearly the most peaked. These peaks lead to the constraints on most transportation networks (e.g., congestion: utilization of available capacity on a facility) thus warranting the attention it earns. E-commerce and electronic mail is relaxing this constraint slightly by no longer requiring real-time interactions (between 8 and 5 during the day), but it is hard to say how strong the impact will prove to be.

It is in large part because of the other competitors using the transportation network during rush hours that we see so much travel pursued outside of these time windows. Where and how people get to activities that *do not* involve their work destination comprises an increasingly important and growing dimension of travel. [9] Part of the problem is that these trips (85 percent of the total) are widely dispersed in character, ranging from doctor's visits to visiting friends to getting a quart of milk. One could, for example, separate non-work activities into two distinct categories: non-work trips that are relatively habitual in nature, and those less regular. The former may vary in frequency but usually involve going to the same destination (e.g., grocery, doctor, health club). The latter group is less regular and (one might hypothesize) more susceptible to being influenced by information sources, advertisements, word of mouth, etc. . . . (e.g., hardware store, clothes, appliances).

Although the longstanding practice in transportation research is to independently consider each place to which people travel, this practice has its limitations. Analyzing individual trips in isolation from the previous and subsequent trips to which they may be connected tells researchers and policy analysts an incomplete story. This is especially true because an estimated 44 percent of all travel away from home combines two or more trips [10] and a reported 27 percent of all commutes combine multiple trips. [11] Looking at the larger pattern of linked trips is to be preferred because linked trips—in combination with basic forces—are what determine the nature of an individual's travel. Travel decisions are often a product of where people were before and where they plan to go—phenomena that go unrealized in a strict trip analysis.

While the idea of a multi-stop journey is straightforward, breaking the concept into different categories proves more difficult. Two decades of research

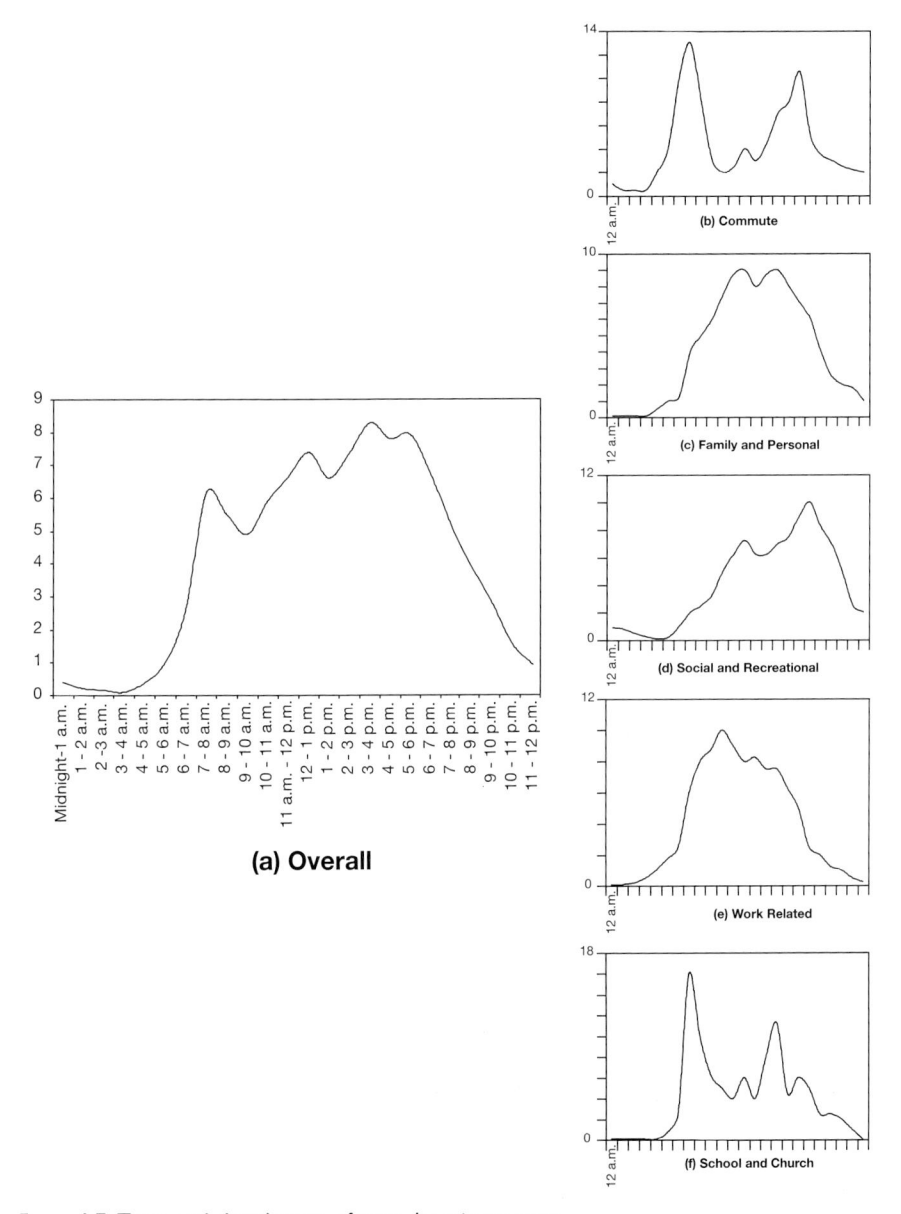

Figure 6.7 Temporal distribution of trips, by trip purpose

suggest strategies to circumvent what have been referred to as the "isolated trip approach." [12] Examining multi-stop journeys recognizes that travel is a function of many factors including types of destinations, previous destinations, subsequent destinations, travel mode, and household and individual characteristics. Multi-stop journeys provide an intuitive way to grasp the interrelated decision process of linked trips and therefore represent a cornerstone of current work in activity-based transportation modeling.

By convention, the literature most often labels multi-stop journeys as tours (sometimes called trip chains) and defines them in terms of the home-to-home loop. Tours are most commonly analyzed by the number of component trips (i.e., stops). Simple tours contain two trips (e.g., home to work and then work to home) and complex tours contain more than two trips. Analyzing the nature and frequency of simple versus complex tours, however, only considers one dimension of the tour: number of stops. It does not do justice to how a separate dimension of travel—purpose—influences the nature of tours; this is illustrated in the following thought experiment described in Box 6.4.

Using tours as a unit of analysis prompts an important challenge—how to assign a single purpose to what is often a multi-trip/multi-purpose tour? To better capture how different purposes of travel—a nominal variable—interact with trips, classification emerges as a preferred strategy. As the lowest form of measurement, classification allows many variables to be considered simultaneously (e.g., the purpose and number of trips on a tour).

Several depictions of travel have been used in prior classification schemes. For example, some research developed a similarity index of travel activity to identify single types of travel for a person over a day. [13] Other approaches group similar types of activities, but allow greater flexibility in how tours are coded. [14] Some strategies develop an elaborate typology of tour-types analyzing the transitions between activities. [15, 16] Others simply look at work-related trips to discern if there is a stop along the way. [17, 18]. Common themes emerge from each of these strategies.

First, the sequence of consecutive trip links that begin and end at home form the predominant way to classify a tour. Second, many strategies use a simple binary system—work versus non-work—to differentiate between travel purposes within a tour; others specify more detailed non-work trip purposes. All approaches provide a separate category for simple tours, yet they all differ in terms of how they deal with the combinations and permutations present in more complex tours.

Any classification scheme depends on the particular purpose of the study or application. Detailed coding schemes have the advantage of more precisely tracking a sequence of detailed travel activities with different purposes. Although even 20 classifications of tour type do not capture all possible trip-purpose combinations, the enormous number of tour combinations produced by matching a more modest set of eight trip purposes with number of trips requires such complex bookkeeping that it is difficult to put into practice. Simple coding schemes, on the other hand, are limited because they do not

Box 6.4 Does travel have positive utility?

The previous chapters introduced and explained the concept of utility as a measure of the happiness or satisfaction gained consuming good and services. The prevailing thought in the transportation industry is that time in travel has a negative utility; we typically want to consume less of it. Much of the discussion in this chapter is based on the premise that destination characteristics strongly influence trip characteristics. The mantra goes: travel is a derived demand, and therefore we only do it when we want to get somewhere else or when necessary.

These premises, however, are increasingly being challenged, and the evidence behind the challenges appears to have merit. For example, in looking at distance traveled, Patricia Mokhtarian and colleagues [19] discovered that subjective variables such as a liking for travel, the adventure-seeker personality trait, the travel stress attitudinal factor, and the excess travel indicator added considerable explanatory power to the standard variables traditionally used in the analytic models used to predict travel behavior. Lest one think such findings apply only to non-work travel, Bruce Hamilton [20] suggests that, compared to model-predicted values for shortest commutes, people's actual commutes were on the order of eight times longer.

Can the positive utility of travel be shown through such measures? Although certainly not confirmatory, it is worth considering that there is likely a threshold separation distance (or time) between home and work. This may have anthropological or other roots. [21] Many people prefer a certain degree of separation, a fact that increases the difficulty of implementing planning initiatives intended to bring travel destinations (e.g., employment) within close proximity to home sites.

If there is a positive utility to travel, it likely plays out differently by time of day, mode, and purpose of travel (the importance of promptness, for example, may be supressed for leisure travel). Figure 6.8 presents different depictions of the utility of travel. Figure 6.8a presents hypothetical values representing the conventional view of the negative utility of travel (taken to an extreme, the assumption being that zero travel has the greatest utility). Figure 6.8b shows that, for many people's automobile commutes, there may be a positive relationship that peaks around 15 minutes; 6.8c illustrates a similar, relationship peaking at 20 minutes if walking. Alternatively, 6.8d shows a longer postive relationship for leisure travel such as the proverbial Sunday drive. Those peaks may have something to do with the presence of opportunities given where people live (I may live in a community with no shops within ten minutes, so my minimum travel is ten minutes), however, it should be noted, people choose to live in those communities. The overall shapes of the different curves, however, depend on the purpose, preferences or mode of travel of the individual.

(a) Conventional wisdom

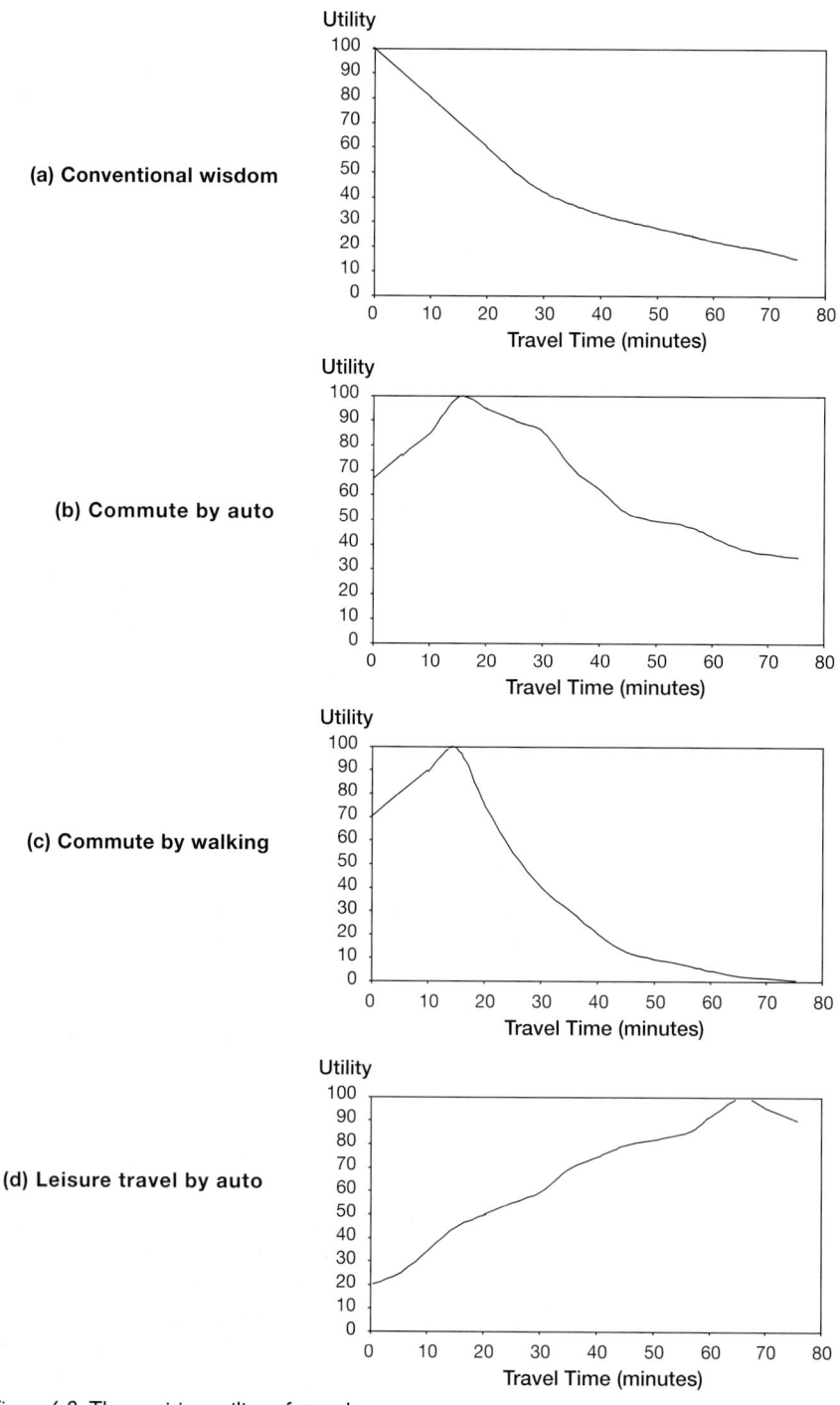

(b) Commute by auto

(c) Commute by walking

(d) Leisure travel by auto

Figure 6.8 The positive utility of travel

Table 6.2 Classifying trip tours

Type #	Tour type	Coding
1	Simple work	H-W-H
2	Simple maintenance	H-M-H
3	Simple discretionary	H-D-H
4	Complex work only	H-W-W- ... -H
5	Complex maintenance only or	H-M-M- ... -H
	Complex discretionary only	H-D-D- ... -H
6	Complex work + maintenance only	H-W-M- ... -H*
7	Complex work + discretionary only	H-W-D- ... -H*
8	Complex maintenance + discretionary only	H-M-D- ... -H*
9	Complex work + maintenance + discretionary	H-W-M-D-H*

Note: * Tripmaking could take place in any order

differentiate between types of non-work activities—activities that may have very different travel characteristics. To be useful and practical, a taxonomy has to be simple and clear, yet travel is so complex that any classification scheme is limited in the incremental advancement it provides. Aggregating trip types into the three groups (subsistence, maintenance, and discretionary trips) provides a basis to code and analyze different combinations of tours that is more economical than using eight different activity types, but more detailed than the simple work/non-work dichotomy. The end result is nine different types of tours capturing complexity and trip purpose as shown in Table 6.2.

Scheduling time wrap up

Like long and medium-term transportation and location decisions, short-term decisions related to what activities to pursue, how often to travel and in what manner are shaped by complementors and competitors who, in turn, affect the constraints and chances of an individual. Complementors make opportunities available, influence how long it takes to complete activities or trips, and may even join us. Competitors sit at the table we want and make us take longer to get there.

Our own decisions are the product of these offsetting forces and of our own underlying preferences, which are satisfied by working/producing (e.g., changing time and effort into money) and shopping/consuming (exchanging money for someone else's time and effort). This basic process is so fundamental that it survives changes in economic systems (from feudalism to capitalism or communism). Policy can influence this process, but certainly not transform it into regulating, for example, when people shop for groceries.

Our task as transportationists is to understand the travel component of the transactions described above—a task for which there is no shortage of potentially relevant statistics. The most common measures come in the form of distance and number of trips. More robust measures consider the mode of travel. Measures can be averaged at the level of a single individual, summed at the level of a household, or aggregated to other units of analysis (e.g., transportation analysis zones or even entire metropolitan regions).

Where to travel, when, by what mode and using what route—all these factors come together as we consider the sequencing, timing, and subsequently, patterns of trips people complete. The net result is a difficult-to-predict concoction of behaviors that may be perfectly rational from the viewpoint of one individual but entirely irrational for another. Understanding why and how people sequence and time their trips leads to a deeper understanding of decisions about mode or route choice. The convenience of a given place to one's home or workplace location it is part of the equation. However, passenger travel in metropolitan areas is affected by a variety of factors—a mixture of demographics, habits, culture, and preferences, with a sprinkle of geography—but mostly a healthy portion of magical dust. There is a good reason why even the most robust multiple regression models designed to predict travel distance rarely explain more than 30 percent of the observed variation. Robert Levine's research, highlighted in the introduction to this chapter, noted cultural factors as one source of variation. The other sources of variation are more difficult to pin down but no less important.

Note

a   Strictly subscribing to the travel time hypothesis, one could suggest that such additional time in activity would be translated back into additional travel ... until a person's so called "budget" is fulfilled.

References

[1] Levine, R., *A Geography of Time: The Temporal Misadventures of a Social Psychologist, or How Every Culture Keeps Time Just a Little Bit Differently*, New York: Basic Books, 1997.
[2] Zahavi, Y.A. and Ryan, J.M., 'Stability of Travel Components over Time', *Transportation Research Record*, 1980, vol. 750: 19–26.
[3] Mokhtarian, P.L. and Chen, C., 'TTB or not TTB, That is the Question: A Review and Analysis of the Empirical Literature on Travel Time (and Money) Budgets', *Transportation Research Part A-Policy and Practice*, 2004, vol. 38 (9–10): 643–675.
[4] Weiner, E., *Urban Transportation Planning in the United States: An Historical Overview*, Washington, DC: United States Department of Transportation, 1997.
[5] United States Department of Commerce, *Manual of Procedures for Home Interview Traffic Studies*, Washington, DC: US Government Printing Office, 1944.
[6] Reichman, S., 'Travel Adjustments and Life Styles: A Behavioral Approach', in P.R. Stopher and A.H. Meyburg (eds.), *Behavioral Travel-Demand Models*, Lexington, MA: Lexington Books, 1976.
[7] Hagerstrand, T., 'What About People in Regional Science?' *Papers of the Regional Science Association*, 1970, vol. 24: 7–21.

[8] Hu, P.S. and Reuscher, T.R., *Summary of Travel Trends: 2001 National Household Travel Survey*, Washington, DC: United States Department of Transportation, 2004.

[9] Handy, S.L., DeGarmo, A., and Clifton, K.J., *Understanding the Growth in Nonwork VMT*, College Station, TX: Southwest University Transportation Center, 2002.

[10] Krizek, K.J., 'Neighborhood Services, Trip Purpose, and Tour-based Travel', *Transportation*, 2003, vol. 30: 387–410.

[11] McGuckin, N., Zmud, J., and Nakamoto, Y., *Trip Chaining Trends in the US—Understanding Travel Behavior for Policy Making*, presented at 84th Annual Meeting of the Transportation Research Board, Washington, DC, 2005.

[12] Damm, D., 'Parameters of Activity Behavior for Use in Travel Analysis', *Transportation Research*, 1982, vol. 16A (2): 135–148.

[13] Pas, E., 'Analytically Derived Classifications of Daily Travel-Activity Behavior: Description, Evaluation, and Interpretation', *Transportation Research Record*, 1982, vol. 879: 9–15.

[14] Bowman, J. and Cambridge Systematics, *A System of Activity-based Models for Portland*, 1998.

[15] Golob, T., 'A Nonlinear Canonical Correlation Analysis of Weekly Trip Chaining Behavior', *Transportation Research A*, 1986, vol. 20 (5): 385–399.

[16] Southworth, F., 'Multi-destination, Multi-purpose Trip Chaining and its Implications for Locational Accessibility: A Simulation Approach', *Papers of the Regional Science Association*, 1985, vol. 57: 108–123.

[17] Ewing, R., Haliyur, P., and Page, G., 'Getting Around a Traditional City, a Suburban Planned Unit Development, and Everything in between', *Transportation Research Record*, 1994, vol. 1466: 53–62.

[18] Hanson, S., 'The Importance of the Multi-purpose Journey to Work in Urban Travel Behavior', *Transportation*, 1980, vol. 9: 229–248.

[19] Mokhtarian, P.L., Salomon, I., and Redmond, L.S., 'Understanding the Demand for Travel: It's not Purely "Derived"', *Innovation: The European Journal of Social Science Research*, 2001, vol. 14 (4): 355–380.

[20] Hamilton, B.W., 'Wasteful Commuting', *Journal of Political Economy*, 1982, vol. 90 (5): 1035–1053.

[21] Marchetti, C., 'Anthropological Invariants in Travel Behavior', *Technological Forecasting and Social Change*, 1994, vol. 47: 75–88.

Chapter 7

Diamond of Exchange

"I only do business with the people I do business with. The people I do business with find out I do business with the people I don't do business with . . . I can't do business with you."

The character, Fred, in the film
Atlantic City, when speaking to
a young drug dealer

Kevin, an avid cyclist, is in the market for a new bicycle (the seven he owns are insufficient). He could legally purchase one at many places: a neighborhood bike shop, a high-end regional bicycle store, a used-bike vendor, a newspaper's classified advertisement, eBay, or even police auctions of found and recovered bicycles.

Kevin selects the high-end store, Grand Performance Bicycles, and purchases a bike. He provides $2,000 to the store and they provide him with the bicycle (he actually gives them his credit card, and they give him a bike, and then Kevin pays his credit card bill).

Where did the store get the bicycle? In what shape was it? Grand Performance Bicycles received a package from their distributor containing a disassembled bike, manufactured by Orbea. In addition to the supplied parts in the main package, they fitted the bicycle with tires from Michelin and components from Shimano. The store also provided several value-added elements such as handlebar tape and water bottle cages. They made the bike easily available to the buyer, allowing him to test ride it. They assembled the vehicle, saving on shipping costs for the manufacturer. They customized elements of the bike, matching the color of the frame to the tires (and to the handlebar tape as well).

The distributor receives bikes and parts from several manufacturers and delivers them to thousands of shops across the country. The bike is transported from the distributor to the store via truck. The shipment required a driver, fuel, as well as a truck and all of its constituent elements. Each of those elements is fed by a network of its own.

Shimano, founded in 1921 as Shimano Iron Works, is a leading Japanese manufacturer of bicycle components (and fishing tackle), and for a long time has held a dominant market share in components for derailleur-equipped bikes (a derailleur is the device to move a bicycle chain from one exposed gear to another). Shimano has its own supply chain, which includes raw materials such as steel, for which it bids against numerous other companies. The rise of steel prices in Asia (due to a building boom in China) drives up the price of Shimano products. The fact that bicycle purchasers know the name Shimano is also interesting, as most products are known by the name of their final packager (an Apple Macintosh, a Ford Mustang, a Maytag Neptune washing machine, Kellogg's Frosted Flakes), rather than by the names of component makers.

Manufacturers like Shimano, based in Sakai, Japan (known as "bicycle city"), near Osaka, ship bikes and components to distributors. However this journey is more complicated. Most bicycle manufacturers are in East Asia, so such shipments typically need to cross an ocean to reach markets in the Americas or Europe. Bicycles from a manufacturer are placed into a standard shipping container at the factory and trucked by a private carrier to a port (say Osaka, Japan's second largest city and port) where the container is lifted from the truck to a cargo ship. The ship carries the bike to a port in the United States (say Los Angeles-Long Beach, the largest port on the US West Coast), where it is off-loaded from the ship and transferred to a short-distance truck, which takes it inland for reloading onto a long-distance truck, which takes it to a distribution warehouse. (Interestingly, Shimano's United States headquarters is in Irvine, California, a suburb of Los Angeles.) The manufacturer makes the bicycle out of raw commodities—steel, titanium, aluminum, plastics, etc. Each one of those commodities it purchases in the marketplace. A similar story takes place for each transaction, however minute.

This chain of events—together with the multiple parties involved—begs the question of the underlying motivation behind why firms—be they office-based, industrial, or retail—decide to locate where they do. How is location affected by supply chains? Why do retailers choose the size and number of stores that they do? When are retail chains successful compared to locally grown businesses? Why do developer actions seemingly contradict what many planners would define as more sustainable planning?

Part 1 of *Planning for Place and Plexus* examined the motivations and decision-making processes of individuals (households) over long, medium, and short-term durations. We discussed how various theories, games, constraints and other factors influenced their overall utility, which in turn influenced individual choices. We now embark on Part 2 to discuss how such factors play out with respect to different agents—private firms—who flex their muscle to shape the urban landscape.

Firms come in various shapes and sizes. One breakdown would differentiate between developers (those responsible for making decisions of where to build, what to build, and how much floor space to build) and locators (those responsible for deciding where to place their business to receive maximum return). Developers, however, respond to and anticipate the needs of locators (occupants who would locate in the buildings that are developed). The importance of both developers and locators has been underestimated and undervalued in the typical understanding of how cities are formed.

After first describing criteria to understand the general behavior of developers, our focus in this chapter quickly turns to locators and their needs. Among those who are locating (removing, of course, household locators, already discussed), the business world can be divided into non-retailers and retailers (Chapters 8 and 9, respectively). In terms of their impact on the transportation sector, both make location decisions based in part on the location of suppliers (including potential employees) and customers. Many non-retailers "sell" goods, and all retailers employ people; however, non-retailers (including office, industrial, and other types of organizations) make different calculations than retailers.

Developer behavior

Developers are ultimately the ones providing space for both retailers and non-retailers. Developer behavior, like that of individuals, is inherently complex. The topic enjoys considerable study by practitioners of regional science, applied economics, business and management, geography and other disciplines. The topic may be addressed at the global, state, or municipal levels; we focus the discussion on factors central to the land use-transportation planning environment within metropolitan areas, describing how elements of place and plexus affect development decisions. Although impossible to ascribe a series of principles to every development project, enough study on the subject hones in on the following ten factors that affect the location and rate of development (in roughly the following order) [1]:

1 market velocity (how active is the specified market?);
2 price of land;
3 availability of hard infrastructure (capabilities related to roads, water, sewers);
4 access choices (intersections, frequency of existing transit services, parking);
5 human infrastructure (education of workforce, nearby school quality, housing, day care);
6 physical character (quality surrounding district, vitality, views and vistas);
7 environmental quality (healthy air and water);
8 predictability (no dramatic changes in zoning or character, appropriate capital improvement plan);
9 amenities (parks, restaurants);
10 available financing.

By this point in the text, readers of *Planning for Place and Plexus* should not be surprised that each of the above criteria in some way relate to accessibility (some more directly than other). *Hard infrastructure* and *access choices* are transportation issues. Firms desire transportation to connect with their suppliers, workers, and customers. Among their supplies are water and sewer services, for which specialized pipeline transportation systems have been created. Water pipes simply move water from its source to its users, and sewers move waste from its source to somewhere downstream (where hopefully it is treated). *Human infrastructure* is also an accessibility issue—can the firm find an appropriate nearby workforce so it doesn't have to pay higher salaries to compensate for long commutes? For many industries, this question affects not only where in the city a firm locates, but what city the firm chooses in the first place. If a firm specializes in designing chips for computer network routers, it is more likely to choose San Jose in California, where there are many engineers who will have a short learning curve, than San Jose in Costa Rica, which despite the presence of an Intel manufacturing plant, has a population with a different skill set. *Physical character* and *environmental quality* are also accessibility issues, based on the desirability of access to aesthetically pleasing surroundings and access to clean air and water. Finally, *amenities* are simply access to recreational and other non-work destinations. *Price* too is an accessibility issue. Price is determined by the value of land, and the value of land is simply the market's assessment of the value of its accessibility to all of the other things it cares about.

Although the above list includes ten different factors, one macro-dimension—accessibility—manifests itself throughout each. However, the weights associated with different types of accessibility (e.g., access to workforce, materials, customers, amenities, friendly government, etc.) are different for each developer and each firm. Some firms just need to be close to a single supplier whereas others seek an entire district with many potential opportunities available; some firms cater to a single customer whereas others serve a variety, and so on.

Although money markets are global, money chases the highest rate of return at the lowest risk. The rate of return and risk vary by location. Places that markets see as growing will attract capital at lower interest rates than places with low growth (and thus a high risk that the investment will not pay off). Spatial differences in *financing*, thus, too, depend on accessibility. *Market velocity* is a measure of growth. Areas which are growing have a higher turnover than those that are stagnant, making it easier to enter or leave; an investor in a growing area is less likely to be stuck with a long-term lease that cannot be broken. *Predictability* concerns local government. The appropriate capital investment plan enables the provision of accessibility in the future. The stability that zoning often provides helps guarantee that the services that are accessible today will remain so tomorrow (with the negative side effect that future opportunities are being foreclosed). In other words, what firms really care about are different notions related to accessibility.

Why firms? Why markets?

Developers usually construct buildings with a specific occupant or type of occupant in mind. These occupants are in many respects the real agent of interest and are usually firms. Ronald Coase, in his 1937 essay 'The Nature of the Firm', [2] and Oliver Williamson, in 1975's *Markets and Hierarchies* [3] asked the question "Why do firms exist?" If markets are so good, every transaction can take place in the marketplace, yet many take place within firms (the office has a copy machine, rather than everyone going to a copy store to make copies). The answer is that firms reduce transaction costs compared with markets, even if the other costs may be higher due to a lack of scale economies. In the case of the office photocopy machine, less time is required to make copies within the building than to go out; in addition to the savings in terms of travel time, no purchase order or reimbursement form is required to make copies within the office, all of which reduces costs. (Though having to key in an eleven digit number to use the copier, as required by some bureaucratic organizations, increases those costs.) The firm may reduce costs even though a single office copier is likely to break down, whereas at the copy store, the presence of multiple machines reduces the likelihood that all machines will be simultaneously broken.

On the other hand, we can ask the converse question: "Why do markets exist?" If hierarchical firms are so efficient, all business could be done within a single firm (or the state), yet many take place in markets. The answer to this question is that the savings in transactions costs may be outweighed by the gains from specialization and economies of scale. A small business cannot justify its own copy machine for the purpose of making two copies a day; even large businesses (e.g., automakers), cannot justify making their own steel (though once, Ford, in the pinnacle of vertical integration, did so at its River Rouge plant in Michigan, turning lumber, coal, and iron which entered the plant on one end of the assembly line into autos driving out the other end).

Box 7.1 presents some challenge to traditional notions of for-profit production that are found in software, and considers open systems. Box 7.2 looks at alternatives to "for-profit" organization as retailers organize themselves.

Diamond of Exchange

This part of *Planning for Place and Plexus* situates a firm's decision of where to locate within a context similar to the diamond individuals faced in previous chapters. Two of the dimensions—complementors and competitors—represent similar phenomena.

Complementors

Firms have complementors (e.g., Shimano has Orbea), who are in separate sectors but help the firm go about its business and achieve agglomeration economies. The classic example of complements in economics is the left shoe

Box 7.1 Firm behavior as increasingly open systems

In his influential 1997 essay *The Cathedral and the Bazaar,* Eric Raymond [4] raised a point similar to that put forward by Coase and Williamson in a completely different context. Raymond addressed the question of whether the best software is developed under monolithic and proprietary systems that centralize all development tasks under the control of a single firm, or whether the "open-source" model, in which users contribute directly to development, is superior; Raymond termed these two systems the "Cathedral" and the "Bazaar" respectively. The classic examples of the two models are Microsoft Windows and the Linux operating system. Windows was of course developed by a single company, and although it issues security patches on an almost weekly basis, Microsoft only releases major operating system changes every five years or so. Linux is much more nimble—a "self-correcting system" in the words of Raymond—and its advocates argue that as a result, it is more secure. Raymond uses in his analogy two instances in the essay that map very tightly onto Williamson *Markets and Hierarchies.*

Some argue that Raymond's Bazaar is distinct from the market, and there are certainly differences. For one, contributors to open-source projects often get paid in social capital and good feeling, which makes the open-source world more like a non-profit organization than a for-profit marketplace. (Many contributors work for other companies, which allow them to do open source work on the side or pay them to do so during business hours to help steer the community in a valuable direction. These companies, like Novell and IBM, market open-source products for profit.) On the other hand, the notion of distributed work and virtual organizations is more similar to a market rather than to a hierarchical firm, church, or government. Still, the open-source projects that Raymond praises require a coordinator (or a group of coordinators), if only to host the source code. No one has yet put forward a fully open and anarchic process for developing products. The difference is the degree of organization and control in a voluntary project compared with a corporate one.

Kevin's bicycle from this chapter's introduction is an open system; he can attach products from many vendors to the carbon fiber frame, and in principle, remove his add-ons (i.e., components) to put them on a different bicycle frame. The standardization, interchangeable parts, and many other benefits from the industrial revolution make this possible. It is also because of this the bicycle truly is more a hacker's machine than the modern automobile, with its sophisticated electronic systems. The evolution of the automobile over the past several decades may have improved reliability, but has made hacking the car (customizing it with specialty parts) much more difficult. Hot rods are exactly what they used to be, but have not made much progress since the time of George Lucas's film *American Graffiti.*

Box 7.2 Cooperation amongst competitors

When we think about corporations, whether publicly or privately held, large or small, we think of profit as their primary motivation. An advantage that large companies have over small is economies of scale—the ability to produce larger quantities at lower costs, the ability to get volume discounts from their suppliers, etc. In response to the emergence of chain stores in the early twentieth century, many small retail businesses in the United States attempted to band together to obtain some of the advantages of the larger chains. The idea seems to have begun on the west coast for grocery stores, and in the Midwest for hardware stores, and spread to other regions.

These organizations are cooperatives. These are not the consumer cooperatives that one may typically consider (described in Chapter 2) when shopping at the local natural food co-op for organic eggs, wheat germ, and hormone-free milk. Rather, these are cooperatives of businesses, and can be better compared to franchises or corporate ownership.

In the franchise model, an entrepreneur or firm comes along, develops a format, and sells the right to franchisees to use that format. Examples include McDonald's (and most other fast food chains). All franchises look very similar to each other, even if they have local proprietorship, since it is that similarity that the franchisee seeks when paying to join the system. In the corporate model, those who developed the format also own all of the local stores (e.g., Wal-Mart), and the stores are even more identical. Retailers' cooperatives (or purchasing cooperatives) are, in contrast, organized from the bottom up. Typically some retailers in an industry in a particular region saw a threat (competing chain stores) and an opportunity (obtaining lower prices from suppliers). They organized a cooperative, a non-profit organization that they collectively owned. The cooperative would typically operate a distribution center and negotiate with suppliers to get lower costs. In addition, the cooperative might offer marketing services. A single hardware store couldn't afford national, or even citywide advertising, but a collective could. Together, the cooperative could arrange for there to be store brands (putting the name of the store or some other unique brand on a product so that there would some exclusivity in the market), which any single store might find too expensive to organize.

Retailers' cooperatives have probably been most successful in the hardware business. Most of today's neighborhood hardware stores are affiliated with a national cooperative, in which they own a stake. Every weekend during the American football season, one can see television (or hear radio) advertisements for Ace Hardware or True Value Hardware, featuring well-known celebrities John Madden (for Ace) and Madden's long-time announcer booth colleague Pat Summerall (for True Value). Although the average size of a neighborhood hardware store has been growing, those stores (and the cooperative networks to which they belong) remain much smaller than "big box" chain competitors like Home Depot or Lowe's that follow the corporate model.

> The lack of an expiration date on hardware has allowed the hardware industry to organize into several national competing cooperatives that also compete with national chains. Many local grocery stores are affiliated with each other as well, although in a less obvious way. The need for food distribution to be based out of regional distribution centers (so that the food arrives fresh) has slowed the emergence of national retailers' cooperatives that are found in hardware. The largest grocery cooperative is IGA, with 4,000 affiliates, which provides marketing and calls itself a "voluntary supermarket chain." However, IGA does not distribute food to its members itself, but works with 37 distributors worldwide. Thus independent supermarkets have formed regional food distribution cooperatives. These regional distribution cooperatives have affiliated as well (into cooperatives of cooperatives) and distribute merchandise, including the Shurfine and Western Family store brands. (And confusingly, some IGA stores carry Shurfine products, whereas others do not.)
>
> As planners and designers seek to create true neighborhoods with local merchants, they need to understand the economics of that sector, and how it is positioned in the marketplace. To have successful locally owned neighborhood hardware stores and neighborhood grocery stores (and neighborhood pharmacies, for that matter) requires having businesses that are affiliated with regional and national organizations to obtain the economies of scale necessary to survive against the deep pockets and broad market reach of large supermarkets, hardware stores, and drug stores. The cooperative model has given locally owned stores a chance to stay in business and even expand, though evidence suggests that in several key sectors, more and more market share is going to the large national chains.

and right shoe; typical two-footed individuals are unlikely to consume one without the other. Industry, however, is not organized so that there are separate firms manufacturing left shoes and right shoes. That is, a customer does not shop at Barney's Left Shoe store and then walk down the street to the Fred's Right Shoe store. There is a natural tendency for many complementary goods to be packaged together, or sold by the same firms. Bicycle frames and bicycle helmets, for instance, will be sold in the same store, but made by separate firms. Many complementors are managed under separate roofs. Consider a trip from your house to Shimano's headquarters in Osaka. Unless you are living in Japan, you are likely to take both an airplane and a taxi or two. The airport taxi service is complementary to the airline, but you pay different vendors (unless you have arranged a package tour).

Many products are complementary, some more obviously so than others. Movies showing at the multiplex may seem to be competitors, but on the other hand, they may serve different audiences within the family. Mom can drag dad to Sophia Coppola's *Lost in Translation* while the teenagers watch the latest remake of *Godzilla* and the young children can see Miyazaki's *Princess Mononoke*. They are all films; for some individuals they are substitutes, other people would go to one and never consider going to another. However, the family can go together to the multiplex, which is competing against different theatres elsewhere in town. A one-screen cinema is at an inherent disadvantage, as it can only satisfy part of the audience, and may be unappealing to the other. At best it can show different films during different times of the day (including a midnight showing of *The Rocky Horror Picture Show*.)

Competitors

But firms also have competitors (for Shimano, one is SRAM, based in Chicago, Illinois; another is Campagnolo, based in Vicenza, Italy), who use many of the same suppliers and compete to serve the same customers. Although most firms choose not to do business with their competitors, they are at least aware of their competitors' strategies and location criteria in their own decision-making. In fact, common lore suggests Hugh McColl, Jr., former CEO of NationsBank, maintained a list in his pocket of so-called "enemies" who were his primary competitors.[a]

Complementors and competitors

Many products are complements and competitors at the same time, making the dynamics of competitor/complementor relationships all the more complicated. Non-intuitively, firms that compete may be more successful when they co-locate. Complementarity and competition are complex phenomena. In many cities, firms are located by district. You might find the bookstore district, the shoe district, the electronics district, the car repair district, the paint district, or even the plastic food district of Kappabashi-dori in Tokyo (where there are multiple firms selling plastic mock-ups of sushi for display in the window of your favorite Japanese restaurant, as well as other restaurant supplies). All the firms in the district are competitors, so why would they locate together? Surely they would be closer to their customers if they spread out like gas stations or coffee shops.

By locating together, firms tap into a critical mass of consumers, who can go to one district and compare product quality and prices across stores. Given the choice between going to several different stores all around town to compare goods for sale, with the resulting high transportation costs, or going to one district and walking around, it is more economical to go to a single place, even if it is farther from your starting point. The higher transportation costs of going farther for a single district are less important than the higher retail costs one finds by going to a neighborhood shop without competition. The buyer can be assured prices will be lower, and quality higher, if he can simply go next door to the competitor. The plastic sushi "chef" has to go to the plastic food district to sell his plastic *toro* or plastic *amaebi*, since that is where all of his competitors, and thus all of his potential customers, can be found.

Not only are plastic sushi shops complements, sushi restaurants may be as well, for a somewhat different reason. Suppose you are the owner of one of these purveyors of raw fish. Every sushi restaurant in town is competing for people who want to eat raw fish. But they also act as an advertisement for raw fish in general. Every time you lay your eyes on a sushi restaurant, you are reminded that you like raw fish (or dislike it, but that is unimportant since then you wouldn't be considering it for very long). When thinking about what to eat for dinner, sushi acquires more mind share against other options (e.g., steak, fowl, etc.), which helps the sushi market in general (as well as each restaurant in particular). Box 7.3 provides some economic definitions of competitive and complementary goods.

Connectors

In the Diamond of Action, individuals were modeled as having a "surface" of chances limited by constraints. Because firms often operate in a more constrained environment, where competitive markets are less important than negotiated relationships, we model companies as operating on a network rather than over space (see Figure 7.1).[b]

Considering how goods and services might move across geography (from place to place) implies simple, isolated markets. Shimano is competing with SRAM and Campagnolo for customers' business, but it is also competing with office-building developers in Shanghai for raw materials. Shimano is complementary to Orbea for bicycles, but Orbea is not in the fishing tackle business; Shimano is. Orbea and Shimano are, to a small extent, competitors for raw materials: some of the same materials used in the manufacture of Orbea's "carbon fiber composite" bicycle frames are found in Shimano's products as well. But most would deem such competition insignificant.

Box 7.3 Defining complementarity and competition

Economics has a clear measure for defining whether two goods are complements or substitutes: the cross-elasticity of demand. Cross-elasticity simply measures what happens to the demand for one good when the price of another good changes.

$$cross-elasticity = \frac{\Delta Q_2}{\Delta P_1}$$

If the demand (Q) of good 2 rises when the price (P) of good 1 rises, the two goods are considered substitutes. If the demand of good 2 falls when the price of good 1 rises, the goods are complements. As is suggested by the name: cross-elasticity, it is highly related to the notion of elasticity as further discussed in Chapter 12.

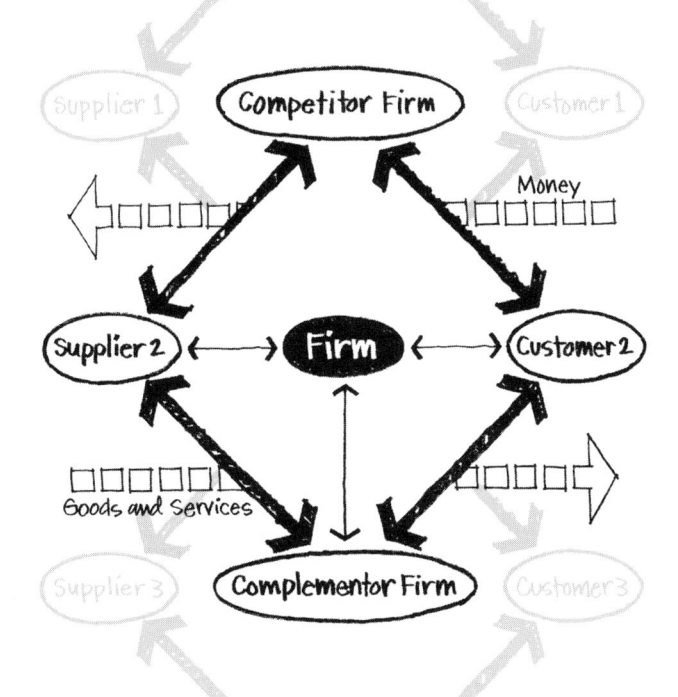

Figure 7.1 The Diamond of Exchange

Firms respond in a similar manner to two of the Cs introduced in the Diamond of Action. *Complementors* help the firm get about its business and achieve agglomeration economies, even if the complementing firm is in a separate sector. *Competitors* may use many of the same suppliers and compete to serve the same customers. However, other "Cs" also come into play. *Connectors,* upstream of the flow of money, are suppliers who provide the material, energy, ideas, transportation, land, and labor that the locator-company combines into goods and services. *Customers* are downstream and would purchase products from the firm or use their services. The shape of the network of goods and services flowing in one direction (from multiple suppliers to multiple customers), and money (or bartered goods and services) in the opposite direction (from multiple customers to multiple suppliers), gives rise to the Diamond of Exchange.

Network analysis of the economy

Approaching the concept of firms as operating within a network gives rise to a broader context of the Diamond of Exchange to develop a network analysis of the economy (see Figure 7.2). For example, one could consider different elements—production, exchange, and transportation—to present them in terms of a network of goods. Network nodes connect via links (and links connect via nodes) that reinforce each other. These links can be physical in nature (e.g., threads, wires, beams, highways, rails, pipes) or socio-economic (e.g., kinship, social, or exchange relationships). The market, on the other hand, is a place (real or virtual) where exchange takes place. An economic network may comprise multiple markets, but there are three main elements:

Figure 7.2 Network model of the economy

(1) the site of production/consumption (material transformation), (2) the site of exchange (ownership transformation), and (3) the connection between the two (spatio-temporal transformation).

Although each of these elements is modeled as a link or node, it should be remembered that each could be expanded to form a subnetwork if there were a desire to increase the detail or resolution of the analysis. A production/consumption agent in an economic network has both suppliers and customers, and can be modeled as an "agent node" on a network. Because production and consumption are two sides of the same coin, they are referred to together: every process consumes inputs to produce outputs. The "exchange nodes" are defined by the convergence of "connection links," and are analogous to markets. The agent nodes are connected to exchange nodes by special "connection links." Connection links account for transportation or communication costs in the production system. The flows in one direction are goods and services that are input into the production process, transformed, and output as refined goods. The flows in the other direction represent money (or a monetary equivalent) that is paid for the goods.

In Figure 7.2, an agent (firm or individual) purchases goods in an input market (Stage 1), and may be supplied by any or all firms in that input market. The goods are brought to the "factory," (the term is used here loosely), transformed (Stage 2), and sold in the output market to any or all customers (Stage 3). The firm is complementary to any firm supplying its input markets and to its customers, while it is competitive with parallel and unconnected nodes.

Clearly even this situation is idealized. Some firms may have different degrees of vertical integration, that is, they may internalize what is represented here as an input market or the output market. However, this figure does reflect the fact that a production process may have economies of scope, so that a single firm produces for more than one output market, as is shown in Figure 7.2 between Stage 2 and Stage 3. In the illustration, there are three stages (labeled 1, 2, 3 from left to right) several markets in each stage (for instance a market for capital and a market for labor) and multiple firms in each market. Extending the chain far enough to the left and to the right, and incorporating enough of the economy, the markets connect with each other again, as the ultimate final consuming agent is the individual consuming goods and an ultimate input agent is that same individual producing labor.

Figure 7.2 is a snapshot describing the processes and relationships at a certain point in time. Over time, links and nodes are added and deleted as the economy grows and contracts, markets change, and innovation occurs in response to entrepreneurship and invention. The purpose of this analysis is to provide a tool to examine how networks and relationships in general do "happen," and thereby to provide a mental model of the economy.

We might extend the idea that flows take least-costly paths to the model. Then "final" customers on the right side purchase a bundle of goods that provides the highest value or lowest cost, profit-seeking production/consumption agents in the middle will act as efficient customers for the initial producers on the left, and efficient producers/transformers in their own right. The network will generate welfare-maximizing flows under the strong assumptions from microeconomics: that property rights are well-defined, externalities are absent, links are competitive, costs rise with outputs, etc. The interesting cases, ones which suggest that the free market is imperfect and will not maximize welfare, arise in the absence of one or more of those conditions.

Framework for Part 2

This chapter introduces the notion of businesses as locators. Although businesses ultimately locate where developers provide space (noting that developers merely respond to the perceived desires of locators), locators can be divided into retailers and non-retailers, each with slightly different requirements.

Each party influences urban development patterns and the decision-making processes of each are typically given inadequate consideration in land use-transportation policy decisions. These three non-public entities—in concert

---

Box 7.4 Networks in the economy

A market may sell the right to use, or the ownership of, physical networks. People buy and sell shares in telephone companies, and bandwidth is also auctioned. To apply the network economy model to a conventional transportation network, think of a roadway link as a composite of the "agent node" and the "connection link." For each link on a highway, there is only one input market and one output market, each identified with a single node (an intersection), which makes the graphic representation and analysis simpler as the agent nodes are unnecessary because the transformation is only spatial, not material. Although there is "conservation of flow" in the network, flows can be one-way, the link moves traffic in one direction with nothing in return. As part of a larger system, the link (more precisely, an agent such as the Department of Transportation, Turnpike Authority, or private firm acting on behalf of the link) receives revenue from the government or users, revenue which is used to maintain the link.

In one sense, the link is selling the right to be traveled on and is compensated by users or government for this right. If it is not paid in money, it deteriorates over time and the payment comes from the link's own capital stock which is dissipated. This means the link deteriorates, it gets potholes that go unrepaired, it cracks and remains unpatched, bridges weaken and begin to crumble. Furthermore, if there is no charge, there is overconsumption (congestion), so that users pay in time rather than money for using a link. Imposing road pricing, discussed in Chapter 12, is a partial remedy to these problems.

The more generalized version of a graphed economy subsumes the transportation network as a special case. The use of this framework serves to incorporate, at least conceptually, financing in the standard highway network analysis, and thereby allows us to identify some pertinent issues.

with public policy and infrastructure decisions—are in large part responsible for the current lay of the land.

The three parties are each interested in the same thing: return on investment. The difference is that they each do so from slightly different perspectives. The return on investment is influenced by the Diamond of Exchange where location decisions are influenced by complementors and competitors as well as connectors (suppliers and customers). In Chapter 2, we explained how complementors and competitors of individuals changed the opportunity surface, increasing chances or imposing constraints. When competitors are complementors, the size of the market measure by number of customers increases. However, competitors beyond a point steal customers more than they draw them. A similar process occurs with suppliers. They have advantages if their customers cluster. So not only is plastic sushi made in the plastic sushi district, the tools and material for making plastic sushi, as well as a labor pool of skilled plastic sushi chefs, can be found in that district. The clustering, however, does not go back forever; there are many uses of plastic besides making sample sashimi.

Notes

a   It is said that McColl once threatened to "launch his missiles" at an acquisition candidate to push a particular deal forward. [5] This attitude toward competitors was supposedly embraced by the small bank executives, who through the years demonstrated such vigor as part of industry leadership.

b   Readers may notice the Diamond of Exchange bears some resemblance to the famous "Diamond of Advantage" put forth by Harvard professor Michael Porter (1986). However, it differs in important ways. First, we have laid the groundwork for thinking about a firm positioned within a network structure. Firm structure and rivalry (competitors), related and supporting industry (complementors), factor conditions (suppliers), and demand conditions (customers) are no longer just amorphous bubbles, but rather chains within the flow of money, goods, and ideas that make up the economy. These ideas are concretized into specific agents. This simple diamond model helps think about the world and where firms locate and how they behave; as with all models, however, it requires elaboration.

References

[1]   Fleissig, W., *Policy Mechanisms to Influence Location Choices. A Developer's Perspective: How Transit, Infrastructure and Amenities Influence Private Investor Decisions*, Prepared for the ICMT/OECD Workshop on Land Use Planning for Sustainable Urban Transport: Implementing Change, Linz, Austria, 1998.

[2]   Coase, R., 'The Nature of the Firm', in S.A. Boulding (ed.), *Readings in Price Theory*, Chicago, IL: R.D. Irwin, 1952.

[3]   Williamson, O., *Markets and Hierarchies: Analysis and Antitrust Implications*, New York: The Free Press, 1975.

[4]   Raymond, E.S., *The Cathedral and the Bazaar*, Sebastopol, CA: O'Reilly, 1999.

[5]   Wikipedia contributors, *Hugh McColl*. Available at: http://en.wikipedia.org/wiki/Hugh_McColl (accessed June 5, 2006).

Chapter 8

Siting

"Take away my people, but leave my factories, and soon grass will grow on the factory floors. Take away my factories, but leave my people, and soon we will have a new and better factory."
Andrew Carnegie

In January 2000, Best Buy, the United States' largest retailer of consumer electronics, announced plans to relocate their corporate headquarters to Richfield, Minnesota. Their aim was to consolidate ten scattered offices from the nearby suburb of Eden Prairie onto a single 1.6 million square foot (150,000 m^2) complex that would employ up to 7,500 people. The City of Richfield, having watched as the Minneapolis-St. Paul International Airport and the I-494 and I-35W freeways gobbled up the city property tax base, and having suffered cuts in funding by the state government, already had seen one redevelopment deal fall through for that site. When word leaked that Best Buy was shopping for a location for their new corporate headquarters, Richfield's Housing and Redevelopment Authority (HRA) wooed the firm through Tax Increment Financing (TIF).

TIF allows municipalities to pay for redevelopment through projected increased property tax revenue as a result of the project. Essentially, the community is borrowing money to support a development (land acquisition, infrastructure, etc.) to be paid back by future property tax revenue generated by that development. A TIF district can capture only taxes collected above and beyond the original value of the land. A common funding mechanism permits municipalities to sell bonds that are paid back by the increases in property taxes over time. Another method is for the private developer of a project to pay for the up-front costs, with future property taxes abated to reimburse the developer. A third and final possibility is for cities to borrow the up-front costs from their reserves.

Developers are attracted to TIF because it allows them to build projects without paying for many of the costs associated with traditional development, such as land acquisition, building demolition, hazardous materials abatement, and infrastructure improvements. Cities like TIF because they can use it to attract development and increase their tax base over the long term.[a] Under the deal struck between Richfield and Best Buy, the city was responsible for acquiring the houses and businesses that were on the site. Best Buy was responsible for much of the cost of redevelopment, including a $23 million portion to be paid from revenues generated by the TIF over a 24-year period. The company was also required to provide $7 million to help fund the construction of a new Penn Avenue bridge spanning Interstate Highway 494. An additional $7 million from Best Buy's property taxes would be used for this purpose, while another $7 million in taxes from the site would go towards Richfield's Housing Fund to help finance new housing construction and rehabilitation. It was estimated that the redeveloped site would generate between $7 million and $8.4 million in annual property taxes, compared with $768,000 produced in the area before the planned development.

Over the months that followed, the HRA and Best Buy acquired 82 of the 84 parcels in the area to clear them for development—the final two parcels, belonging to the Walser Automotive Group, a car dealership, were acquired through condemnation. Walser sued, claiming that the TIF district did not meet the legal requirements of serving a public purpose (private property was taken and given to benefit another private entity).

The case went before the Minnesota Court of Appeals and the Minnesota Supreme Court, who ruled that Richfield's HRA failed to follow all of the legal requirements in establishing the taxing district. The ruling had the potential to undermine Richfield's plans to use TIF; the city therefore entered into mediation with Walser to reach a settlement. That settlement, approved in March of 2003, required Richfield to pay Walser $18.5 million for their two properties, plus relocation costs. Richfield had intended to give them just over $9 million, but Best Buy made up the bulk of the difference by contributing an additional $9 million. The court-approved resolution also directed Richfield to alter how they established redevelopment districts. The ruling affects the standards applied when local governments in Minnesota evaluate property to certify TIF districts.

The issue of takings went to the United States Supreme Court in a different case: *Kelo* et al. v. *City of New London* et al., in which the court upheld the city's taking of private land for economic development purposes in accord with a plan, where the economic growth is benefiting one set of private parties at the cost of a different set. [2, 3]

Several issues motivated Best Buy's relocation decision. First, there is the pool of talent in the retail industry. Best Buy, being a retail company and the

Twin Cities region being home to several others, most notable among them Target Corporation, creates some type of pooling of labor force in the retail management sector. Both Target and Best Buy were founded in the same metropolitan region, and have kept their headquarters there because the costs of relocation outweighed benefits. That said, Best Buy did move down the road, but with the relatively short distances involved, doing so probably cost very few workers—so second, there was the pull factor of convenience for their existing workforce.

Third, they moved to a site on the I-494 beltway, in the booming southern suburbs of the Twin Cities, just a stone's throw from the Mall of America, providing superior access to "Hard infrastructure" and "Access choices." The Twin Cities has relatively clean air and water compared to many cities, being near the headwaters of the Mississippi River and on the Northern Hemisphere's Jet Stream, so Minnesota's used water flows down the river and air pollution is swiftly carried across the border to Wisconsin. The park system in the area is also quite strong (though, despite *haute cuisine* food companies like General Mills and Hormel (producers of Spam), it would be hard to argue that Minnesota dining is a world beater). And finally, the factor of financing, and the subsidy from local government, reduced the price of their new site sufficiently that it caused the firm to choose Richfield rather than the competing suburb of Eden Prairie.

Chapter 7 introduced the idea that where firms choose to locate in metropolitan areas is influenced by four factors: suppliers, competitors, complementors, and customers. These four factors help define the economy as a network and a relevant part of that network is the supply chain—"a network of facilities and distribution options that procures materials, transforms the materials into intermediate and finished products, and distributes the finished products to customers." [4] Specifically, the firm's particular supply chain, including both materials and labor, often governs its location decisions. In other words, a business's location (particularly those firms *not* oriented towards selling goods to people) is dominated by factors related to the proximity of both labor and material supply.

Literature stemming from regional economics, geography, and real estate studies provides extremely rich descriptions of the history, nature, and evolution of business location theory—more than can be discussed in these pages. A foundation of the approach presented in this book is that firms are one of the three primary agents who flex their muscle over the urban landscape. Surprisingly, an analysis of the behavior of private firms has been absent from most writings on land use and transportation.

This chapter, therefore, describes how decisions involving the siting of firms plays out in three respects. Given that there are extensive accounts of firm location behavior, this chapter offers only thumbnail descriptions to highlight important elements of dominant theories. It first describes central tenets of industrial location theory. It then introduces the concept of agglomeration economies. The third aspect of siting decisions analyzed here is a firm's proximity to the workforce, a concept known in the land use-transportation literature as jobs-housing balance.

Traditional industrial location theory

Alfred Weber[b] initially made the case that industries locate to minimize transportation costs. Industries transform some raw material into a final product. To analyze their location, he introduced the Material Index, which is the ratio of weight of raw materials requiring transport to the weight of the final product, and tells whether a good is *weight-gaining* or *weight-losing*. A weight-losing good has heavy raw materials relative to the final product, and so it is most efficient to process that good near where the raw materials are obtained; an example is copper mining. In contrast, if the good is weight-gaining, so that the raw materials can be added to some local product, it is best to process it near final consumption; an extreme example is fountain sodas, where the syrup is combined with water and ice in the fountain, which is dispensed for the customer.

The concept of weight-gaining or weight-losing can be understood mathematically or physically. Either way, imagine a schematic such as the one presented in Figure 8.1, which aims to minimize the transportation costs of materials moving from *C1* and *C2* through some point *P* to a final market *M*. Weber introduced the concept of *isodapanes*, points of equal transportation costs around the minimum cost point (in Figure 8.1, point P). That is, *P* is not merely the centroid of the triangle. Subsequently, Weber introduced the Varignon Frame (Figure 8.2) to solve the problem with two raw material inputs and one output, where the weights on the frame represented the weights of the input and output goods.[c]

Weber's model, however, also considered two other factors which are increasingly important in the location decisions of firms: agglomeration and labor. Agglomeration, clustering and linkages across firms may justify locating in a specific place. Low cost labor may justify higher transportation costs of goods.

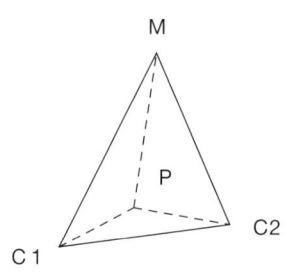

Figure 8.1 Locational triangle

Figure 8.2 Varignon frame (taken from Alfred Weber's *Theory of the Location of Industries*, 1909, available at: www.csiss.org/classics/content/51)

Agglomeration economies

The term "*economies of agglomeration*" represents the gains accruing to firms located near one another. Fujita and Krugman [10] argue that concentrating production and economies of agglomeration are the driving force of city formation. We could add the concentration of exchange activities to that mix, as agglomeration lowers not only the costs of production, but also the costs of exchange. The notion of exploiting economies of agglomeration is at odds with the jobs–housing balance because residences and firms are competing for the same land. The market will allocate scarce resources (land) to the highest bidder, which in business districts tends to be business.

Because there are economies of agglomeration at some level, it does not necessarily make sense to want the ratio of jobs to employed residents to equal 1 in every city or town. Yet the closer the ratio is to 1, the shorter the total commuting time will be. By allowing markets to create imbalances, we are collectively gaining from trade by increasing commuting time.

The previous chapter discussed Williamson's *Markets and Hierarchies*. One of the implications of a spatially based transactions costs model is that firms that locate in business districts (be they central business districts like the traditional downtown, or regions like Silicon Valley) can be smaller, more specialized, and less integrated, because they can purchase more services in the marketplace with lower transaction costs. And on the flip-side, their customers can find them, their competitors, and their complementors all in one place, lowering transactions costs, and thus increasing overall economic activity. Box 8.1 considers the "new economic geography," which builds models accounting for agglomeration.

Proximity to the workforce

Although weight gaining/losing and agglomeration considerations may, depending on the industry, still be important in firm location, as the epigraph to this chapter suggests, the employees themselves deserve special attention. After all, employees represent the labor that is consumed by the employers, a factor that is especially important in white-collar work.

The relationship between jobs and workers is central in the urban and regional economics literature, though that literature is still far from reaching a consensus on which came first. Do people follow newly created jobs into regions, or do jobs follow newly arrived migrants? The primary reason the debate has not been settled lies in the endogeneity of both workers and employment. Firms locate in part to ensure that there is a supply of cost-effective, quality labor, a necessary input for production. A firm that requires low-skilled labor does not want to pay a premium for an educated workforce, while a firm that requires skilled workers cannot locate where those skills cannot be found. Regions experiencing rapid job creation are likely to attract new residents, while regions experiencing an influx of new migrants are likely to experience increases in jobs. [11] Untangling these causal effects is an interesting exercise, but equally germane to the work of land use-transportation planning is the fact that jobs are plentiful in some areas while residential opportunities (housing) can be found in abundance in other areas, leading to an imbalance between job-rich/housing-poor areas and job-poor/housing-rich areas.

Jobs–housing balance

The uncertainty associated with endogeneity, in part, gives rise to questions about many policy initiatives. Consider, for example, one of the more well-known policy goals that emerged in the late 1980s and took off in the 1990s: the effort by cities and metropolitan areas to balance their employment and housing opportunities. The debate over the effectiveness of job-worker (or job-housing) balance policy, and the impact this policy has on travel demand, is one of deep ideological difference.

Given a large enough geographic region (e.g., the world), all things are balanced. At the metropolitan level, by definition, jobs and workers are *largely* in balance, as metropolitan areas are defined as cohesive markets. More obvious issues arise at the sub-metropolitan scale. Although some claim it is just a matter of defining the "right" geographic scale, most policy attention focuses on the sub-metropolitan scale because that is where many *im*balances come into play—leading, many claim, to excessive auto travel.

Box 8.1 The "new" economic geography

Prior to his career as a *New York Times* columnist, Paul Krugman was an economist who helped make major advances in international trade and spatial economics. Among those advances was building a model illustrating how the proper balance of agglomeration (center-seeking) and dispersion (center-fleeing) forces can lead to the formation of a city with multiple centers. This model, dubbed Edge City Dynamics, was presented in his book *The Self-Organizing Economy*. [12]

The agricultural model of von Thünen was extended by Alonso as a model of location of residences relative to the central business district, and later formalized by Beckmann [13] as the New Urban Economics (NUE). The assumptions of that model were quite restrictive, [14] including: (1) the central city is located on an isotropic plane, (2) all employment is located at the center of the city in its central business district, and (3) the city is connected by a densely packed network of radial transport links that facilitate transportation from the hinterland into the Central Business District (CBD). The network allows for any individual at the same radial distance from the CBD to traverse the distance to the center equally well. Land rents are thus based solely on distance from the CBD.

Clearly these assumptions are unrealistic, with many cities having fewer than 10 percent of their workers taking jobs in the central city. A city with multiple centers, or edge cities, is clearly more realistic than the more classical models. With the advent of the New Economic Geography (NEG) (which is newer than the New Urban Economics), the analytic tools to construct a somewhat more realistic model were at hand.

Alfred Marshall proposed three reasons for agglomeration economies. [15] First, industries concentrate because knowledge is concentrated, and knowledge from one industry spills over onto related industries: "the mysteries of the trade become no mysteries, but are as it were in the air." Second, labor gets more and more specialized (and thus efficient) in larger and larger markets. Third, large markets have strong "backward" and "forward" linkages, that is, they are embedded in supply chains that can be spatially concentrated to reduce transportation costs.

The New Economic Geography formally developed a mechanism for specifying the agglomeration forces, that is, a way of capturing increasing returns. Goods are differentiated into many varieties, each of which has an increasing returns technology with labor as a sole input, called D-Goods. [16] Because of scale economies, each good is produced at only one plant, but because of transportation costs, producers prefer to locate nearer their consumers. Because consumers are also laborers, these forces lead to concentration.

There are also goods that are immobile and tied to the land, such as agriculture, called H-Goods. Because some consumers will be located with these goods, some producers will find it advantageous to locate near these consumers, producing the dispersion or center-fleeing forces. In an urban context, housing takes space, and so

> is tied to the land, and could be a dispersive force, attracting firms that would like to pay less for labor and be nearer consumers.
>
> As much of an improvement that the NEG is over the NUE, it still has problems. These models assume the "iceberg" model of transportation, where goods melt away with increasing distance, and ignore networks. Though this is acknowledged as a weakness, the best that analytical economics can do is to assume cities as hubs (e.g., a port) and note that connections occur between cities. Complex structures like networks may require simulation to fully capture. The models really don't treat firms (or individuals) as agents. There have been attempts to reformulate Krugman's Edge City Dynamics model as an agent-based model [17] which produces somewhat different results.

It is not surprising that the concept of institutional market intervention, designed to encourage the co-location of residences and firms, gained popularity during the late 1980s—the same period that spawned volumes of literature on road congestion and its associated environmental impacts. Since this time, a number of regional policies have been adopted across the United States with the aim of building "more livable communities." But without a clear understanding of who leads and who follows (jobs or residents) it is still a difficult issue to resolve. Specifically, the role played by firms in this system remains unclear.

In Minnesota, for example, the 1995 Livable Communities Act aims to fill in existing residential-rich environments with jobs, rewarding communities that do so through tax credits. New Urbanist designs often favor the placement of housing or some sort of professional employment opportunities atop retail stores. Community development corporations, in large part, aim to enhance the skill set of their communities and, in turn, attract additional employment opportunities. In both cases, the implicit assumption is that neighborhood residents will work close to home. Policies to encourage local job-worker balance rely on the assumption that doing so will serve to reduce the amount of daily travel required by individuals, thus tempering road congestion and environmental degradation. The issue of choice is often framed in terms of the question of job-worker (or job-housing) balance. In its simplest version, the idea is to identify specific areas (i.e., municipalities), and find the ratio of the number of jobs to the number of workers.

To illustrate the concept, consider the two communities in Scenario 1: one exclusively residential in character (Community A) and another with nothing but businesses (Community B). Both communities are unbalanced, clearly requiring high travel to connect them. To remedy the situation, jobs-housing policy suggests it would be best to inject Community A with 100 jobs and Community B with 100 workers. The communities would then be in perfect balance, in theory, assuming that the types of jobs match the skills and preferences of available workers. At a minimum, some of the workers will work in the same town in which they live.

	Community A	Community B	Community A	Community B
Scenario 1	100 workers 0 jobs	0 workers 100 jobs	J-W ratio: 0	J-W ratio: infinity
Scenario 2	50 workers 50 jobs	50 jobs 50 workers	J-W ratio: 1.0	J-W ratio: 1.0

Such an example is simplistic, however, as it assumes only two municipalities, and does not take into account boundary effects, jurisdictional size, and other factors. Even the most pristine of scenarios (such as Scenario 2) requires us to question such a ratio. How are we to interpret such a ratio? Some areas, such as downtowns, clearly have a surplus of jobs, leaving other areas with a surplus of workers. In many cases, the jobs available are not appealing to the people who live there, or are not at the appropriate skill level. In these cases, "perfectly balanced" towns still have to import workers from outside their boundaries.

The idea of balance suggests alternative ways of measurement that are not so naïve. Rather than simply looking at the number of jobs and workers in a place, we can look at the accessibility to jobs and workers from a point. Recall that accessibility describes the ease of reaching destinations. We might say that if we can reach just as many jobs as workers in the same time window, the point in question is "in balance." Individuals who live in areas with high access to jobs have shorter commutes, those who live in areas with lots of competing workers have longer commutes. Levinson [18] found that a 1 percent increase in origin jobs accessibility (opportunities) for auto commuters decreases commute durations by 0.22 percent. Furthermore, the fact that congestion is still a nuisance during peak periods of the day suggests that there are more than a fair share of people commuting to and from work. Employment decisions and locations are clearly an important element of people's travel and will continue to be in the future. The advantage of this measure is that it includes all of a region's workers and jobs, measured at a point, weighted by their importance (nearer jobs are more valuable than farther ones, nearer workers are more significant competitors for jobs than farther ones).

Questioning balance

The bulk of existing congestion in most metropolitan areas is a result of the peaking nature of the work commute. Reducing travel distances would help. The focus on jobs-housing balance, however, likely receives more attention than it deserves for at least two primary reasons. First, less than half of the population is employed.[d] Thus, the concept applies to only a subset of the population, although it is the important breadwinning subset.

Second, the increasing role of social networks (a concept described in Chapter 4) suggests that the once-prominent role of geography (people sought employment opportunities close to their home), may have decreased in importance. Furthermore, assorted claims describe the changing nature of geography and commute patterns. Giuliano [19] argues that the relationship between job and housing location is complex, and where people choose to live "may have little to do with job access considerations," leading to the conclusion that spatial location is increasingly irrelevant. Taken to an extreme, this even implies that information technology will lead to "the Death of Distance." [20] We believe space still matters, distance has not died, and geography is not at an "end." The jobs that people hold tend to be within some fixed area around the home; how close, however, we do not really know. People who have plenty of opportunities within close proximity to their homes travel less to reach jobs; people with fewer opportunities must travel farther, and must be more organized about it.

Balance skeptics suggest that regional polices attempting to strike a synthetic balance between jobs and workers have little or no impact on minimizing total travel (distance) in the short term, as a result of implicit policies and personal preferences that fundamentally oppose spatial balance in favor of functional segregation. However, recent writings suggest that the market has failed to achieve balance in three out of four metropolitan regions. [21] Providing case studies to support such findings, Weitz concludes that increases in housing costs tend to be more gradual in areas with a jobs-housing balance. Weitz counters the skeptics and points to those actions planners can take to help bring appropriate housing, jobs, and workforces together, resulting in overall community improvements. As a policy question, it might be worth asking larger questions such as whether self-containment is good, and whether the city or the metropolitan area should be balanced. If not, what deviation from balance is reasonable?

Imbalance by design?

The market may not provide jobs-worker balance because the market does not want balance—at least not perfect balance. The central business district is a *business* district for a reason; businesses prefer to be nearer to other businesses (customers and suppliers, competitors and complementors) than to their workforce. Just as many firms want to be open for business when others are (giving rise to the eight-to-five workday and the morning and afternoon rush "hours") to enable temporal coordination, they want to locate near each other to enable spatial coordination. If I have business to conduct, it is easier to go across the hall or down the street than across town or down south (up north, out west, back east). Firms would rather avoid paying its employees for work-related travel when they could be productive (but fail to pay their employees for their unproductive commute time, at least not directly).

A consequence of imbalance may be spatial mismatch, wherein the workforce may not be compatible with local jobs (and it is a problem if a low-skilled workforce can only find appropriate jobs far away). The consequences in one particular case are discussed in Box 8.2

Wrap up for siting a business

The standard urban economics model of employment location is derived from basic principles of classic land economics, bid-rent theory, and traditional land use/transportation interactions first introduced by Johann Heinrich von Thünen almost two centuries ago (discussed in Chapter 3). A farmer's profit at a given location depended on two factors: how much people in the city were willing to pay for different crops, and how much it would cost to transport those crops to the market. The two key variables are the cost of the goods and the cost of the transport. This early work later stimulated theories including William Alonso's [22] bid-rent curve in a monocentric city, Walter Christaller's central place theory (described in Chapter 11) and George Zipf's [23] Law about the hierarchy of places, as well as the work of August Losch, [24] who theorized about firms' location decisions and the spatial competition between them. These early contributions provided conceptual foundations to understanding how sites with higher accessibility spark more competitive bidding (producing declining land rent gradients from high access locations) and the spatial separation of firms competing for market share.

These frameworks, however, fail to explain the edge cities, suburban activity centers or secondary business districts that have been on the rise since the deployment of freeways. Employment does not only cluster in, around, and near central business districts. This observation suggests that there are additional factors at work in explaining clusters of employment. These factors are often referred to as "agglomeration economies" and have to do with the inter-firm externalities that come as a result of spatial proximity—externalities that come in the form of information spillovers, local non-traded inputs, and a locally skilled labor pool. In addition to such agglomeration effects, however, one must not lose sight of the role of longstanding factors, such as those related to transportation costs, as well as Tiebout's theories that people "vote with their feet" when choosing where to be live based on "bundles" of amenities, governmental services and taxes.

Firms', employers', and retailers' behavior behave as profit maximizers—a maxim that usually falls deaf to social, environmental, and other concerns. Firms such as Best Buy seek their long-term greatest reward, whether that involves maximizing benefits by locating in an urban core or minimizing costs by getting large subsidies from local governments. However, to do that, they must operate within the networks in which they find themselves embedded. The networks include their suppliers, and their suppliers include workers. So how firms locate relative to the labor market they employ is an important question. This is related to the notion of job-worker balance. If jobs and workers were both

Box 8.2 Spatial mismatch and the banlieue

The riots of November 2005 in the *banlieues* (suburbs) of France may seem strange to Americans accustomed to suburbs as places of middle-class satiation and tranquility. The suburbs around Paris are quite different from those around most cities in the United States, and many (though certainly not all) are more analogous to what are referred to as the Projects in the US. In recent decades, the word *banlieue* has taken on the meaning of low-income high-rise housing projects (*cités*) inhabited by immigrants, especially those from Africa. Modernist architect Le Corbusier designed some of the more famous *banlieue* high rises, although others were designed in his style. His idea of the house as a "machine for living in" gives a sense of the modernist design guidelines, aiming for maximum efficiency in the storage of people. The *banlieue* provided a blank slate on which modern architects could more easily impose their vision than in the already developed central city.

In Box 4.5, we ask why the rich commute longer than the poor, if their value of time is higher. In the United States, the wealthy migrated to the suburbs many decades ago, while minorities live disproportionately in the center city. Because jobs are spread throughout the metropolitan area, the wealthy use automobiles to reach them, while the very poor, without cars, can only reach jobs by transit, which provides much less flexibility. This suggests the possibility of spatial mismatch, in which low-skilled urban workers cannot reach low-skilled jobs located in the suburbs.

In France, the poorer and less-skilled workers live disproportionately in the suburbs, and yet face even more severe unemployment problems than similar populations in central cities of the United States. The mobility problems are similar: the poor lack cars, and public transit, although better in France than in the US, still is insufficient. However, the mismatch is not simply a transportation-land use problem, but a problem of human capital (as well as financial capital). The skills possessed by workers do not match the skills needed by employers, and a lack of ready cash, along with a different culture and regulatory regime, inhibits private investment in these communities.

One violent incident usually provides a spark, but the spark only inflames the populace if it is on edge. The reasons for the unrest in France, like the causes of riots in US cities in the 1960s, or the Rodney King riots of 1992 in Los Angeles, or the 2001 Cincinnati riots, or many others, are multiple. *The Economist* magazine blames French unemployment as the cause of the riots there, and the source of the problem on "35-hour week, a high minimum wage, and tough hiring and firing rules." It also notes "there are no black or brown mainland members of the [French] National Assembly," which may provide additional explanation for the neglect of the issues facing the *banlieue*. One cannot exclude the brutalist nature of modern architecture as a contributory factor, though high-rises populated by the wealthy do not have the same social problems as high-rises for the poor. And one cannot rule out spatial mismatch, as unemployment is even worse for the poor in the suburbs of France than for the poor in the center of US cities.

spread uniformly, they would be in perfect balance. But firms must also locate with respect to material suppliers, customers, and their complementors and competitors. For those reasons, firms often cluster with each other (and thus often times away from labor), leading to job-worker imbalances. The degree to which policies can and should aim to reduce economies of agglomeration to achieve economies of commuting is an important debate.

When firms cluster, their labor pool may cluster as well, causing networks that "take off." These are cities, and the more stuff a city has, the more it will get. Network externalities and other spillovers are critical to creating places. Developers will follow this to a point, but ultimately cities may be self-limiting in size. That limit though is dynamic, and changes with both the economy and technology. The megacities that are emerging in recent years would be impossible without modern technologies: not only transportation, but also water, sewer, electricity, and communications, though there remain vast areas where some or all of these services are lacking.

Notes

a However, the incentives to both businesses and municipalities offered by TIF have the potential to create competition among cities for development projects. For example, Bryne [1] tested this hypothesis in the Chicago area, finding that a municipality's decision to adopt TIF is likely influenced by nearby cities' adoption of TIF. Internal factors influence the decision as well, including a city's tax rate and influx of new residents. Because some revenue captured by TIF can be used in a city's general fund, cash-strapped cities are more likely to employ it.
b Alfred Weber (1868-1958), who lived in Germany during the Nazi period, was a critic of Hitler, and lost his position as a professor at the University of Heidelberg from 1933 until the end of World War II in 1945. His brother, sociologist Max Weber, was famous for *The Protestant Ethic and the Spirit of Capitalism*.
c Notwithstanding Weber's contributions, critics of Weber's model note that, like von Thünen's and Alonso's models, it lacks a network and thus fails to consider distance and scale economies. However, these are common criticisms of any abstract model. Other critics note that some of Alfred Weber's ideas originate with the obscure Carl Launhardt [5, 6, 7, 8] and Francis Greulich. [9]
d For example, of the 281,421,906 recorded population from the US census in 2000, only 128,279,228 (46 percent) recorded doing any work for pay or profit.

References

[1] Bryne, P.F., *Tax Increment Financing: Determinants of Property Value Growth*, Institute of Government and Public Affairs Working Paper 102, Chicago, IL: University of Illinois, 2002.
[2] Epstein, R.A., 'Supreme Folly', *Wall Street Journal (Eastern Edition)*, June 27, 2005: A14.
[3] Legal Information Institute, *Kelo v. New London*. Available at: http://straylight.law.cornell.edu/supct/html/04108.ZS.html (accessed July 10, 2005).
[4] Harrison, T.P. and Ganeshan, R., *An Introduction to Supply Chain Management*, 2005. Available at: http://lcm.csa.iisc.ernet.in/scm/supply_chain_intro.html (accessed May 15, 2007).

[5] Launhardt, C.W.F., *The Theory of the Trace: Being a Discussion of the Principles of Location*, 1872.
[6] Launhardt, C.W.F., *Die Bestimmung des Zweckmaissigsten Standortes einer Gewerblichen Anlage: Zeitschrift des Bereines Deutscher Ingenieure*, 1882.
[7] Launhardt, C.W.F., *Mathematische Begrundung der Volkswirtschaftslehre*, 1885.
[8] Launhardt, C.W.F., *Theory of Network Planning*, 1888.
[9] Greulich, F., 'Transportation Networks in Forest Harvesting: Early Development of the Theory', in *Proc. Int. Seminar on New Roles of Plantation Forestry Requiring Appropriate Tending and Harvesting Operations*, 2002.
[10] Fujita, M. and Krugman, P., 'When is the Economy Monocentric? Von Thünen and Chamberlin Unified', *Regional Science and Urban Economics*, 1995, vol. 25: 505–528.
[11] Patridge, M.D. and Rickman, D.S., 'The Waxing and Waning of Regional Economies: The Chicken-egg Question of Jobs versus People,' *Journal of Urban Economics*, 2003, vol. 53: 76–97.
[12] Krugman, P., *The Self-Organizing Economy*, Cambridge, MA and Oxford: Blackwell Publishers, 1996.
[13] Beckmann, M., 'On the Distribution of Urban Rent and Residential Density', *Journal of Economic Theory*, 1969, vol. 1: 60–67.
[14] Button, K., *Where Did the 'New Urban Economics' Go After 25 Years?* 38th Congress of the European Regional Science Association, Vienna, 1998.
[15] Fujita, M., Krugman, P., and Venables, A.J., *The Spatial Economy: Cities, Regions, and International Trade*, Cambridge, MA: MIT Press.
[16] Fujita, M. and Mori, T., *Transport Development and the Evolution of Economic Geography*, Institute of Developing Economies Discussion Paper No. 21, 2005.
[17] MAML, *Evolution of Central Places: Urban Morphongenesis*, n.d.
[18] Levinson, D.M., 'Accessibility and the Journey to Work', *Journal of Transport Geography*, 1998, vol. 6 (1): 11–21.
[19] Giuliano, G., 'Is Job-housing Balance a Transportation Issue?' *Transportation Research Record*, 1993, vol. 1305: 305–312.
[20] Cairncross, F., *The Death of Distance: How the Communications Revolution is Changing our Lives*, Cambridge, MA: Harvard Business School Press, 1997.
[21] Weitz, J., *Jobs-housing Balance* (PAS 516), Planning Advisory Service, American Planning Association, 2001.
[22] Alonso, W., 'A Theory of the Urban Land Market', *Papers and Proceedings of the Regional Science Association 6*, 1960: 149–157.
[23] Zipf, G.K., *Human Behaviour and the Principle of Least Effort*, Reading, MA: Addison-Wesley, 1949.
[24] Losch, A. *The Economics of Location*, translated from the 2nd revised edition by William Wolgom., New York: John Wiley and Sons Science Editions, 1967.

Chapter 9

Selling

> "I am the world's worst salesman, therefore, I must make it easy for people to buy."
>
> F.W. Woolworth (1852–1919)

On January 31, 2004, Minnesota's second enclosed mall, the Willard Torsen-designed Apache Plaza Shopping Center, closed (Minnesota's and America's first mall was the famous, Victor Gruen-planned Southdale, which was still in operation, as of this writing). On March 20, 2004, a "bulldozer bash," the mall's last event, was held, allowing the community to bid farewell to its center.

Shortly after, Apache Plaza was razed to the ground to make way for a New Urbanist mixed-use development dubbed Silver Lake Village. Silver Lake Village contains about 219,300 sq. ft. (20,000 m²) of retail, 25,000 sq. ft. (2,300 m²) of office space, 676 apartments, including "market rate," "urban flats," and senior units, 26 townhouses, and a 5.6-acre (2.3-ha) park. New tenants along this main street-like retail center include Ficocello's Salon, Papa Murphy's pizza, Wireless World, Applebee's, Cold Stone Creamery, EB games, Caribou Coffee and Chipotle Mexican Grill, St. Anthony Village Wine and Spirits (a municipally owned liquor store), and America's favorite store, Wal-Mart. Only the Cub supermarket remains from the Apache days. That these stores are all chains hints at some of the changes affecting retail.

Apache Plaza and now Silver Lake Village are located in the City of Saint Anthony, a first-ring suburb of 8,400 residents to the north of the Twin Cities of Minneapolis and Saint Paul and straddling Hennepin and Ramsey Counties. Apache Plaza opened on October 19, 1961, five years after Southdale, and was located on the opposite (northern) side of Minneapolis, initially giving it a large marketshed. At its peak, it was home to 530,000 sq. ft. (49,000 m^2) of retail, occupied by 68 stores, among them (at various times) JC Penney, Herberger's, Montgomery Ward's, Rothschild Young Quinlan, G.C. Murphy, Van Arsdell's and Woolworth's (only Herberger's and JC Penney are still in business).

Despite containing architecturally interesting elements, including a sunken garden, terrazzo floors, and ten 3-inch (7.6 cm) thick hyperbolic roof structures, each measuring 65 × 71 feet (20 × 22 m), that lifted the roof and provided space below its perimeter for abstract, stained glass clerestories symbolizing American Indian themes, Apache never achieved landmark status. Apache Plaza soon faced competition. In 1972, the larger Rosedale (one of the Twin Cities "dales"—a local suffix for shopping malls) opened nearby to the east, with better highway access. By 1979, consideration was being given to closing Apache Plaza, but the owners chose to invest and remodel the center. Just before the remodeling was complete in 1984, a tornado in a few moments undid the years of effort, destroying the stained glass. A new round of repairs were undertaken, but merchants left and were not replaced. The mall retrenched, with part being torn down to make way for a grocery store.

In 1992, a tornado of another sort hit, when the 4.2 million sq. ft. (390,000 m^2) Mall of America opened, and the deathwatch on Apache Plaza reached full force. Ideas about what to do with the abandoned mall proliferated, including serious suggestions about turning it into a warehouse or a high-tech incubator. Ultimately, Silver Lake Village was decided upon. See Box 9.1 "The Abandoned Mall Blues," for more color.

Box 9.1 The abandoned mall blues

Apache Plaza never reached the fame of Harvey, Illinois' Dixie Square Mall. Dixie Square was abandoned for over 25 years from 1978. Opened in 1966 with a Montgomery Ward and J.C. Penney's, success lasted all of four years before crime discouraged shoppers and merchants. However, it became most famous for the car chase scene in the 1980 comedy *The Blues Brothers*, where cars drove through (the now abandoned) mall and destroyed it further. Vandals later wrecked what John Belushi and Dan Aykroyd didn't. In 2005, a plan to bring some new retailers (Costco, Kohl's, Old Navy) to the site was put in place. The website www.deadmalls.com is dedicated to the phenomenon of abandoned malls.

Full circle?

At least in the United States, retailing in the past century has now come full circle—from main streets and business districts, to open-air shopping plazas, to enclosed pedestrian realms, to open-air (but centrally managed) main streets where parking is in front of shops and pedestrians are again exposed to the elements. Any number of questions arise, including why the shift in the first place and why some places go "back to the future."

To understand the changes in predominant venues for selling goods, one must understand some of the changes in transportation. The previous chapter considered where non-retailing firms locate so that they can be optimally embedded in their network of suppliers (including labor) and consumers, complementors and competitors. This chapter focuses on places for selling goods. Although retailers demonstrate many of the same macro-preferences as non-retailers, the way such preferences play out is markedly different.

First, retailers directly interact with people as consumers, whereas the businesses considered in Chapter 8 (including the headquarters of retail firms such as Best Buy) conduct transactions primarily with other firms, and deal with people as suppliers (labor). The differences are important. People are interested in many attributes when they (as economists put it) "maximize utility." Firms, on the other hand, are seemingly simpler actors who "maximize profits." Of course, firms are comprised of people acting as their agents (e.g., buyers representatives), who are as likely to be swayed by marketing as any other consumer, hence the trinkets distributed at trade shows and other marketing ploys when firms deal with each other. Still, retailing takes this to a higher level.

Second, the size of the transactions most retailers deal with pales in comparison to business-to-business transactions. Most retail exchanges are not accompanied by contracts (automobiles are one exception). Moreover, in the course of a single shopping excursion, consumers may visit (and buy from) multiple vendors. So for consumers, there are benefits to be found when retailers co-locate, especially when competitors co-locate so that shoppers can compare price and quality.

We offer Table 9.1 to depict the potential interactions that can take place between buyers (customers) and sellers (vendors) depending on their relative locations. In general, vendors and customers conduct business with each other from fixed locations (customers from home or work; vendors from a physical location), mobile locations, or increasingly, virtual locations.

For example, both customers and vendors can remain fixed and communicate through telephone and advertising: catalog sales, "as seen on TV," advertisements, and telemarketing campaigns follow this strategy. The buyer can be mobile and go to a fixed store: traditional retailing. The buyer can remain home and have the sales force come to them: door-to-door sales. Or the buyer can meet with a seller at a third location, such as a street fair or a trade show convention. In scenarios involving mobile vendors or customers, there are transport costs to one of the parties. Alternatively, the buyer can enter the

Table 9.1 A matrix of exchange

		Vendor		
		Fixed (store)	Mobile	Virtual (electronic)
Customer	Fixed (at home/ work, not using a computer)	Telemarketing, Catalog sales (third party delivery)	Traveling salesman (vendor delivers goods)	
	Mobile	Traditional shopping (customer pickup)	Trade show, fair etc. (customer take-away or third party delivery)	
	Virtual (electronic)			e-Commerce (third party delivery)

Source: www.icsc.org/srch/rsrch/scope/current/gla.pdf

virtual world and connect with a seller there. In these cases, which are increasing in frequency, a third party is responsible for bearing the transport costs. There are many hybrid combinations, including online research and offline purchase (or vice versa) that are becoming popular.

This chapter traces the range of retail transactions in turn. We open by describing the evolution of traditional retail, the onset of shopping centers and malls, and relevant trends that affect these establishments. We then turn to explaining the rise and fall of door-to-door sales and then to door-to-door delivery. We outline salient dimensions of retail location strategies, and close by positing a relatively simple model of retail location to suggest some of the complexity that is involved in such considerations.

Evolution of retail

The first-ever sales transaction has not been recorded, and probably went untaxed. As the first transaction, it certainly did not occur in a shopping center, and almost certainly did not take place in a store. It also likely did not involve money, which had yet to be invented, so it really was more of a bartering procedure. This first trade was an invention on the order of the wheel. Some scientists have even speculated that the ability of modern humans to trade is what advantaged us over the Neanderthals (who were physically larger and stronger than we, and already settled) and motivated the evolutionary push to put that competing species out-of-business, so to speak. [1]

Moving forward tens of thousands of years from the extinction of the Neanderthals, the Greek Agora, which acted both as a marketplace and public gathering area, was developed, probably some time around 1000 BCE. The

Agora in Athens became a public area during the administration of Solon (638–558 BCE). The Roman Forum, too, was a marketplace and public venue, and in the City of Rome, several specialized markets developed, such as the Forum Boarium, dedicated to the cattle trade. One reference identifies over 30 fora in Ancient Rome. [2] Rome, at this time, traded with places as far away as China along the Silk Road and with islands in the Indian Ocean, indicating that a similarly evolved culture of markets and exchange was in operation throughout much of the world. Market squares later took their place in medieval Europe.

Shopping streets were, of course, common in urban areas. Enterprising retailers could transform the traffic brought by roads and bridges into customers. Even expensive transportation facilities would find themselves choked with stores, slowing transportation while quickening commerce. The medieval London Bridge is one of the more famous examples of a bridge that was covered with buildings serving as retail stores, residences, and even churches. Opened during the reign of King John in 1209, it had taken 33 years to build. Eventually, buildings up to seven stories high were constructed on this prime real estate. This bridge lasted until 1831, after a new bridge (without buildings) was constructed. The example of the London Bridge illustrates the conflicts between two functions of roads: movement and access, which will be discussed in more depth in Chapter 11.

Permanent marketplaces were supplemented by temporary and traveling fairs. The first fairs have been dated to 500 BCE, and may have occurred earlier. Fairs were events where foreign traders could show their wares, and were often coupled with religious festivals, taking place at and around temples. The fair changed over many centuries, evolving into several different types of activities, ranging from the World's Fairs to state and county fairs to conventions and trade shows. They are now less a place for purchasing than for information exchange. In fact, the International Association of Fairs and Expositions (IAFE), which specializes in agricultural events (like state fairs), itself has an annual convention and trade show in Las Vegas.

Markets were enclosed at least as early as 1786, when a member of the French royal family rented gardens to create the wooden *Galeries de bois du Palais Royal*. In 1800, a warehouse was transformed into a Bazaar in London. In later decades, at other locations throughout Europe, streets were covered with metal and glass roofs, such as Saint Hubert Gallery in Brussels, opened in 1847, which survives to this day. These are clearly early predecessors of modern shopping centers that have management under a single organization.

In 1823, Alexander Stewart opened a dry-goods store in New York. Dry-goods stores were certainly not rare, and what distinguishes department stores is basically size and scope. In 1846, Stewart opened the Marble Dry-Goods Palace on New York's Broadway, which by 1862 had taken over a full city block and was the largest retail establishment in the world. Department stores were, as the name suggests, stores with individual departments, which were often contracted out. Paris' Bon Marché opened in 1838 and expanded in

1852, and Macy's opened in 1858. John Wanamaker entered the Philadelphia retail sector in 1861 and by 1876, in time for the Centennial Exposition in Philadelphia, had constructed a department store that was the largest space in the world devoted to retail selling on a single floor. [3] There is thus some controversy over which department store was first, because it is unclear where a regular store ends and a department store begins.

The shopping center followed streetcars and citizens to the suburbs. Baltimore, Maryland's Roland Park shopping center (with six shops) opened in 1896 to serve the needs of that new streetcar suburb, and is considered to have been the first to provide off-street parking. [4] Roland Park, designed by Frederick Law Olmsted among others, may also have been the home of the first homeowners' or community association. Since it was built before the enactment of municipal zoning regulations, developer Edward Bouton placed covenants on houses to ensure that the character of the neighborhood would remain unchanged over time. The association collected revenue to support common areas like the neighborhood park. It is now listed on the National Register of Historic Places. Bouton was later commissioned by Bethlehem Steel, owner of the Sparrow's Point steel mill near Baltimore, to create a workingman's Roland Park in Dundalk, a first-ring suburb. [5]

Others credit Country Club Plaza, opened in 1922 in Kansas City, Missouri, and developed by Jesse Clyde Nichols, with being the first shopping center. [6] It certainly was the first to fully adapt to the automobile; early plans called for eight gas stations at the center, with plenty of off-street parking. In a sense, Country Club Plaza is a planned extension of the city; the shops are at street level, and automobile streets transect the complex. But it is a part of the city designed solely for shopping, and it is at a much larger scale than anything that came before. It has also kept up with the times, remaining a functioning and successful center with over 100 stores, in a way that many later enclosed malls, such as Apache Plaza, have not. Table 9.2 summarizes the evolution of the retailing hierarchy.

Table 9.2 Evolution of the retailing hierarchy

	1900–1950		1962–2000		2000+	
	High end	*Low end*	*High end*	*Low end*	*High end*	*Low end*
Regional	Downtown/ Dept. store	Downtown/ Five and Dime	Enclosed mall/Dept. store	Big-Box- Center/ -Mart	Life-style center	Warehouse stores
Village		Grocery	Supermarket		Supercenters	
Neighborhood		Drug store	Convenience store			

Source: Food Marketing Industry Speaks 1993–2003 Key Industry Facts—Prepared by FMI Information Service June 2003 www.fmi.org/facts_figs/keyfacts/storesize.htm

Shopping malls were discussed briefly in the introduction. A "mall" is distinct from a "center" in that the shops in a mall open inward onto a pedestrian realm, rather than outward onto a street. There are also pedestrian or auto-free streets that could be classed as malls. The key is that a mall is a pedestrian-oriented area lined with buildings. Use of the word "mall" to refer to a shopping area derives from "pall-mall," which was an alley where a game with a croquet mallet was played, and pall-mall comes from the Italian for "Ball-mallet." After games played in long alleys stopped being fashionable, that same street became a shopping area.

An enclosed, climate-controlled mall is a variant of the mall. As shown in Table 9.3, of the some 47,700 shopping centers in the United States, only 2.4 percent, or about 1,130 are enclosed malls, but these malls have a much larger footprint; the largest 0.8 percent of such centers (the 421 centers larger than 1 million sq. ft. (93,000 m²) in area) comprise 7 percent of total floor space. The Mall of America alone contains just under 0.1 percent of total

Table 9.3 Profile of retail in the United States

Type	Number in United States
Retail establishments	1,400,000[a]
Supermarkets	127,000[b]
Convenience	138,205[c]
Sears	870[d]
K-Mart	1,479[e]
Wal-Mart discount stores	1,354[f]
Supercenters	1,713[g]
Sam's Club	551[h]
Shopping malls and centers	47,718[i]
Enclosed malls	1,130[j]
Lifestyle centers	130[k]
Factory outlet centers	278[l]

Notes:
a National Rail Federation
b www.meatnews.com/mp/northamerican/dsp_particle_mp.cfm?artNum=364 (May 2005)
c National Association of Convenience Stores (2005)
d www.aboutsears.com
e www.kmartcorp.com/corp/story/general/kmart_glance.stm
f www.walmartstores.com/Files/2005AnnualReport.pdf
g www.walmartstores.com/Files/2005AnnualReport.pdf
h www.walmartstores.com/Files/2005AnnualReport.pdf
i www.icsc.org/srch/about/impactofshoppingcenters/Brief_History.pdf
j www.icsc.org/srch/about/impactofshoppingcenters/Did_You_Know.pdf
k www.iscs.org/srch/about/impactofshoppingcenters/Did_You_Know.pdf
l http://history.sandiego.edu/gen/soc/shoppingcenter.html (2000)

shopping center floorspace in the United States. Many malls are now owned not by developers, but by Real Estate Investment Trusts (REITs). REITs own or have an interest in half of 1,130 malls. [7] Leasable area has steadily increased (Figure 9.1).

While malls are losing their dominance, they are not done yet, as the continuing growth (on the order of 1 percent per year) suggests. Still new forms are emerging, among them festival market places, power centers (containing what even the International Council of Shopping Centers refers to shamelessly as "big box" retail and "category killers"), and the new fashion, lifestyle centers, which are much more like Country Club Plaza than a modern shopping mall. Table 9.4 classifies shopping centers today.

Among the transitions currently affecting retail is the elimination of free-standing department stores, which have been rounded up into multi-store centers. In 1993, there were 281 single-location department stores in the United States. In July of 2002, there were only 53 single-location department stores remaining, according to Chain Store Guide. [8]

Figure 9.2 shows the increase in size of the average supermarket food store, a trend that has continued steadily since the first Piggly Wiggly store was opened by Clarence Saunders in 1916 in Memphis, Tennesee. By 1923, San Francisco's Crystal Palace store was already 68,000 square feet, [9] though that was atypical. Yet, over time, grocery stores and later supermarkets have steadily grown in size, at the expense of more specialized stores and at the expense

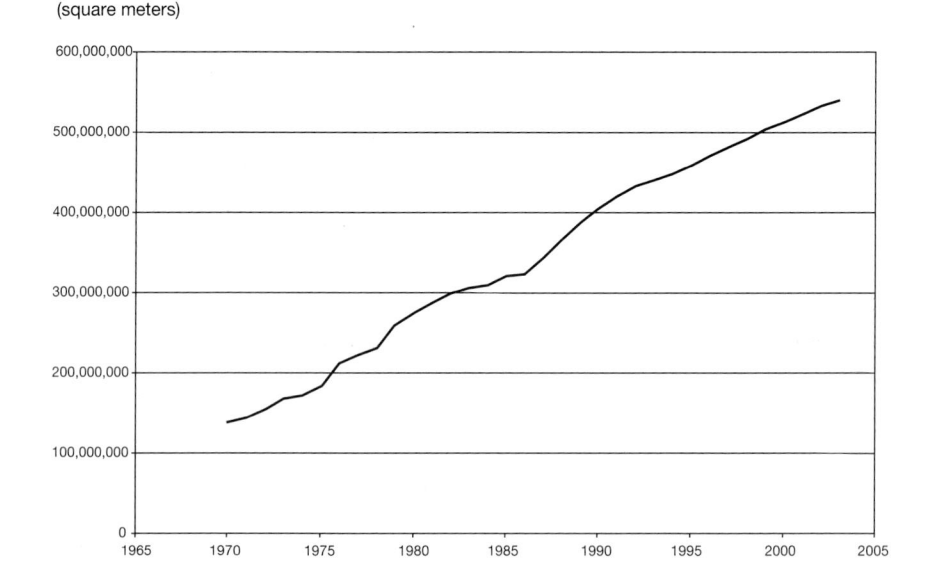

Figure 9.1 Leasable retail area of US shopping centers (1970–2003)

Table 9.4 Shopping center classifications

Center Type	Retail Sq.Ft.	Typical Anchor	Primary Trade Area
Neighborhood	30,000–150,000	Supermarket	3 miles
Community	100,000–350,000	Discount dept. store, Supermarket, Home Improvement, Large specialty/Discount apparel	3–6 miles
Regional	400,000–800,000	Department store, Jr. dept. store,	5–15 miles
Super-Regional	800,000 +	Department store, Jr. dept. store,	5–25 miles
Fashion-Specialty	80,000–250,000	Fashion	5–15 miles
Lifestyle	150,000–500,000	Large book store, Sporting goods, Home furnishings, Family apparel, Multiplex, dept. stores	5–8 miles
Power	250,000–600,000	Category killer, Home, Discount dept. store, Warehouse club, Off price	5–10 miles
Theme/Festival	80,000–250,000	Restaurants, Entertainment	N/A
Outlet	50,000–400,000	Big Box Retail	25–75 miles

Source: International Council on Shopping Centers
www.icsc.org/srch/about/impactofshoppingcenters/ShopCentDef.pdf

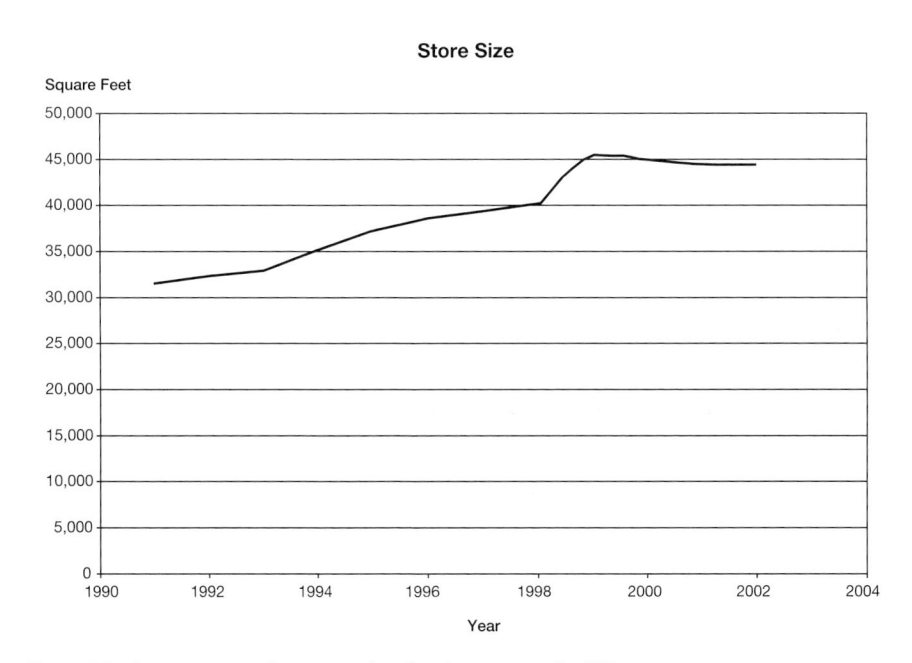

Figure 9.2 Average size of supermarket food stores in the US

of the number of grocery stores, which have declined for a number of years. Thus, fewer but bigger stores conveyed food to Americans, who could shop less frequently and store more food in their larger refrigerators in their larger houses.

Evidence for the continuing consolidation of retail stores comes from Figure 9.3, which shows that the total number of pharmacies is declining, while the number of consolidated stores selling drugs (mass merchandisers and supermarkets) is increasing. Clearly, people are buying more at larger more general stores, and less at smaller specialty retailers. This phenomenon is intertwined with their trip making, as it reduces the number of trips required, but increases the distance of travel to make those trips.

The rise and fall of door-to-door sales

Foxell writes of Metro-land [10], the idyllic north London suburbs built by the Metropolitan railway in the early twentieth century, "This service economy is illustrated by the variety of tradesmen that called at our home: the milkman twice a day, with a horse-drawn cart; the baker once a day, with a large upright barrow on two wheels, the handles of which lifted him off the ground when going down hill; the postman thrice; the butcher's boy by bicycle twice a week; and the grover twice a week. Others like the coalman or the Gas, Light & Coke Co. in their steam-powered Sentinel Lorry also made their regular deliveries. Over a longer period, visits could be expected from the men from the Prudental (insurance), Hoover (vacuum cleaners), Singer (sewing machines)

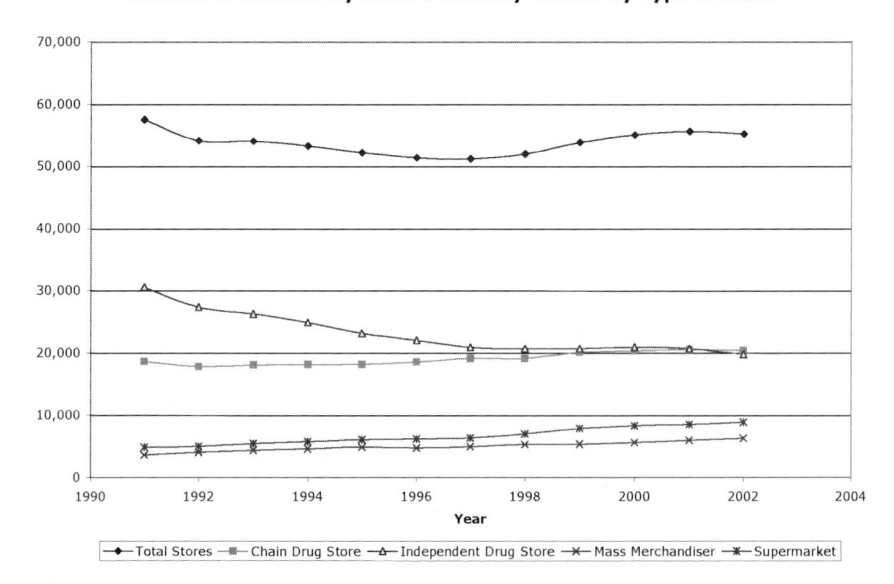

Figure 9.3 Number of community retail pharmacy outlets, by type of store
Source: www.nacds.org/wmspage.cfm?parm1=506 NACDS estimates based on IMS HEALTH, NCPDP, and American Business Information data. Franchise operations such as Medicine Shoppe are included as chains.

and the like – all using a service call to take the opportunity to sell new products. There was something reassuring about seeing such familiar faces and catching up with the latest gossip. In addition there were itinerant callers such as the Walls Ice Cream man on his tricycle as well as the French onion sellers, gypsies with pegs and posies, rag and bone men, tinkers (metalsmiths), and the knife-sharpeners with their pedal-driving grinding wheels."

When we were young and still at home during the daytime, we remember from time to time vendors knocking on our doors. For 20 years, David owned (and properly maintained) a hairbrush his mother purchased from the Fuller Brush Man. The Fuller Brush Company, founded in 1906 by Nova Scotian Alfred Fuller, became famous for its sales force, knocking on doors and opening up suitcases full of brushes and cleaning equipment. Fuller was no novice in door-to-door brush sales, having worked for a company that did just that before striking out on his own. Fuller, however, thought he could make better brushes and thus sell more, and apparently he was right as the company grew very quickly. It soon became so iconic that Disney dressed the Big Bad Wolf as a Fuller Brush Man in its cartoon *Three Little Pigs*. Later, Red Skelton starred in a movie called *The Fuller Brush Man*. One Fuller salesman, Frank Stanley Beveridge, founded Stanley Home Products in 1931, and used door-to-door sales, combined with Stanley parties to peddle cleaning products. One of Stanley's saleswomen was Brownie Wise, who soon became Vice President and General Manager of Tupperware, helping to bring Tupperware parties to America. Mary Kay Ash also gained training from Stanley before establishing her cosmetics empire and famous fleet of pink Cadillacs. Today, Stanley is co-marketed with Fuller Brush.

There are, of course, many other home marketers: the expression "Avon Calling" came from Avon Products, which was founded in 1886 by David McConnell as the California Perfume Company (based, logically enough, in New York). It is now the world's largest direct seller, with offices from California to Kazakhstan, and almost five million sales representatives. Others are multi-level marketers, like Amway, which at its base has sales in people's homes and through catalogs, but those individuals report to distributors, and so on, in a pyramid-like fashion. Sometimes, the customer had to come outside, as when the Good Humor, Jack and Jill, or Mister Softee Ice Cream truck clanged its bells.

Singer was one of the earliest transnational consumer products companies, ranging from the United States to Czarist Russia to Latin America, where it employed salesman such as Eugene O'Neill in Argentina. Singer Sewing Machines rose to such prominence with door-to-door sales that in the late 1800s it was the world's largest employer of salesmen. [11] This business success allowed it to build the world's tallest building in New York in 1908, a distinction the Singer Building held only until 1909 when it was surpassed by the Metropolitan Life Insurance Company Tower. The Singer Tower came down in 1969 to make way for the World Trade Center. The company filed for bankruptcy in 1999.

Other door-to-door businesses of yesteryear included encyclopedia sales. Sales of bound encyclopedias have fallen for a number of reasons, foremost among them the rise of the Internet as well as electronic versions of encyclopedias.

Box 9.2 Ground floor retail everywhere

It seems the New Urbanist new building ideal is apartments with ground floor retail. Numerous infill developments bear this out, such as the Excelsior and Grand project in St. Louis Park, Minnesota, (a first ring suburb of Minneapolis) which claims to offer "[t]he best of urban and suburban living" and to exist "on the cutting edge of urban planning"; one wonders whether this type of urban form will remain an exception, or whether it can become the rule. [12]

Bear Stearns has estimated United States retail space per capita at 5.3 sq. ft. (0.49 m^2) per person in 1964 and 19 sq. ft. (1.77 m^2) in 1996. This change itself is remarkable—a near quadrupling of retail space in just over 30 years—and certainly deserves further study.

Downtown Hong Kong has ground floor retail everywhere, and Hong Kong as a whole has a population density of 2,415 persons per square mile (6,254 persons per km^2). The average flat size in Hong Kong is 650 sq. ft. (60 m^2) and the average household size is 3.4. That results in 191 ft^2 (17.7 m^2) per resident.

The average space for residents in the United States is a just bit more than in Hong Kong. The average new house in the US, as shown in Figure 3.10, has risen from 983 sq. ft. (91.3 m^2) in 1950 to 2,350 sq. ft. (218.3 m^2) in 2002. In the United States, the average household size is 2.61 persons (Census 2003). Since not everyone lives in a new home, and many live in apartments, we have to make an estimate, suppose that the average residential unit is 1,500 sq. ft. (139.4 m^2), which gives an area per person of 575 sq. ft. (53.4 m^2).

How many stories of residential development would be required to support ground floor retail in residential buildings? This excludes the ground floor of commercial buildings, though if office buildings did have ground floor retail as well, then residential buildings would need to be taller.

A simple equation gives us this result:

$$H = \frac{L}{R}$$

where:

H = height above the ground floor
L = living space (Per person: 191 sq. ft. (17.7 m^2) in Hong Kong, 575 sq. ft. (53.4 m^2) in the US)
R = retail area (Per person: 19 sq. ft., (1.77 m^2))

In other words, in Hong Kong, a building ten stories above the ground floor (floor zero, following the American floor-numbering convention) would have enough residents to support ground floor retail at US levels of consumption. Applying US residential retail space per capita implies a building 30 stories above the ground floor to support ground floor retail. Although Hong Kong does have many apartment buildings ten or more stories in height, they are still are a rare sight in the United States (excepting Manhattan and a few other city centers), and 30-story apartment buildings are scarcer still. Even the ten-story height is generally not reached in New Urbanist developments such as Excelsior and Grand.

The job of encyclopedia salesman was most famously held by Warner Erhard before he went on to found EST Therapy, and satirized on comedies such as *Monty Python's Flying Circus*, *Happy Days*, and *Friends*. Vacuum-cleaner salesmen have also been held in low esteem, yet to show the merits of a vacuum cleaner, bringing it to people's homes might be effective. Door-to-door salesmen sold about 70 percent of vacuums in 1938. Clearly that number has dropped.

Door-to-door marketers, especially those who do "cold calls" to potential customers who have not previously expressed an interest in their product, are disappearing from the retail landscape. And the blame for the fall of door-to-door sales can be laid on the doorstep of increasing participation by women in the labor force, as discussed in Chapter 4. The more women in the workforce, the fewer at home, and the more time wasted knocking on the doors of empty houses. Perhaps one of the reasons for the rise in retail area per capita in the United States (as discussed in Box 9.2) is that door-to-door sales have declined. When cosmetics, vacuums, encyclopedias, plastic kitchenwares, and other home products are bought outside the home, more space in shopping centers is needed to sell them.

The rise (and fall) and rise (and fall?) of door-to-door delivery

Door-to-door delivery, however, differs from door-to-door sales. The delivery requires only a catalog (be it paper or electronic) and some way of getting the order and finances from the consumer to the manufacturer and the goods from the manufacturer back to the consumer.

The enabler for this type of exchange was the United States Postal Service's Rural Free Delivery (RFD). [13] The need for RFD lay in several factors. The remoteness of rural America meant 30 million residents had to travel to town to pick up their mail. The poor quality of roads made this difficult. Postmaster General (and department store founder) John Wanamaker pushed for RFD, which began in the 1890s, and after experimentation it was finally inaugurated in 1896 in West Virginia and later expanded to 29 states. By 1901, Congress made RFD permanent. RFD had several effects. One was to give added weight to federal involvement in the good roads movement. Article 1, Section 8 of the US Constitution gives Congress the power "To establish Post Offices and post Roads"; though federal aid for state roads did not really begin until 1913, and was not a significant force until 1916.

A second effect is that retailers such as Montgomery Ward, L.L. Bean, Charles Tiffany, W.A. Burpee, and of course Sears, Roebuck & Company took advantage of RFD. Especially with the addition of parcel service to traditional postal service, the mail order catalog business took off. Sears, which had been publishing specialty catalogs since 1888, issued its first general merchandise catalog, the "Big Book," in 1896; the Christmas edition came to be known as the "Wish Book." The catalog truly was general merchandise, selling automobiles by catalog from 1909 to 1913 and bungalow houses from 1908 until the Great Depression. In fact, Sears didn't open its first retail store until 1925, and the general Big Book catalog was discontinued in 1993, well before the widespread adoption of the World Wide Web.

By the time Sears was scaling back its catalog business, mail order, along with-toll free numbers, had become a booming industry. The emerging Internet saw the rise of numerous e-commerce vendors. Amazon.com (founded 1994) and eBay (founded 1995) relied both on the Postal Service, as well as on express carriers such as Federal Express (founded 1971) and United Parcel Service (founded 1907). Jupiter estimates US online sales at $65 billion, and projects that such sales will grow to $117 billion by 2008, at which time they will amount to about 5 percent of all retail sales, although the online sector is growing faster than traditional retailing.

Online research influences a great deal of offline purchasing, but missing from online sales are goods that are widely consumed without much research, such as supermarket food items, as well as items such as gasoline that are impractical to deliver. Many have tried to extend the reach of online purchasing to replace the supermarket, recalling the milkman of yore, but companies such as Webvan failed to succeed. Webvan, which attracted more venture capital than any Internet retailer except Amazon.com, delivered food to customers in seven cities, and established a new warehouse distribution system (paying $1 billion to Bechtel for this) in each of those cities. It acquired rival startup Home Grocer, but wound up spending money faster than it could earn it for long enough that it had to declare bankruptcy July 10, 2001, after the peak in the stock market bubble (but before 9/11). Even more ambitious, Kozmo.com, which also served seven cities, promised free one-hour delivery for a variety of goods—from videos to coffee and ice cream—ordered online. Unlike Webvan, Kozmo.com never went public, lasting from 1998 to April 2001. Webvan-like services (Peapod and Simon Delivers, among others) do remain, with lower capital costs. Whether these are profitable remains to be seen.

Product differentiation, undifferentiation, and markets

In sales transactions involving mobile customers, the location of the store plays an important role. There are competing theories suggesting how and why certain locations might have advantages over other locations. For example, some theories suggest it is important to differentiate oneself from competing products; alternatively, other theories call for making products as similar as possible in certain markets.

An example of the latter is provided by Hotelling's Law, also known as the principle of minimum differentiation. Imagine, for example, a beach 1 km long; it is summer, and you are in the ice cream business. Where are you going to set up your cart? Assume further that you are the only vendor, that customers will go to the nearest vendor (provided the vendor is no more than ½ km away), and that customers are uniformly spread out across the beach. The answer is fairly simple: to maximize your profits, you should locate in the middle. This is because if you locate to one side or the other, you will lose some business as people at the more distant end of the beach don't want to walk more than ½ km to get to your cart.

Your rival, Hagen, decides that your beach is so profitable he should open a second cart. Where should he set it up? The answer to that is that he will set it up in the middle as well. The reasoning is the same: if he locates to one side, say the left, he gets all the business from his left, half of the business between himself and you, but none from the right; while if he locates in the center he theoretically accrues half the business on each side, which is a much better outcome. (A socially preferred outcome would be for you to locate at 250 meters and Hagen at 750 meters from one end of the beach, which would reduce the maximum travel costs.)

A third vendor, Baskin, decides to locate his cart on the beach as well. Now Baskin could also locate in the middle, and split all of the business three ways, but it is more profitable to locate just to the left, and get 100 percent of the 49 percent of the beach to your left (and 50 percent of 1 percent of the beach between him and you and Hagen) instead of 33 percent of 100 percent of the beach. Hagen will also move if Baskin is stealing business, this time (say) 1 percent to the right, also garnering 100 percent of 49 percent of the beach (and 50 percent of 1 percent of the beach). This leaves you with 50 percent of 2 percent of the beach. So you decide to move, just to the left of Baskin, garnering 100 percent of 48 percent of the beach. So maybe Baskin hops over you, and you hop over him, and so on—but that wouldn't be a good solution either, since as you and Baskin move to the left, Hagen slowly moves his cart leftward as well, ensuring he has all of the business on his right. So instead of hopping to the left of Baskin, it may be better to move to the right of Hagen. There is a potentially stable end-point to all of this hopping, and that is Baskin at 167 meters, Hagen at 833 meters, and you at the half-way point (500 meters). At this point, everyone equalizes profits. If you move to left, you cannot improve your profits, but Hagen gets richer at the expense of Baskin, prompting Baskin to move again.

There are several things to note about this example: vendors are free to locate, so that there are no costs associated with moving (since the vendors have pushcarts); there are different customers each time (so former customers' memory of your location is unimportant); and entry is free, so in theory, another player could come in without cost to himself.

Real markets are much more complicated than Hotelling's Beach, developed in a 1929 paper, and there are other strategies a new entrant could take; instead of differentiating themselves spatially, they could come in and offer a different product (higher quality and more expensive, lower quality and cheaper). Product differentiation (like spatial differentiation) usually occurs if there are more than two vendors, or if there are demand curves.

However, insights like this are valuable. Sometimes vendors want to be as close to customers as possible, and try to establish as large a market area as possible for themselves, and thereby separate from competitors. Sometimes vendors would rather be near their competitors, and split a larger market with them, particularly if there are spillovers, which explains why similar stores are often near each other.

Take, for instance, Walgreens, as an example of a corporation with varying motivations from a location standpoint. The leading pharmacy retailer in the United States currently explores strategies to place stores in close proximity to one another—and to competitors—in order to maximize profits. Real estate planners employed by Walgreens are constantly on the lookout for locations that will yield the highest probability for approval by the Real Estate Committee. According to information freely available on the Walgreens website, such sites would contain some of the following characteristics: (a) a freestanding location at signalized intersection of two main streets with significant traffic counts; (b) direct access to service the site; (c) roughly 75,000 square feet of land to accommodate parking for 70+ cars and a pharmacy drive-through; (d) at least 14,560 square feet of building space; (e) a trade area population of 20,000; and, finally, (f) a readerboard pylon sign.

Furthermore, stores like Walgreens, by their very nature sell so many products that some of their offerings compete with each other while simultaneously being complementary. Most supermarkets sell Coke, Pepsi, and RC Cola, which are to the untrained tastebud quite similar products, so they are competing with themselves, but collectively supermarkets find stocking the competing varieties are complementary because there are some families that have been bitterly divided by the "cola wars" and buy multiple brands, and others who simply go for the lowest price. Moreover, cola goes with many other food products internal to the store.[a] And stores can be complementary to each other, so a hair salon locates adjacent to the supermarket, since both are places to which people tend to travel, on average, between once a week and once a month. So, once a month after getting their hair done, a shopper can walk to the supermarket, and save a trip.

The role of information

Just as William Whyte [14] found that people tend to sit where there are places to sit, we assert that people go where they know they can go. The central issue to understand (as transportationists) then relates to how people learn of such locations and the impact such information has in influencing travel. Learning and knowing where to travel is a tremendously underestimated, terribly undervalued, and poorly understood dimension of travel.

Consider the case of a fictional Eddie Haskell. Eddie is 18 years old and is shopping for colleges. He knows about Northwestern University because that is his father's *alma mater*. He knows about the University of Maryland because his mother went there. His problem is that he wants a highly selective university but wishes to leave the East Coast and does not like big cities. For information, he turns to the *US News and World Report's America's Best Colleges*. This compendium, together with other college reference manuals (and of course other anecdotal information from neighbors and friends) provides a plethora of information about where to look, where to visit, prices, and locations. The set of rankings (and other sources) inform our soon-to-be collegian's decision about where to apply. While Eddie is consulting such information to make a four-year decision, such a process (though perhaps less formal and systematic) is not unlike what most individuals employ when deciding where to relax, shop, eat, or visit (and thus where to travel).

Similar to the tenets discussed in Chapter 4 (looking at employment decisions), people receive inputs via two means: media networks and social networks. Any self-respecting hotel or tour group operator will quickly aver that word of mouth communication is the most successful means of recruiting new clients. Furthermore, there is a reason why businesses continue to use direct mail coupons, and why information directories continue to make money for the companies that publish them.

Zagat is a popular restaurant guide that publishes an annual guide to restaurants by city. The restaurants are rated on food, price, atmosphere, etc. These ratings are not simply a one-way dictat from owners Tim and Nina Zagat, but are collated from the opinions of thousands of *Zagat* members. It is a network in two ways. First, like a publisher or broadcaster, the *Zagat* guide conveys information to its consumers (readers or viewers), who can then act on such information, representing a one-to-many network. But *Zagat*'s guide is more than a one-to-many network. It also takes feedback from those same consumers (who are members of the network) and incorporates that in the next edition.[b]

The value of *Zagat* is that there is a huge sample of restaurants; each rating is not simply the product of a single reviewer and a single meal, but rather dozens of raters of each restaurant with many meals. Because it is a valuable network to join, more reviewers/diners join it. There are, of course, other restaurant guides and rating services, but as with other network externalities we have seen, there is power in numbers. If your taste differs from that of the patrons of *Zagat*, you can try others; but you need to steer away from ones that allow paid promotions.

This is only one of the many variations for such a product. There are countless other informational avenues that are available on the Internet. Take, for example, many trips (for shopping trips, movies, or other) that tend to be precipitated by knowledge of when, where, how much, etc. As we write, the authors are consulting the Internet to explore possible venues for New Year's activities.

Although we may have historically thought of media as a one-to-many system, there have always been feedback mechanisms. Whereas once these were slow and infrequent (like the letter to the editor), with the Internet, the media can become a truly many-to-many system.

More widely disseminating information serves to increase the number of opportunities we feel comfortable seeking. Instead of being confined to the restaurants in their own neighborhoods or a few that are reviewed in a newspaper, diners now view restaurants in every neighborhood of their city as potential destinations. They can pursue the quest for the perfect "Juicy Lucy" (a cheese-infused hamburger) by searching online,[3] and then travel outside their neighborhoods rather than staying put.

But while formal media has its role, socially conveyed information, as discussed in Chapter 4 in the context of job-seeking, is also significant. The impact of word of mouth through colleagues, friends, neighbors or others is still a powerful means of information exchange.

Those annoying direct mail coupons and advertisements that appear in the mailbox exemplify traditional media. Although they don't affect all travel, they do call attention to the range of retail, food, and entertainment services within driving distance. These ads appear in the mailboxes of everyone who lives in a particular postal zone, or shares certain particular demographic characteristics, or has shopped at a certain store in the past. Radio, television, newspaper, magazine, and Internet ads are driven by similar considerations.

The Internet not only lets people do traditional things better (e.g., find out where to shop, acting as an electronic yellow pages); it permits one to do some modestly new things (actually make the orders and pay for them online, like an advanced, automated version of the Sears catalog), and do more radically new things (sell things online to the highest bidder to strangers across the world), which far exceeds the capacity of newspaper classified advertisements. Indeed, Internet shopping (like catalog shopping) has taken consumer retail to a new level that may be slowly chipping away at the relevance of geography. But, as much press and attention as this notion receives, the reality is that place-based retail still dominates—by a wide margin. There remains considerable uncertainty as to how the phenomenon of online spending will fully manifest itself, particularly as it relates to travel. Past failures of electronic home shopping have been documented. [15] One need look no further than the experience of Kozmo.com (the now defunct online delivery service for sundries, which ceased operations after only three years) or WebVan (a grocery delivery service), to learn of the relative uncertainty associated with this line of service, and the demand and cost-structure required for these services to become profitable.

Collectively, new technologies provide information that not only extends the range over which consuming takes place—enabling and encouraging consumers to travel farther—they permit consumers to purchase without traveling at all. The goods may be shipped by traditional carriers (i.e., the Postal Service, United Parcel Service, Federal Express) or if possible, they may be distributed electronically (e-tickets, downloadable music, movies, and books). Although some items will not lend themselves to electronic transmission (clothes, food, cars), many can be acquired without physical presence. There will of course be a backlash, and few will be willing to purchase things online without having a good sense of what they are getting, but as bandwidth (the capacity of communications networks) rises, the amount of information that can be exchanged rises concomitantly, and the need for travel drops. The extent to which chat rooms replace "visiting friends," videoconferences replace "work," or "telemedicine" replaces "doctors" remains to be seen. However, if history is a guide, replacing one kind of trip may simply permit another.

Selling wrap up

The modern marketplace provides consumers considerable flexibility that was unavailable in a convenient fashion just a quarter-century ago. The variety and choice offered by the marketplace allows consumers to undertake a particular activity at several different locations. By extending the hours and days they are open, retail stores accommodate diverse personal schedules. Below, we identify some additional factors affecting commerce [16]:

- *Bargain Hunting*—Competition attracts price-conscious consumers to trade higher travel and time costs for lower cost merchandise. This is particularly the case when consumers are familiar with the goods they seek and are responding to regional advertising. Newspapers typically carry advertising inserts describing low-priced goods available only in big-box stores and off-price retailer establishments sited throughout a metropolitan region.
- *Comparison Shopping*—Although all stores generate some measure of non-local trips, many business activities (e.g., stores selling furniture, major appliances, or automobiles) generate higher levels of longer, "comparison" shopping trips. In these cases, customers will bypass other similar stores to shop there.
- *Preference for Variety*—People also travel farther to find variety or a unique shopping experience. For some, shopping is a recreational activity, and "satisfaction" is a large component. Malls that include food courts, multi-screen cinemas, amusement rides, electronic game parlors, concert stages, traveling festivals, and fashion, automobile, hobby, and craft shows play to this preference.
- *Schedule Flexibility*—Consumers exhibit considerable flexibility in the time scheduling of trips to retail centers, often made possible by extended store hours. Non-work trips are combined with trips to and from work, and they originate from work sites. Tours involving several non-work activities typically occur after work hours and on weekends. When visiting parts of Europe (Netherlands, Germany), the authors were puzzled how society functioned when households with two working members found stores that closed at the same time as other businesses. When does one shop? (The answer seems to be, when one is supposed to be at work, which may help explain the lower productivity found in Europe).

Years of analysis from the fields of urban economics and real estate provide sound explanations for the location of retailers in the aggregate. From the early foundations of Walter Christaller's Central Place Theory (described in more detail in Chapter 11) to more contemporary theories, there seems to be a mantra with four choruses. Retailers choose to:

1   locate near pools of potential shoppers;
2   locate within the range of wholesale distribution;
3   locate on highly trafficked streets to ensure high visibility; and,
4   locate in clusters of complementary stores to exploit network effects by maximizing attractiveness to customers.

To add to the old adage, it appears as if it boils down to four considerations: Location, location, location, oh, and location. Apache Plaza ultimately failed on these dimensions, particularly once the pool of shoppers moved to a better location and a newer mall.

The commercial market prefers store sites determined by the needs of developers and owners to succeed financially. Key retail site selection criteria and the market forces currently shaping a metropolitan region's retail structure help understand these trends. These key retail location decision criteria have been highlighted by others and can be explained by wanting to (a) take advantage of agglomeration economies and scale economies, (b) maximize visibility, access, and parking, and (c) minimize environmental impacts and complications within zoning and other resistance issues from the public.

The above factors lead to a tendency of stores wanting to cluster together, usually at locations of high regional and local access or visibility (e.g., major intersections or freeway ramps). They do so to achieve a market advantage. In addition, stores of all kinds are getting larger, both in floor and market area, taking advantage of economies of scale. The location criteria combined with consumer characteristics have produced the pronounced retail structural trends seen in metro regions across the United States. These trends include:

1   planned shopping centers dominating the retail market;
2   smaller malls (still auto-reliant) clustering around major malls;
3   a growing share of "Big Box" superstores in the retail market;
4   dominant chains preferring stand alone sites so as not to advantage competitors with spillover traffic;
5   dining out continuing as a major factor in site design, as does;
6   the convenience of services for driving to and through establishments.

American society (and resulting development) has put itself in a situation where trips to the local, close, and convenient establishment (e.g., the corner store/bakery/hardware store) get passed over in favor of other (most often larger) establishments. Consumers have long expressed strong preferences to buy cheaply, to compare competing products, and to experience variety. It is important for land use and transportation professionals to recognize that residents appear willing to travel often and farther than their neighborhood commercial center to find them.

Home and shop have seen changing relationships with technology, as consumption takes place in a much larger market, retailers serve more consumers, and consumers have a greater choice of retailers. The consequence of these larger markets is a decline in local shopping opportunities. Critics

Box 9.3 Starbucks at the end of the universe

"... and if you walk to the end of the block, there sits a Starbucks. And directly across the street—in the exact same building as that Starbucks—there is ... another Starbucks. There is a Starbucks across the street from a Starbucks! And ladies and gentlemen, THAT is the end of the universe." Lewis Black

Comedian Lewis Black has said he discovered the end of the universe in Houston, where two Starbucks (the largest chain of coffee shops in the United States) are located across the street from each other. The audience may have thought he was joking, but it is true,[a] and there are websites devoted to finding these sites.[b] Why would a vendor do this? Virginia Postrel interviewed the company's director of business development, Brooke McCurdy:

> People are amazed that we have stores across the street from each other. But they're different stores. [If the color of one store] reminds you of something from your childhood that you intensely dislike, you can go three stores down and say "I like this better. I just feel better here."

Postrel argues the company is differentiating its product through design elements of the environment in which coffee drinkers sip. Others have noted that some Starbucks are exclusively non-smoking, while others allow patrons to light up in places where smoking indoors at restaurants is still legal. Starbucks also co-locates with the bookstore chain Barnes and Noble as well as a number of grocery chains (and co-locates its sister chain, Seattle's Best, with Borders), so there may be more than just two Starbucks in close proximity. Also, if a particular Starbucks has so much business that it can't be handled at a single standard size store, the chain may feel it best to open a second location rather than expand the first, just to better manage the facilities. Customers might feel odd if a Starbucks got to be as large as a grocery store. Howard Schultz, chief executive of the company, said of a location in Vancouver, British Columbia where one Starbucks allowed smoking and one was no-smoking initially (both are now non-smoking): "We kept looking at it and looking at it, and finally we said, let's take the other side of the street. My board members thought I was out of my mind,"[c] Although the first store did see a small (10–15%) diminution of sales for the first 18 months, eventually both resumed growth and had different customer bases. "We've repeated that strategy countless times." Site selection processes are now highly analytical, using Geographic Information Systems tools including detailed socio-economic and demographic data.

Nancy McGuckin, a travel behavior analyst at the United States Department of Transportation, looking at the National Household Travel Survey, identified the "Starbucks effect," wherein 1.6 million travelers a day attach errands to their morning trips. Until recently, those trips involved getting out of the car, but Starbucks is rapidly deploying drive-through locations throughout its system; as of 2005, drive-through outlets represented 15 percent of the chain's locations.[d] The travel survey

188 Selling

> data don't reveal to what extent that added errand is a visit to a Starbucks, nor how far away it is, but with over 9,500 locations worldwide (over 6,800 in the US)[e] and $5.3 billion in revenues annually, Starbucks is serving many cups; competitor Caribou Coffee serves 65% of its cups before 10 am.[f]
>
> You can find your local Starbucks density by going to their website[g] and entering your postal code, which gives the number of Starbucks within five miles (8 km) of your location. As of this writing, our count stands at 18, which is pretty low compared to some places. The rapid proliferation of Starbucks locations prompts one to consider an interesting question that merges tenets of social networks and commute minimization (issues discussed in Chapter 4). Presumably, the type of work in each Starbucks location is reasonably similar across each franchise (a phenomenon shared by employees of banks, fast food establishments, and fire fighters). What if all Starbucks employees worked at the establishment nearest their home? How much travel would be saved? Considering that reports suggest that only about 20% of employees work at their nearest Starbucks, the savings could be big.[h] However, considering that many employees got their job at the Starbucks through some sort of social network, it is a difficult suggestion to see materializing.
>
> [a] Gogoi, Pallavi. The New Science of Siting Stores. *BusinessWeek* July 6 2005 www.businessweek.com/technology/content/jul2005/tc2005076_7033.htm?campaign_id=rss_techn
> [b] www.gergltd.com/users/isaac.gerg/starbucks/ The End of the Universe – Lewis Black Skit! The Proof!
> [c] Kirbyson, Geoff. Howard Schultz: Not Your Average Joe www.brandchannel.com/careers_profile.asp?cr_id=47 Aug. 30, 2004.
> [d] Starbucks sees growing demand for drive-thru coffee, by Elizabeth M. Gillespie, www.usatoday.com/money/industries/food/2005-12-24-starbucksdrivethru_x.htm.
> [e] Starbucks: Starbucks Announces Strong June, Revenues; www.starbucks.com/aboutus/pressdesc.asp?id=521.
> [f] Fuhrman, Elizabeth, The Last Drop: The Starbucks Effect. Beverage Industry www.bevindustry.com/content.php?s=BI/2005/05&p=23.
> [g] www.starbucks.com/retail/locator/default.aspx.
> [h] Scigliano, Eric (2002) Trading PLaces: It's a homegrown, low-cost alternative to commuting via highways and transit. Just add software. Seattle Weekly, November 13, 2002 www.seattleweekly.com/2002-11-13/news/trading-places.php

bemoan this loss of "Third Space," places that people can meet and socialize outside home and work, places such as bars, cafes, and other hangouts, [17] but it has been lost consciously. People choose to move to neighborhoods without these features, and fail to create them for themselves. Despite the rise of Starbucks (see Box 9.3), the large American house (see Chapter 3) means that most Americans, unlike many urban Europeans, can relax at home.

Notes

a Speaking of cool refreshing pop, think about where you find them on the supermarket shelf. Most larger stores have an aisle devoted to soft drinks. The sellers pay for placement and stock the shelves, Coke on one end, Pepsi on the other and RC and its related brands (Sunkist, Canada Dry, Dr. Pepper, Seven Up) in the lower

rent middle of the aisle. The middle is less desirable because it has lower accessibility: shoppers in a hurry can dart from their cart on the cross-rows to the aisle and back to pick a single item from the shelf more easily for products at the end than for those in the middle. This exemplifies place and plexus writ small.
b   The next edition is published the next year in the published version, sooner for the electronic version, which can be accessed online by subscribers or via a handheld personal digital assistant such as a Palm.
c   To aficionados, a Juicy Lucy is not simply a slice of cheese melted between two patties, rather the cheese is cooked within the raw meat.

References

[1] Horan, R., Bulte, E., and Shogren, J., 'How Trade Saved Humanity from Biological Exclusion: An Economic Theory of Neanderthal Extinction', *Journal of Economic Behavior and Organization*, 2005, vol. 58: 1–29.
[2] Platner, S.B., *A Topographical Dictionary of Ancient Rome*, London: Oxford University Press, 1929.
[3] Schoenherr, S.E., *Evolution of the Department Store*, 2006. Available at: http://history.sandiego.edu/gen/soc/shoppingcenter4.html (accessed September 15, 2007).
[4] Live Baltimore Home Center, *Neighborhood History: History of Roland Park*, 2005. Available at: www.livebaltimore.com/nb/list/rlndprk/history (accessed September 7, 2007).
[5] Mause, T., *Dundalk's Timeline*, 2005. Available at: www.DundalkEagle.com (accessed April 27, 2007).
[6] CCP, *One Man's Vision Shapes the City: Urban Historians Recognize the Country Club Plaza as One of the Most Successful Commercial Developments in the Country*, 2005. Available at: www.countryclubplaza.com/plaza.aspx?pgID=893&newsID=1&exCompID=45 (accessed Sept. 5, 2007).
[7] Gladwell, M., 'The Terrazzo Jungle', *The New Yorker*, 15 March 2004: 120–127.
[8] Newsweaver, 2002. Available at: http://newsweaver.co.uk/kogod/index000025582.cfm (accessed September 5, 2007).
[9] Hardwick, M.J., *Mall Maker: Victor Gruen, Architect of an American Dream*, Philadelphia, PA: University of Pennsylvania Press, 2004.
[10] Ideafinder, *Vacuum Cleaner*, 2005. Available at: http://ideafinder.com/history/inventions/story042.htm (accessed October 1, 2007).
[11] Martin, H.C. *Singer Memories, Singer Door-to-door Salesmen*, 2005. Available at: www.singermemories.com/cast-of-characters-salesmen.html (accessed June 10, 2005).
[12] Excelsior & Grand, 2006. Available at: www.excelsiorandgrand.com (accessed July 7, 2007).
[13] Ridgway, D. *Postal History: Sharing the Story of 100 Years of Rural Free Delivery*, Smithsonian Institute Research Report 89, 1997.
[14] Whyte, W.H., *The Social Life of Small Urban Spaces: Project for Public Spaces*, New York: The Conservation Foundation, 1980.
[15] Talarzyk, W. and Widing, R., 'Direct Marketing and Online Consumer Information Services: Implications and Challenges', *Journal of Direct Marketing*, 1994, vol. 8 (4): 6–17.
[16] Nelson, D. and Niles, J.S., *Market Dynamics and Nonwork Travel Patterns: Obstacles to Transit Oriented Development*, Proceedings of the Transportation Research Board, Washington, DC, 1998.
[17] Oldenberg, R., *The Great Good Place*, New York: Paragon House, 1989.

Chapter 10

Diamond of Evaluation

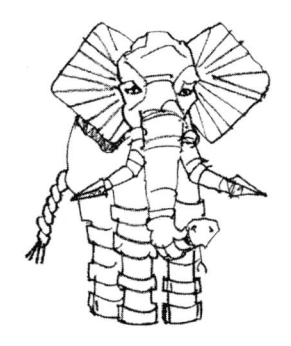

It was six men of Indostan
To learning much inclined,
Who went to see the Elephant
(Though all of them were blind),
That each by observation
Might satisfy his mind

The First approached the Elephant,
And happening to fall
Against his broad and sturdy side,
At once began to bawl:
"God bless me! but the Elephant
Is very like a wall!"

The Second, feeling of the tusk,
Cried, "Ho! what have we here
So very round and smooth and sharp?
To me 'tis mighty clear
This wonder of an Elephant
Is very like a spear!"

The Third approached the animal,
And happening to take
The squirming trunk within his hands,
Thus boldly up and spake:
"I see," quoth he, "the Elephant
Is very like a snake!"

The Fourth reached out an eager hand,
And felt about the knee.
"What most this wondrous beast is like
Is mighty plain," quoth he;
"'Tis clear enough the Elephant
Is very like a tree!"

The Fifth, who chanced to touch the ear,
Said: "E'en the blindest man
Can tell what this resembles most;
Deny the fact who can
This marvel of an Elephant
Is very like a fan!"

The Sixth no sooner had begun
About the beast to grope,
Than, seizing on the swinging tail
That fell within his scope,
"I see," quoth he, "the Elephant
Is very like a rope!"

And so these men of Indostan
Disputed loud and long,
Each in his own opinion
Exceeding stiff and strong,
Though each was partly in the right,
And all were in the wrong!

Moral:
So oft in theologic wars,
The disputants, I ween,
Rail on in utter ignorance
Of what each other mean,
And prate about an Elephant
Not one of them has seen!

The Blind Men and the Elephant,
by John Godfrey Saxe (1816–1887)
(based on an Indian fable)

Few planning issues incite such emotion and are as contentious as bridge crossings. In fact, in every community where the authors have served in professional capacities as planners, there was always a contentious bridge project lingering in public debate: an additional crossing of the Snake River in Jackson Hole, Wyoming; an additional roadway over Lake Washington in the Puget Sound; the need for increased capacity connecting Oakland to San Francisco; the reconstruction of the Woodrow Wilson Bridge on Capital Beltway surrounding the District of Columbia; and the proposed Stillwater Bridge spanning the scenic St. Croix River separating Minnesota from Wisconsin east of the Twin Cities.

The last case, for example, is now into its second decade as a topic for debate. The battle lines are drawn, the stakes are high (the price tag for the new bridge estimated at a minimum of $150 million), and not surprisingly, there is no consensus on the right solution. The lack of resolution in this particular scenario can be attributed, in part, to the distinct issues under consideration, each with differing "optimal" solutions.

Many area residents (the majority of whom live on the Wisconsin side of the river) claim it is unrealistic to expect them to travel eight miles south to cross the river via Interstate Highway 94, the main east-west highway in the area (see Figure 10.1). People have a right to get to the Twin Cities in a timely way, they say, and it is inefficient to prohibit this; furthermore, it is unfair to remove the mobility on which they currently rely if the existing bridge is closed and no replacement constructed.

Some business owners and other economic interests suggest that not providing for a viable bridge crossing creates issues of inequity. They argue that they would suffer economic hardship without a bridge because traffic would bypass the area. Other business owners, in contrast, believe that cut-through traffic in downtown Stillwater, where the existing bridge from the east terminates on town streets, creates an unhealthy environment and unpleasant experience for shoppers (most of whom come from the Twin Cities, west of Stillwater) visiting the historic and now tourist-oriented old downtown that features many antique dealers, bookstores, and restaurants.

The St. Croix River is subject to the federal Wild and Scenic Rivers Act, and thus any construction along it evokes a strong environmental ethos. Among the arguments motivated by environmental concerns are that constructing an additional bridge will damage the scenic bluffs at the river's edge, and that building new bridge pillars will harm the water quality of one of the country's cleanest rivers. Additionally, there is concern that additional bridge capacity will further encourage automobile-based transportation and spur additional development of the rural Wisconsin landscape where farms and woodlands predominate today.

Finally, there are the preservationists who oppose removing the iconic steel truss lift-bridge built in 1931 on the grounds that its absence would seriously diminish the overall historic experience for anyone visiting the area. The existing bridge, however, is old, too low to withstand floods, and requires major repairs.

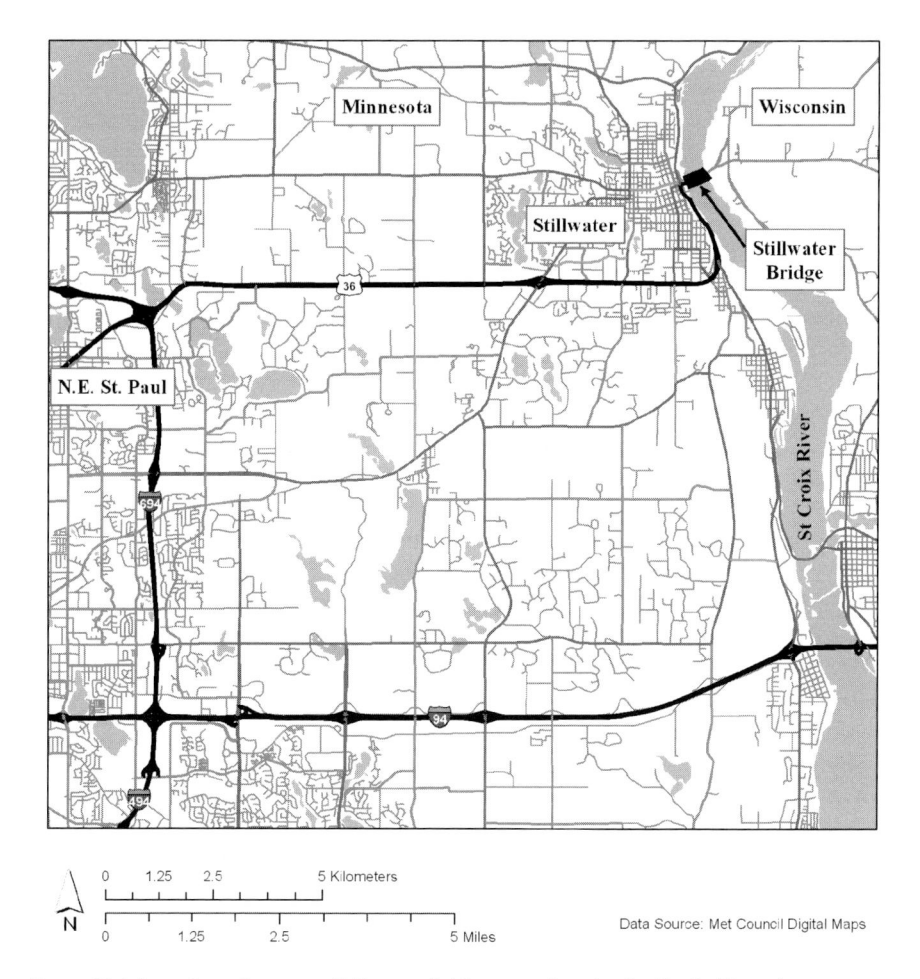

Figure 10.1 Location of current Stillwater Bridge spanning the St. Croix River (some proposals call for a crossing south of the current crossing and south of town, roughly extending MN-36 eastward)

Diamond of Evaluation

Agents putatively working on behalf of the public (e.g., government bodies such as city, regional or state planning departments or departments of transportation) aim to nudge market forces and motivate human behavior to be more consistent with articulated goals and objectives of a community. A problem, however, comes when the differing perspectives, claims, and proposed solutions are in conflict with one another. How can one evaluate proposals changing the place and plexus of a community? What economic interests consider efficient, the environmental planner may deem destructive.

What the equity planner seeks to expand is often at odds with what the historic preservationist wants to maintain.

Such contradictions are blatantly apparent when considering different perspectives on traffic congestion. Most residents and political leaders think of traffic congestion as an evil—traffic-snarled streets or highways that waste countless hours, decrease economic output, and frustrate residents. Few would be interested in an urban area so choked on its forms of mobility that it stifles travel and productivity.

On the other hand, some leading thinkers suggest that congestion is a positive indicator of a region's success. [1, 2] The beauty of congestion, they claim, is that it serves as an equilibrating device between services to be provided and locational/travel preferences. But there comes a point at which increasing levels of traffic challenge the efficient use of time—the point where road capacity can no longer absorb the increase in traffic, resulting in a decrease in average speed. If average speeds get slow enough, the economic growth of the community may stall. If economic growth rates decline (or go negative), the level of congestion will settle into an equilibrium (or dissipate), and it can no longer be blamed (again, Yogi Berra's famed quote, "Nobody goes there anymore, it's too crowded.").

These issues bring to mind the parable of the Blind Men and the Elephant, related at the beginning of this chapter. The story tells of how individuals (or in this case, different types of planners) each comprehend only a tiny portion of the whole picture; they then tend to extrapolate all manner of dogmas from that perspective, thereby launching a reductionist view. To help understand the long lists of competing aims and aspirations that make up many planning initiatives, we coin the Diamond of Evaluation (Figure 10.2), comprised of the 5 "Es."[a] This organizational device encourages a holistic, rather than reductionist, understanding of the factors affecting transportation-land use plans, policies, and initiatives.

Measures of effectiveness

Assuming the five criteria described in the Diamond of Evaluation cover the major goals of land use-transportation planning prompts one to prescribe measures of effectiveness. There is no single measure of effectiveness that adequately describes the transportation or land use systems; likewise, there is no single measure that jointly describes efficiency, equity, experience, environment, or expediency. Planners are therefore forced to undertake the inconvenient but necessary task of using several measures. Even so, some measures are more useful than others.

It is necessary to outline attributes that help define good measures of effectiveness (hereafter referred to as MOEs). Although the below list is by no means definitive, it serves as a starting point for further discussion. MOEs should:

- align with user experience and be understood by those users;
- be measurable, or calculable from available (observable) data;
- be predictable, or able to be forecast;
- be useful in a regulatory or control context (so that the measure can be used as a basis for regulating new development in order to maintain standards, to rank projects so that better projects can be selected, or to help guide operational traffic engineering decisions);

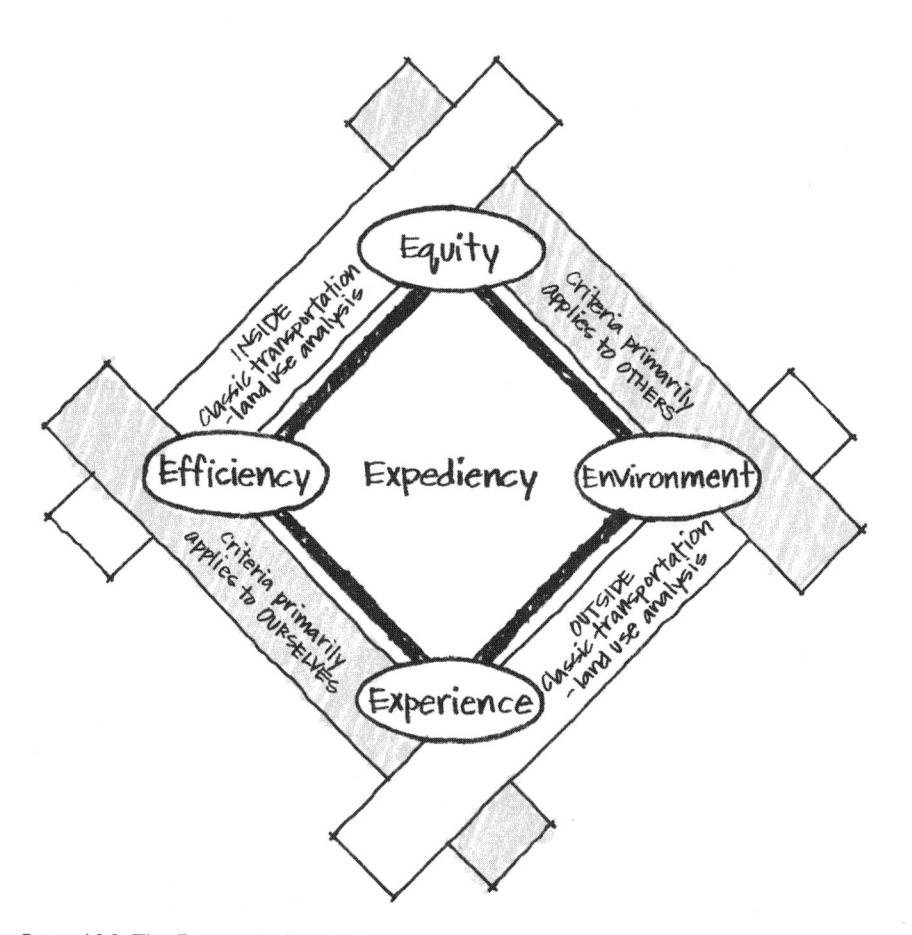

Figure 10.2 The Diamond of Evaluation

The Diamond of Evaluation reduces the range of goals and objectives to five "Es"[a]—efficiency, equity, environment, experience, and expediency—each of which speaks to a different justification for government actions. The first four Es describe basic motivations and goals for a preferred land use-transportation system. A two-by-two matrix organizes planning goals according to whether they primarily apply to ourselves or to others, and the extent to which the relevant issues have been integrated into the concerns of classic land use-transportation models and analysis (as it has been practiced over the past half century). The final criterion, expediency, transects all orientations.

- scale or aggregate well (e.g., it should be possible to combine measures on separate links to obtain a measure for the entire trip); and
- be collectively complete—in a model sense—in that one could combine the MOE's to attain an overall measure.

It is also important to distinguish between the normative (what *should be*) and the positive (what *is*).[b] To say that the speed on a link is 50 km/h tells us nothing about whether that situation is good or bad; it is simply a fact (this measure is positive). Only by comparing the measure to a normative standard (for instance, a speed limit) can one determine if there a problem related to excessive speed (the speed limit is 30 km/h), a congestion problem (the speed limit is 110 km/h), or no problem at all. The following sections introduce evaluative criteria to apply to each MOE in a transportation-land use context and further explain MOEs that may be useful.

Efficiency

> ef·fi·cient *n.* The quality or property of acting or producing effectively with a minimum of waste, expense, or unnecessary effort.[c]

Cities form because they concentrate goods, services, and ideas that can be exchanged in a manner that minimizes transportation costs. Rising levels of traffic and attendant congestion, however, prompt many to question such motivations. Are cities indeed minimizing transportation costs? In the US, the Intermodal Surface Transportation Efficiency Act of 1991 (ISTEA) refocused attention on the non-efficiency aspects of transportation. Although the concept of efficiency was clearly an important part of the bill (after all, it is in the title), the lack of a concise definition of "efficiency" in the legislation left many struggling with its meaning in this context. Does efficiency mean providing services that are cost effective? Providing transportation services that allow people to use their time more efficiently? What does it mean to have a minimum of waste or unnecessary effort?

The ambiguity prompts us to clarify such matters. To do so, we offer hypothetical cases and present differing perspectives on what many consider to be inefficient behavior:

- A poor working mother who takes three buses and nearly three hours to get home from her job;
- A wealthy executive stuck in traffic and about to miss her flight, who would be willing to pay to avoid it, but lacks the opportunity since "we are all stuck in it together;"
- A suburban "soccer mom" who operates "mom's taxi service" shuttling her (and her neighbors') children between soccer practice, piano lessons, study group, and back home;

- An exurbanite who adds 16 miles round trip to her commute because the preferred bridge does not connect her home in Wisconsin to her job in the Twin Cities (referring to the scenario described at the beginning of this chapter);
- A large corporation, its livelihood dependant on transporting goods to the airport for timely, next-day delivery, which loses money because its delivery trucks are sitting in freeway congestion.

Cumulatively, it is easy to see how these examples invoke an element of waste, undesired expense, or unnecessary effort, suggesting that efficiency is one of the hallmarks of a good transportation system, and that efficiency is not always at preferred levels. A problem is that efficiency means different things to different users. It also means different things to analysts, requiring different measures. The reasons for the different measures are that their uses vary: planning, investment, regulation, design, operations, management, and assessment are among the aims.

For purposes of land use-transportation planning, we feel it is useful to approach the concept of efficiency across two dimensions: (1) the overall scope of the analysis, and (2) the degree to which there is a spatial component involved. The second dimension is especially important in land use-transportation research as there are links, subnetworks, trips, and entire networks over which evaluation may be important. These spatial units correspond to the block, neighborhood, corridor, and metropolis of planners. Dividing the dimensions in a binary manner yields four cells to consider (Table 10.1), each with differing perspectives of evaluation.

Furthermore, different professions have different definitions of efficiency. Managers aim to minimize costs and maintain the *productivity* of a given system element. Economists measure *utility* (or consumers' surplus) and try to ensure that benefits exceed costs. Engineers focus on maximizing *mobility* (the speed and capacity of the system) and safety, as the other perspectives are out of their control, but tend to focus on parts of the network rather than on the trip as a whole. Planners looking at the longer term consider the location of places with respect to each other in a measure of *accessibility*.

The professions generally take the *objective* viewpoint of the omniscient central planner (who may be an engineer, manager, or economist) rather than the *subjective* perspective of the travel consumer. Any of these measures, when applied are usually taken at a specific point in time (e.g., the present or a

Table 10.1 Measures of efficiency

	Aspatial	Spatial
Partial	Productivity	Mobility
Comprehensive	Utility	Accessibility

forecast year), and assume all other circumstances are otherwise unchanged. Box 10.2 considers evaluation of the ability of systems to respond to changed circumstances.

We discuss each of these perspectives on efficiency below. The economist's perspective on effectiveness typically revolves around the notion of benefit-cost analysis, a task which involves summing the net present values of benefits and costs.[d] If the benefits exceed the costs, the project provides economic efficiency. Of course this is easier said than done, as measuring benefits and costs is a non-trivial exercise in many cases. First and foremost, such a task requires forecasting—predicting the future—which is always uncertain and contingent.

Planners and engineers have helped economists develop tools for more robust benefit-cost analysis, but although these tools are mathematically sophisticated, they remain crude. They require analysts to make speculative assumptions about human behavior—either that it will remain stable over time, or change in some known way. Benefits to users in public projects are often measured as the sum total of the utility accruing to consumers. Because utility is not directly measurable, the concept of consumers' surplus is often used. This concept is examined in greater detail in Chapter 11. A typical transportation economist argues that the sum of the change in consumers' and producers' surplus is the appropriate benefit measure with which to compare conditions existing before and after road widening, land development, or other non-systemic changes in policy or infrastructure. Consumers' surplus is closely related to the concept of utility, previously discussed.

Productivity is usually presented in terms of output divided by input: the larger this ratio, the more productive the system. But what are the outputs and inputs, and how are they measured? Beginning with the inputs, we have, broadly, capital and labor. Labor includes all the human time required to produce a service. When considering the productivity of transit service, labor inputs are the employees of the transit agency, including bus drivers, mechanics, managers, and accountants, among others. (When considering travel by private automobile, the driver's time must be included as well.) Capital includes all the buildings and equipment needed to operate the service (e.g., buses, garages, offices, computers.) Capital may include land and energy, though those are often separated. Although labor may go into each of the capital components, to the transit agency it is viewed as capital (the labor required to build the bus is considered in the labor productivity of the manufacturer of the bus, but not that of the transit agency which operates it). Labor productivity can be measured by dividing the output measure with hours of labor input. Similarly, capital productivity can be defined as the output measure divided by the capital, in monetary terms, that is required to produce that output. Capital is somewhat trickier than labor because capital is often a stock, whereas output and labor are flows. For example, if it costs one million dollars to build a road section with a multi-year life, it is difficult to measure the productivity of capital as simply annual output divided by that one million dollars. Rather, that stock needs to be converted to a flow, as if the highway department were renting the road. This conversion depends on the interest rate and the life of the facility. Productivity is not a perfect welfare measure, but it indicates whether welfare is increasing or decreasing. As emphasized earlier, other gauges may be required to measure overall welfare. Furthermore, in this section we have only described the productivity of transportation, not the activity system to which all travel belongs.

Next we arrive at Measures of Effectiveness (MOEs) often used by engineers and planners. Every year, for example, major newspapers in the US eagerly await the well-known Urban Mobility Indicators produced by the Texas Transportation Institute (TTI), in order to relay to their readers how well (or in a perverse sense of pride, how poorly) their city is performing [3] according to this annual ranking of metropolitan congestion levels. The indicators measured by TTI include hours of delay, speed of traffic and number of cars experiencing congestion—all of which are considered to be measures of *mobility*. Such measures have attained the status of received wisdom within the transportation industry and are now firmly embedded in planning concepts such as typical Level of Service applications; higher volume-to-capacity ratios, for example, mean slower travel times, less ease of movement, and thus reduced mobility.[e]

Measures of mobility are typically derived from, and restricted to, the impedance component of accessibility measures as described more fully in Chapters 3 and 4; mobility captures how difficult movement is in general. Mobility measures provide a snapshot of only one dimension of transportation system performance: the ability of residents to transport themselves under certain conditions (e.g., free-flow travel times). In their biggest deficiency, measures of mobility fail to say anything about where the traveler is traveling to.

This is where accessibility measures—those more germane to planners—come to the rescue. Measures of accessibility focus on the ends (rather than the means) and on the traveler (rather than the transportation system) (see Box 10.1). Accessibility measures ask: do people have access to the activities that meet their needs or in which they would like to participate?[f] These measures represent the ability to get what one needs, and include both an impedance factor (reflecting the time or cost of reaching destinations) and an attractiveness factor (reflecting the qualities or attractiveness of potential destinations). They measure the ease with which other pieces of land and their associated activities can be reached from a given origin. [4–8] If a transportation or land use change enables someone to reach activities that are more desirable in less time, then the accessibility (and possibly the value) of a parcel of land increases.[g] A central problem in planning applications is that the terms "mobility" and "accessibility" are often used loosely and interchangeably in the land use-transportation dialogue, leading to confusion about how and where these concepts should be employed (See Box 10.1).

Equity

> eq·ui·ty *n.* The state, quality, or ideal of being just, impartial, and fair.

Although democratic political systems are little concerned with efficiency, most are deeply concerned with fairness and justice. Concepts such as fairness and justice are difficult to define, much less ensure, in a planning context. Geographic and socio-economic variations in transportation services are well documented; for example, suburban areas tend not to support transit or pedestrian travel modes, contributing to higher rates of drive-alone travel. This in turn contributes to a pattern of discrimination against the transportation-disadvantaged with no means of auto travel—the physically challenged, developmentally disabled, poor, elderly, or very young.

But, of course, the whole picture is not so simple. The faster and more glamorous forms of public transportation (e.g., heavy rail) have been shown in many markets (e.g., the San Francisco Bay Area, Northern Chicago) to transport white-collar workers to urban cores. Similarly, in contrast with rail, bus riders are, on average, much poorer than the general population, with disproportionate numbers of elderly and minority passengers. Federal and state transit subsidy policies have generally not remained consistent with land use demographic shifts in urban transit use that have been occurring over the past half-century, but instead have tended to support suburban and downtown commuter services such as radial rail transit networks in efforts to lure discretionary customers out of their automobiles. While this trend in funding priorities may have improved the range of options available to suburban commuters, the resulting inattention to local bus services has diminished accessibility for inner-city residents, particularly to employment opportunities. [9]

The term "equity" has both a descriptive (positive) and a normative meaning, describing the distribution of benefits and whether the distribution is for the better or for the worse. *Horizontal equity* refers to equivalent or impartial allocation of benefits and costs among individuals and groups who are similar in terms of wealth and ability. *Vertical equity*, on the other hand, refers to the distribution of benefits and costs across different social strata, such as income groups, or groups and individuals with different physical abilities.

Borrowing from Krumholz, equity planning has perhaps best been defined as providing more "choices to those . . . residents who have few, if any choices." [10] If this mantra is accepted, then it becomes a noble and easily understood goal for transportation service planners since the inequity of such services is often clearly visible. The resulting situation is one in which the allocation of transit services across incomes, races, and jurisdictions is not happenstance. It is connected to social and economic processes that have produced the current racial and economic polarization between suburbs and central cities.

Social welfare comprises both efficiency and equity. Public sector investment decisions are made in non-market forums that often suffer from a short-term

Box 10.1 Manhattan or Manitoba: access versus mobility

Scan the "Goals and Objectives" section from any transportation plan across the globe. You will likely find some mention of accessibility, access, and/or mobility. For example:

- One of the four themes in creating the Transportation Strategic Plan for Seattle (Washington) was to "Provide mobility and access through transportation choices;"
- The 2020 Regional Transportation Plan for the Chicago (Illinois) region aims to "provide an integrated and coordinated transportation system that maximizes accessibility and includes a variety of mobility options that serve the needs of residents and businesses in the region;"
- From Europe, the central objective of the transport section of the strategic plan for Torino (Italy) is to "complete projects involving the system of accessibility and mobility by increasing the network;"
- And, even the Minnesota (US) Department of Transportation is on board with their mission statement: "to improve access to markets, jobs, goods, and services and improve mobility by focusing on priority transportation improvement and investments that help Minnesotans travel safer, smarter, and more efficiently."

A central issue, however, is that these planning applications fail to let the citizens know what they mean when using these terms—an issue that is important to iron out because of the confusion that can arise from misunderstanding the relationships between them. Mobility is concerned with the impedance component of accessibility—in other words, how difficult it is to travel. Specific strategies to enhance mobility will *usually* increase accessibility as well, by making it easier to reach destinations. But the reverse is not always true; a city can have good accessibility with poor mobility.

Consider, for example, Hong Kong or Manhattan. Residents of these cities usually endure severe traffic congestion, but they also live within a short distance of all needed and desired destinations; thus, because their destinations are close to one another, the travel times between destinations are relatively short, even if travel speeds are low. These places have good accessibility despite having poor mobility; furthermore, this example shows that accessibility does not depend on good mobility (though it does depend on having some mobility by one mode or another).

A place can also have good mobility but poor accessibility. Think of rural Manitoba in Canada. This area has ample roads, low levels of congestion, and (perhaps) even high speeds of travel, but relatively few destinations for shopping. Thus, good mobility is neither a sufficient nor a necessary condition for good accessibility.

Planning for mobility has taken on the meaning of making it easier to get around. In most planning applications—especially considering the fact that automobile travel constitutes the overwhelming majority travel in most cities—planning for mobility has come to mean making it easier to drive around. This focus on the ease of traveling along the transportation network itself (rather than focusing on the ease of reaching destinations) has aligned well with modern planning paradigms; this is especially true in the United States, where road building has been the most popular solution to congestion. These paradigms prize the planning-for-mobility perspective because it accommodates growing levels of travel and increases the potential for movement.

In theory, mobility planning can comport with accessibility planning. But at least in the United States, the practice of planning for mobility has deteriorated levels of accessibility. According to Susan Handy:

> as a result of this emphasis, accessibility in the US is largely mobility-dependent, and mobility in the US is largely car-dependent. In the suburban areas of metropolitan regions, transit service is relatively sparse and destinations are generally beyond walking distance, leaving residents with no option but to drive. The result is a lower level of accessibility, at least for those who need or would like to travel by modes other than the automobile. But even for those residents who prefer to drive, accessibility is threatened. As traffic levels invariably increase in these areas, getting around by car becomes harder, and accessibility ultimately declines.

Planning for accessibility, in contrast, means making it easier for people to get where they want to go. Land use policies designed to bring destinations within walking distance of residential areas are one example of this paradigm. But planning for access may not even require retrofitting neighborhoods. For example, transit services that link specific groups of users to their desired destinations, such as reverse commute programs and other client-based transportation services, are examples of planning for accessibility. Efforts like these reduce the need to drive, although they don't necessarily reduce actual driving. [11]

viewpoint, and from the dominance of select individuals. To achieve "objectivity," public sector investments generally rely on previously described benefit-cost calculations to compare various proposals. Yet, using benefit-cost analysis as a decision-making tool in public choices results in considering equity separately from efficiency. In most cases, the efficiency criteria employed by decision makers for a project override equity considerations.

A situation is considered Pareto Efficient (or Pareto Optimal) if there is no way to make all agents better off, that is, if it is impossible to improve the outcome for Person Y without worsening the outcome for Person Z. As a criterion for decision making, there are two problems with Pareto Efficiency in this context. First, some things, such as time, are not fungible, making exchange difficult (it is difficult to hand out ten minutes worth of time). Second, the exchange does not actually occur. Therefore, while the Pareto criterion is important from an efficiency point of view, it is unhelpful in trying to understand equity. We must recognize that every new transportation project and policy creates both winners and losers. A project that appears equitable to the decision maker may not appear so to an individual affected by the project.

Two additional concepts are important in understanding equity issues. *Equality of opportunity*, or *process equity*, is concerned with equal access to the planning and decision-making process. In contrast, *equality of outcome*, or *result equity*, examines the consequences of the product or policy. The Constitution of the United States, for example, enshrines the former (the Declaration of Independence fails to go quite as far, only positing the *right to pursue happiness*, not *happiness* itself). The actual equality of outcome is extremely difficult to ensure. In contrast with the utilitarian aim of maximizing total welfare, the egalitarian view would maximize the welfare (or opportunities) of the least advantaged member of society, and thus move society toward greater equity, as championed by the ideology of the Environmental Justice movement. Compared with the wealthy, the poor spend a larger portion of their income on transportation (as well as on a variety of other necessary goods). Furthermore, the poor and disadvantaged have historically borne the burden of transportation investments and improvements that are often sited in their neighborhoods.

Executive Order 12898, signed during the Clinton administration, calls for "Environmental Justice" requiring "fair treatment for people of all races, cultures, and incomes" in the development of environmental laws and policies. This document thus only examines environmental outcomes, and only addresses a few socio-economic strata. To be blunt, this is insufficient.

There are a variety of ways to segment populations in order to examine the equitability of the distribution of gains and losses from a transportation project to specific sub-populations (for example, see [12, 13]). Different segmentation strategies will result in different assessments of a project's fairness. Because there is no right way of defining sub-populations, multiple groupings should be considered. Towards this end, transportation evaluation could include an *Equity Impact Statement*. Doing so would specifically consider the winners and losers by specifying sub-groups. Outcomes of the project (e.g., travel time and delay, accessibility, consumer's surplus, air pollution, noise pollution, accidents) would be assessed for each of the population groups. Although inequities across some dimensions are almost inevitable, it is crucial—both for fairness and for political expediency given the growing environmental justice movement—to acknowledge these inequities and their relative magnitudes before proceeding with a project.

Social equity can only be completely realized when the needs of all groups are adequately represented in the decision-making process. This calls for including an opportunity to participate as a key criterion in an Equity Impact Statement. Thus, the planning process would consider the degree to which each group has had an opportunity to affect the project.[h] An Equity Impact Statement, a summary example of which is provided in Table 10.2, would consider the inputs (the opportunity to participate in decision making) as well as the outcomes (mobility, economic, environmental, health, and others) for transportation projects.

This application merely suggests possible considerations; a more comprehensive application would include all of the important outcomes/indicators that are being considered and help reveal important trade-offs raised by a project or policy. Comprehensively applying such a checklist would require the user to consider each cell (for example, in considering the spatial/opportunity cell, the user could ask a set of questions: Was the opportunity to engage in decision making fair across all jurisdictions (or locales within a jurisdiction)? Did each place have a say in the planning, engineering, public meetings, financing, and final decision process? To what extent were small places given a voice equal to large? To what extent were populous areas given a voice in accord with their population? Furthermore, in situations where a potential solution causes undesirable trade-offs (in relation to equity, the environment, the economy, etc., across stratifications), new alternatives might be considered or "designed." Such a disaggregated approach to decision-making is likely to be more transparent and democratic (if stakeholders are included in the decision-making process) and avoid the inherent problems associated with aggregation. It also highlights areas where alternatives might be considered (or are needed).

Population stratification only looks at the sub-populations of various groups (however they are defined, e.g., socio-demographically) and investigates the distribution of both opportunities to participate and project outcomes; spatial (or jurisdictional) stratification would examine how different areas (ranging from small areas such as census blocks or traffic zones, to larger areas such as census tracts, to entire political jurisdictions or metropolitan areas) are affected by the project. For example, the US Congress has a House of Representatives, whose seats are allocated proportional to population, and a Senate, which has two seats for every state. One ensures population equity, the other a type of spatial equity. Temporal stratification would consider the benefits and losses to current residents in comparison to those of (potential) future residents. Many transportation and land use policies, such as impact fees, have significant temporal effects. Modal equity considers whether users of different modes (e.g., drivers, pedestrians, transit riders) receive different gains or losses from a project and whether they have had equal input into the decision. Generational equity differentiates individuals by age (e.g., do the elderly or middle aged benefit at the expense of the young)? Gender equity

Table 10.2 Equity impact statement checklist

Stratification	Process	Outcomes				
	Opportunity to engage in decision-making	Mobility	Economic	Environ-mental	Health	Other
Population						
Spatial						
Temporal						
Modal						
Generational						
Gender						
Racial						
Ability						
Cultural						
Income						

contrasts men and women. Because there are known differences in the transportation use patterns by gender, distinguishing the effects on the two groups is important. Ability compares the fairness accorded to those without any physical or mental disability to those facing such challenges. Racial and cultural equity considers the effects on different races, ethnic groups, religions, and cultures. To date, insufficient research has examined the transportation uses by different racial and cultural groups; transportation projects are likely to have different impacts on different racial and cultural groups, if only because of historic patterns of spatial segregation. Similarly, some investments that serve certain vehicle types and certain areas will inevitably favor the rich over the poor, an issue addressed by examining income equity.

Collecting such data is sure to be difficult and costly. Some data, such as income, race, or gender, may be gleaned geographically (from census data), but not according to network use (which can only be estimated with models). Further, there will inevitably be the need to forecast when land use changes are anticipated. There may also be privacy concerns when collecting such data. Nevertheless, it is important to make reasonable attempts to estimate this information in a consistent way across alternatives, so that general trends can be assessed.

Identifying equity problems is a first step. Solving them is more complicated. Philosopher John Rawls imagined two individuals shrouded in a "veil of ignorance"—they know what they prefer, but they fail to know things such as their social class. [14] They must agree to divide some spoils (political rights, money, etc.), but do not know which side of the spoils they will get. Rawls asserted that they would come to a fair agreement, because each has an equal possibility of receiving either side of the division. The problem with the Rawlsian solution is that some individuals may be risk takers. A somewhat better solution is the so-called "pie cutter problem." Imagine there is a pie and several (N) people who want to eat it; how can each be assured of receiving an equal share? The solution is to let one person cut the pie into N pieces, but stipulate that the person who cuts the pie will receive the last piece. Assuming he likes pie, he will ensure that the pieces are as close to equal as possible in order to get the largest possible last piece. However, the pie-cutter problem assumes a zero-sum world, where the problem is simply how to divide the spoils (pie), rather than how to increase the amount of spoils (pie) available; in many cases, it is possible for participants to achieve gains through trade.

One of the primary principles of Rawls's Theory of Justice is that the most disadvantaged are made relatively better off under new social arrangements. While the pie cutter solution is simple, it does not directly address the issue of helping the most disadvantaged members of society. Unfortunately, Rawls's theory might be difficult to operationalize.

More practical solutions to equity problems include ideas such as "bundling" improvements together, so that not only is there a net benefit (when all projects are considered together), the number of winners exceeds the number of losers by a significant amount. This is the approach most often taken in transportation appropriations bills. The downside, of course, is that this gets expensive and may not be efficient, and may result in the overproduction of transportation services, e.g., to pay off a project's potential opponents.

Environment

> en·vi·ron·ment *n.* The totality of circumstances
> surrounding an organism or group of organisms.

A third MOE stems from the environmental motivations of a preferred land use-transportation system. The environmental effects of transportation have been noted for some time, since at least the emergence of the railroad, which spewed smoke and sparks onto neighboring lands. Literature from the mid-1800s complains about the effects of the railroad, the fires started by trains passing a dry field, the unnatural speed of the trains, and their disruptive effects on wildlife. The Army Corps of Engineers, which is responsible for managing the lock and dam systems on US inland waterways, applied benefit-cost analysis to waterway projects as early as 1936. [15] Environmental impacts began to be more formally considered in the 1960s as multiple-objective analysis gained acceptance.

A significant problem arises in efforts to quantify environmental effects and prioritize their importance. For example, Matt Kahn asks [16] whether it is important to focus solely on the city's per-capita ecological footprint. What is the role of public health criteria? How about pollution? Clearly, how analysts prioritize different environmental challenges such as local air pollution versus climate change plays a key role in determining which metric is suitable to measure benefits and costs. Since the 1960s, significant effort has gone into measuring the economic cost of environmental externalities so that these costs could be directly included into benefit-cost analyses. If these externalities (costs that are incurred in an economic transaction, but are not borne by parties to that transaction) can be quantified, and charged for, individuals and agencies will take into account their environmental impacts when making decisions. The revenue raised by such charges could in theory be used to remediate damages incurred by the project, or to prevent them in the first place. This strategy has thus far been more theoretical than practical for a variety of reasons.

A second more qualitative strategy has been in place in the United States since the National Environmental Policy Act of 1969 established the Environmental Impact Statement (EIS) as a systematic way to ensure that environmental considerations were taken into account in decision making. Although the EIS is mostly procedural, and only requires provision of information, the clear and public presentation of information makes it difficult for decision makers to capriciously disregard environmental issues. The EIS is prepared by government agencies (or consultants) to assess in detail the expected environmental impacts of a project. Major infrastructure projects are subject to the EIS process before they can be approved. The EIS requires determination of *purpose* and *need* for the project. The EIS compares *alternatives*, including the "no build" or "do nothing" alternative. The EIS evaluates foreseeable direct and indirect effects of the project on the affected environment and the consequences of those effects. The EIS must also note any conflicts with federal, state, or local policies, and incorporate comments and objections from the public, interest groups, and government agencies. Recommendations put forward in the EIS are inputs into the final decision, but the EIS itself does not result in a project being approved or rejected.

A different approach to measuring environmental impacts, *Natural resource consumption*, has been on the agenda since Malthus, and later Jevons, argued that resource constraints would doom growth. The scarcity of resources was a significant geopolitical factor for much of the twentieth century; the need for natural resources was used as a justification by Japan for its actions in Asia in the 1930s and 1940s, and the US embargo of resources to Japan was a factor in the attack on Pearl Harbor. The oil shocks of the 1970s again brought out public fears of scarcity. In the planning literature, the topic was returned to prominence by Newman and Kenworthy's now-infamous curve documenting a reciprocal relationship between average fuel consumption and average development density for a sample of 32 cities. [17] Notwithstanding methodological issues for which it was attacked, the curve focused attention on the relationship between urban form and energy consumption. Automobile-reliant travel produces an urban form that, barring renewable energy sources, creates a relentless demand for fossil fuels. Given current levels of consumption, some estimates, applying what is referred to as the Hubbert Curve,

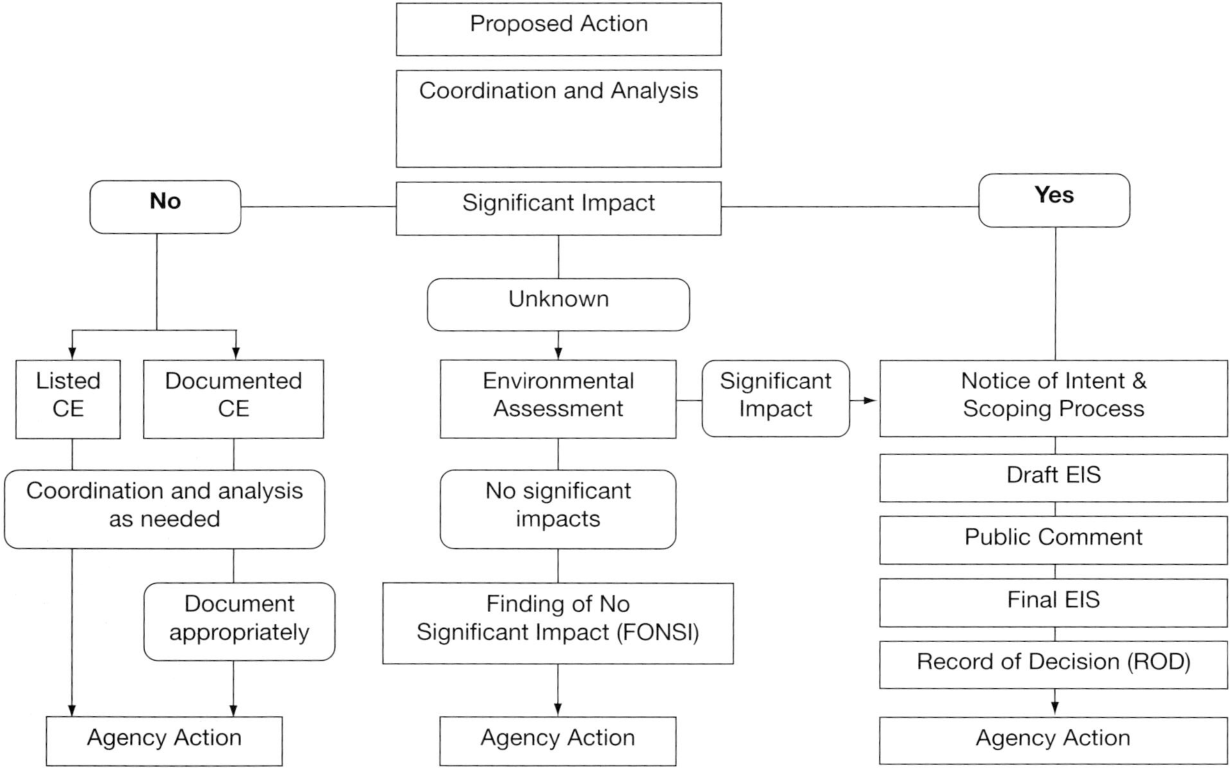

Figure 10.3 Decision process options for NEPA

predict that the world's oil production will peak in the first decade of the twenty-first century. [18] In fact, many claim the automobile (and its attendant primary reliance on non-renewable sources of fuel) is a motivating force for continued involvement in the Middle East by outsiders.

But politics aside, energy consumption is not where the buck stops. Excessive auto reliance has spawned intense attention to air pollution. This concern emerged in the United States during the 1960s, and led to the passage of the first Clear Air Act, signed into law by Richard Nixon in 1970. The Act resulted in pollution standards being set for air quality in urban areas and for automotive tailpipe emissions. While the tailpipe standards were fairly successful, resulting in the deployment of catalytic converters in new cars, the urban area standards were less so. The issue came to a head in the early 1990s with the passage of the Clean Air Act Amendments (1990), which along with ISTEA (1991) tied federal highway funds to metropolitan areas' compliance with air quality standards. These laws spotlighted the effects of tailpipe emissions from private automobiles, commercial trucks, and other transport propelled by the internal combustion engine.

Recent statistics focus on rates of population growth versus rates of development growth. [19] Statistics about the *loss of open space* make the hair on the back of environmentalists' necks stand up, documenting the rapid rates at which wetlands are being developed and the countryside consumed. Some contend that space is plentiful [20] and that there is little or no risk of it being eliminated. Others argue that, although plentiful, open space is not generally accessible. [21]

Experience

> ex·pe·ri·ence *n.* An event or a series of events participated in or lived through.

The conditions described above related to efficiency, equity, and the environment have prompted many concerns about the overall quality of life and its possible decline. Such concerns may be triggered by growth, congestion, evolving economic structures, and the increasing complexity of daily life. In the final analysis, it is probably a combination of each.

Citizens of the developed world today enjoy unprecedented levels of wealth and prosperity; but at the same time, several reports suggest that the complexity of their lives is increasing as well. The relatively simple days of households replicating the lifestyle of the Cleaver family in the popular 1950s television program *Leave it to Beaver* are gone, if they ever really existed. Demands resulting from increased work and consuming aspirations have made daily travel substantially more complex than it was in past decades. It is difficult for families to balance demands for good schools, a convenient work location for multiple workers in the household, safe and attractive neighborhoods, and accessibility to other activities (e.g., soccer or piano lessons, friends, medical care) without spending extraordinary amounts of time in travel. Households are placing greater demands on a transportation system that is becoming more geographically dispersed. The net effect is a dramatic change in their overall *experience* of urban, suburban, and exurban environments on a daily basis.

The characteristics and traits of such an *experiential* component vary widely. Many roll with ideas brought forth in Robert Putnam's book *Bowling Alone*, which argues that some forms of community design (e.g., low density, auto-reliant urban areas) have deleterious effects on the social fabric of communities and on the psychological lives of individuals. Social divisions, lack of neighborly interaction, and the emergence of social outcasts are all outcomes of our existing built form. As an example, in the wake of the tragic 1999 Columbine High School shootings in Colorado, a *New York Times* editorial and other writings went so far as to suggest that conventional suburban housing development was a contributing factor in spawning such deviant behavior.[i] In one sense, we agree with Michael Moore's diagnosis in the documentary film *Bowling for Columbine*—that the "culture of fear" is a problem, and social isolation does not help build trust. These anecdotes, however, are only the tip of the iceberg. The influence of twentieth-century land use-transportation planning and suburban development on the fabric of contemporary society, Although systemically under-researched, has caught the attention of many and has been widely critiqued. [22]

Further, although social capital is conceivably a powerful element of experience, many find the concept of social capital to be quite elusive. An alternative measure often introduced is based on the ability of residents to relate to their community via a means other than the automobile—the ability to get around by walking and bicycling, and the frequency with which these modes are used. The number of people walking or cycling to nearby destinations is in rapid decline. According to the 2001 National Household Travel Survey, nearly one-third of adults in the US did not take any walking trips in the previous week. A seemingly endless list of factors contribute to this phenomenon.[j]

What appears less clear is the central motivation in communities worldwide for wanting to increase levels of walking and bicycling. Some camps hope to promote walking and cycling as highly feasible and attractive modes of transportation that can compete with the automobile in some travel markets. In support of their position, they point to a list of factors that may include the usual suspects: less congestion, reduced consumption of natural resources, and decreased pollution. But some leading practitioners and academics suggest that several of the benefits most often touted for walking and cycling facilities are *not* the benefits that ultimately produce increased walking or cycling. A select few benefits that are more related to the relatively ambiguous goal of "livability" appear to hold more hope for meeting expectations.

For example, a prominent US transportation consultant, after reviewing much of the literature on the benefits of non-motorized modes and discussing the matter with policy officials, argues that:

from a policy perspective, the subject of non-motorized transportation presents a bit of a dilemma. Statistics are spotty and the literature appears to be heavily populated with advocacy. Thus, the overarching policy questions are whether non-motorized transportation, in fact, is a transportation services issue or a lifestyle issue, and is that distinction important. [23]

As Giuliano and Hanson [24] suggest, "building communities with abundant walking and biking opportunities may be more about livability than solving transportation problems." Of course, measuring livability presents its own challenges.

Adding support to the above line of reasoning, a growing number of arguments continue to emerge suggesting that the motivations for many land use-transportation initiatives are slightly misguided. Compact urban forms with mixed uses may be a necessary, but not sufficient condition to address transportation problems and remedy automobile reliance. Although research does show that people living or shopping in more compact built forms tend to drive less frequently or for shorter distances, the relationships have not proven to be nearly as strong as initially anticipated. Therefore, leading practitioners and academics suggest that decreased auto travel should not be the central goal of such initiatives; instead, efforts should be focused on expanding the number of choices available in terms of places to live and travel modes available. [25, 26] (also, see this point in Chapter 14). More adequately providing for residential areas in which walking, cycling, and transit are available would have the effect of increasing the range of choices—and, subsequently, improving the overall *experience*—of many households.

Expedience

ex·pe·di·ent *n.* Something contrived or used to meet an urgent need.

With four-year time horizons and ever-increasing bureaucratic legalese, the projects that rise to the top are those that can be pushed through the process in the most expedient manner. One need look no further than the many initiatives where politicians and others have pushed light rail projects for transportation corridors where the skids were already greased (for example, areas where an Environmental Impact Statement was already completed and approved) and *not* necessarily where light rail would attract the greatest ridership.

Yet because of the artificial constraints we as a society have imposed, we cannot let the perfect be the enemy of the good, and expediency—a clearly qualitative and perhaps unquantifiable measure—is where human judgment is required to select among the solutions that satisfy different goals to different degrees.

Box 10.2 Alternative evaluation paradigms

Places and Plexuses are large-scale, technologically enabled, complex, dynamic, socially interactive systems. Each has a number of properties, whose observed or predicted performance can be evaluated with the Es of efficiency, equity, environment, and experience, mediated by expediency. Systems architecture describes the elements of a system and their inter-relationships. The elements may be modular (one vehicle can be isolated from another vehicle) or integral (the operation of a freeway depends on the interaction of vehicles on coupled links, and cannot be easily decomposed).

Levis [27] describes four types of architectures, which help us understand, design, guide the evolution of, and manage long-lived complex systems:

- The functional architecture (a partially ordered list of activities or functions that are needed to accomplish the system's requirements);
- The physical architecture (at minimum a node-arc representation of physical resources and their interconnections);
- The technical architecture (an elaboration of the physical architecture that comprises a minimal set of rules governing the arrangement, interconnections, and interdependence of the elements, such that the system will achieve the requirements);
- The dynamic operational architecture (a description of how the elements operate and interact over time while achieving the goals).

A key to understanding the long-term performance of the system is not simply how it operates, (assuming all else to be equal, which is how traditional efficiency measures are operationalized) but rather, how it operates when its environment and other circumstances change. A number of system properties are qualities that express those attributes. [28]

- Robustness—"the demonstrated or promised ability of a system to perform under a variety of circumstances, including the ability to deliver desired functions in spite of changes in the environment, uses, or internal variations that are either built-in or emergent;"
- Adaptability—"the ability of a system to change internally to fit changes in its environment," usually by self-modification to the system itself;
- Flexibility—"the property of a system that is capable of undergoing classes of changes with relative ease;"
- Scalability—"the ability of a system to maintain its performance and function, and retain all its desired properties when its scale is increased greatly, without causing a corresponding increase in the system's complexity."

These attributes, which describe the system but are generally not considered "functional requirements" are among the many "ilities," so-called for the suffix that is often attached to their name. They comprise system properties that go beyond more traditional or static measures such as maximizing consumer's surplus. They are often difficult to ascertain before deployment, particularly for new systems.

Ilities include: accessibility, accountability, accuracy, adaptability, administrability, affordability, agility, availability, composability, configurability, customizability, degradability, demonstrability, dependability, deployability, distributability, durability, evolvability, extensibility, fault tolerance, flexibility, footprint, interoperability, maintainability, manageability mobility, nomadicity, openness, performance, portability, predictability, reliability, responsiveness, reusability, robustness, safety, scalability, seamlessness, security, serviceability (a.k.a. supportability), simplicity, stability, survivability, tailorability, timeliness, trust, understandability, and usability.

Evaluating wrap up

Just as Einstein noted that the point of view of the observer shaped the measurement of time, point of view also affects the perception of transportation level of service. Moving towards measures of effectiveness that align with user experience will highlight potential problems before they manifest themselves. Doing so begins by identifying goals based on each of the Es from the Diamond of Evaluation. To operationalize such goals, we need to develop metrics for assessing them. Collecting data and measuring performance are the beginning, not the end. Those performance measures should be used in making decisions about changing (or not changing) the place and plexus that residents inhabit. We would not claim that a simple technocratic process is in general sufficient for decision-making; but since we see inferior solutions being adopted all the time, on both major and minor projects, it is clear that both decision makers and the public lack systematic information about alternatives.

This chapter identified four major classes of efficiency measures: mobility, utility, productivity, and accessibility. Each has strengths and weaknesses that justify its use, but not its exclusive use, as a gauge of transportation system performance. As suggested by the famous metaphor of blind men examining the elephant that opened this chapter, there is no single perspective that can be accurately measured and will correctly and completely describe a transportation-land use system. The idea of addressing land use-transportation problems from different 'perspectives' recalls Söderbaum's idea of positional analysis, [26] in which each conclusion is conditional in relation to each ideological orientation articulated and considered. The idea is to facilitate learning processes and decision-making and not to dictate the 'correct' way of arriving at the best and optimal decision. [29]

Mobility is the traditional measure used by engineers, and has the advantage of ease of measurement. Travel time is a useful measure that aligns with user experience, but users care about trips rather than simply links. Utility might match travelers the best, if only it could be measured. Consumers' surplus is a useful system measure, but the aggregation built into the measure means that it does not match any particular user's experience. Productivity is important to examine when managing the system, but again it is not experientially based. Accessibility provides an overview relationship of transportation, activities, and land uses, but may be hard to explain and is not easily operationalized into policy.

Efficiency looks at the overall outcome, equity considers the distribution of the outcome across individuals and groups. A key difficulty is that subjective perspectives of travelers contrast with the objective views of professionals; but only by considering that subjective perspective as an input into decision making can new decisions be implemented in a political environment. Although at one level everyone understands that change creates winners and losers, at another, only the aggregate net gain is generally considered. Economists hold that so long as the losers can be compensated from the gains of the winners (whether or not they actually are), everything is okay, an idea called Pareto Optimality. The losers, understandably, don't accept that idea. Thus, economic decisions are devolved into the political and legal arenas, where voices are not necessarily weighted equally. Diffuse winners may not expend energy to defeat concentrated losers, despite an overall "net gain." By the economic calculus, society is worse off. Can this outcome be anticipated and avoided?

Winners and losers are created all the time, particularly in transportation projects and even in the simplest of transportation projects. [30] This phenomenon is not just due to the taking of land, or creating pollution effects, but also reducing mobility from the relatively narrow transportation perspective. It is essential to develop measures of effectiveness that identify these issues before they become political problems. Unfortunately, no single measure of effectiveness will capture everything. Complexity implies uncertainty, so any one measure will be incomplete. Yet, the alternative of not doing the analysis is also unacceptable.

The technocratic process is insufficient in part because of at least two considerations. First, many tenets within each of the Es are in conflict within one another. There is often a trade-off between efficiency and equity, between equity and the environment, between the environment and experience, between experience and efficiency, as illustrated with the Stillwater Bridge case. We need to add a fifth E, expedience, to facilitate trade-offs across the other four. Second, there is often overlap between many of the Es. For example, choice is an important element of accessibility; more choices, either in terms of destinations or of travel modes, mean greater accessibility by most definitions. But adding more choices (for whatever purpose) is often used as the rationale that is embedded within experience.

Framework for Part 3

The third part of this book discusses the actions of public agencies in the transportation-land use nexus. We can think of such public agencies and governmental entities as agents, in similar manner to individuals and firms. As discussed above, evaluating the progress of such agencies is much easier said than done, not to mention the difficulties of measurement and operationalization.

Public agencies pursue a number of different strategies to further their aim. Agencies often believe their land use and transportation goals can be achieved by matching infrastructure supply with consumer demand (a perspective that focuses on the operating aspects of the system). But matters are more complicated than this: land use-transportation projects need to be well designed and built when they make sense.

We therefore break the discussion in Part 3 of the book into three chapters: designing (11), building (12), and operating (13). Chapter Eleven introduces four aspects of land use-transportation systems under the rubric of the "Diamond of Design," including hierarchy, morphology, layers, and architectural content. We begin with this chapter because the proper design of systems of place and plexus ultimately sets the stage for all that follows.

Management of scarce resources and attribution of costs associated with building and maintaining facilities.

Assess existing plans, practice, and policies according to their efficiency, equity, environmental consequences, and the experience they engender in users and non-users.

Building new infrastructure, and its consequences.

Design new place and plexus considering hierarchy, morphology, layers, and architecture of those facilities.

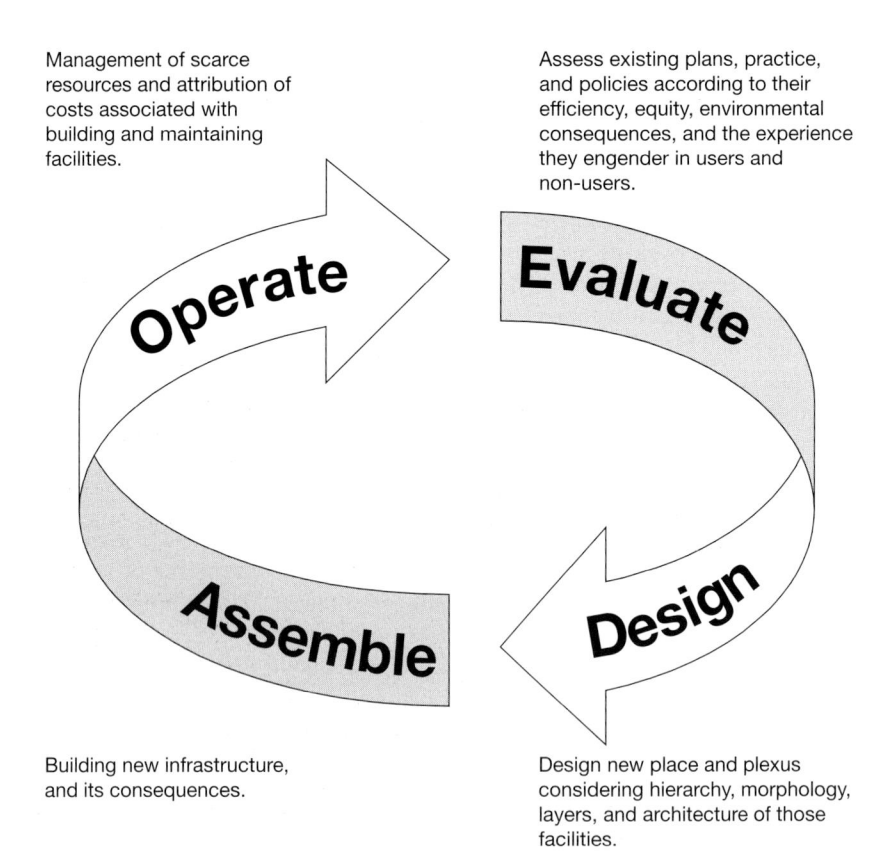

Figure 10.4 The product development cycle

Chapter 12 describes the impacts of building different types of infrastructure; these impacts play out in the "Diamond of Assembly." Finally, Chapter 13 responds with strategies for appropriate operation of land use and transportation systems by allocating demand for these services. Strategies play out over long and short time horizons and commonly rely on policies to adjust the timing or pricing of the availability of land or space on roadways; this two-by-two framework creates the "Diamond of Operation." The chapters in Part 3 can be thought of as a cycle, as illustrated in Figure 10.4.

How do such strategies get implemented? Agencies have different sets of authorities and resources: departments of transportation build and maintain roads, while land use agencies regulate development and the location of attributes. The perils and pitfalls of separating function and authority is well recognized in the planning community. Provided there is sufficient coordination (internal or external) that balances between different initiatives and objectives (which is not always the case), regions can come closer to realizing their stated goals.

Notes

a   "Es" have long been employed in various evaluation frameworks. For example, transportation engineers often reference the four Es: Engineering, Education, (law) Enforcement, and Encouragement. Three Es are almost always referenced in discussion of sustainability: Economy, Environment, and Equity. In contrast, the Es we propose are intended to be positive in nature.

b   For example, in most cases the three Es of sustainability are applied with a normative bent to them: a prosperous economy, a quality environment, social equity (for example, see: www.abag.ca.gov/planning/ smartgrowth/3Esofsustainability.htm).

c   Each of the five definitions (here and the following four sections) came from *The American Heritage Dictionary of the English Language,* Fourth Edition.

d   For example, in determining whether to build a project, select a policy, implement a system, or provide a service, it is possible, with the help of many assumptions, to estimate the net present value of the future stream of profit or welfare using cost/benefit analysis. But because of the required assumptions, benefit/cost analysis may not be sufficient to manage a complex system such as a transportation network on a day-to-day basis. There is a desire to monitor the transportation network on multiple dimensions, to understand how well it is performing (and how accurate were previous projections), and to steer future decisions. Metrics might assess how efficiently labor or capital is employed (to determine where future labor or capital should be employed). They might consider market share against competitors, the state of complementary services (for instance, access to transit or parking in the case of a transit system) or the satisfaction of customers and vendors (to gauge future market share and the price and quality of inputs).

e   Notwithstanding popular culture's obsession with the standard TTI measures, an emerging concern in transportation relates to reliability. The emergence of reliable transportation does not begin with highways; it may end there. Reliable transportation is most widely associated with overnight couriers such as Federal Express, and it has been a critical issue since the beginning of transportation, with rewards (be they tips, bonuses, or continued patronage) for fast service and penalties for slow service being a feature of many modes. Reliability is important both for unusual trips (such as going to the airport), and routine ones.

f   The word accessibility is derived from the words "access" and "ability," thus meaning *ability* to *access*, where "access" is the act of approaching something. The word is derived from the Latin *accedere* "to come" or "to arrive."
g   Hedonic theory, introduced in Chapter 3, suggests that individuals do not purchase goods, but rather the bundle of attributes composing the good. Someone does not buy a house, but rather the qualities of that house: location (accessibility), size, type of construction, appliances, noise from nearby roads, etc. Every house combines the various attributes slightly differently. Hedonic models are used to pull apart these attributes, and develop demand curves for the various attributes (goods or bads). However, these attributes are interrelated, houses with high accessibility will be more expensive, which will lead to more investment in other attributes, leading to better maintenance and more frequent remodeling.
h   This would be raised, for example, by answering questions such as "was the group included among the analysts and decision makers in proportion to its share of the affected population?"
i   However, we think that is a bit of a stretch, and note that historically high-density urban land uses have been blamed for the same thing.
j   For example, even public health officials are increasingly interested in urban planning issues, especially as they relate to increasing rates of obesity in the population as a whole and the poor physical condition of American youth. To increase physical activity, many argue that communities must be designed to facilitate walking and cycling and in general, increase the attractiveness of these modes so they will be used.

References

[1] Downs, A., *Stuck in Traffic: Coping with Peak-hour Congestion*, Washington, DC and Cambridge, MA: Brookings Institution Press and Lincoln Institute of Land Policy, 1992.
[2] Taylor, B.D., 'Rethinking Traffic Congestion', *Access*, 2002, vol. 21 (Fall): 8–16.
[3] Levinson, D., Krizek, K.J., and Gillen, D., 'The Machine for Access', in D. Levinson and K.J. Krizek (eds.), *Access to Destinations*, London: Elsevier, 2005, pp. 1–10.
[4] Hansen, W., 'How Accessibility Shapes Land Use', *Journal of the American Institute of Planners*, 1959, vol. 25: 73–76.
[5] Black, J. and Conroy, M., 'Accessibility Measures and the Social Evaluation of Urban Structure', *Environment and Planning A*, 1977, vol. 9: 1013–1031.
[6] Pirie, G.H., 'The Possibility and Potential of Public Policy on Accessibility', *Transportation Research*, 1981, vol. 15A (5): 377–381.
[7] Morris, J.M., Dumble, P.L., and Wigan, M.R., 'Accessibility Indicators for Transport Planning', *Transportation Research*, 1979, vol. 13A (4): 91–109.
[8] Wachs, M. and Kumagai, T.G., 'Physical Accessibility as a Social Indicator', *Socio-Economic Planning Science*, 1973, vol. 7: 437–456.
[9] Garret, M. and Wachs, M., *Transportation Planning on Trial: The Clean Air Act and Travel Forecasting*, Thousand Oaks, CA: Sage Publications, 1996.
[10] Krumholz, N. and Forester, J., *Making Equity Planning Work: Leadership in the Public Sector*, Philadelphia, PA: Temple University Press, 1990.
[11] Handy, S.L., 'A Cycle of Dependence: Automobiles, Accessibility and the Evolution of the Transportation and Retail Hierarchies', *Berkeley Planning Journal*, 1993, vol. 9: 21–43.
[12] Talen, E., 'Visualizing Fairness: Equity Maps for Planners', *Journal of the American Planning Association*, 1998, vol. 64 (1): 22–38.

[13] Talen, E. and Anselin, L., 'Assessing Spatial Equity: An Evaluation of Measures of Accessibility to Public Playgrounds', *Environment and Planning A*, 1998, vol. 30 (4): 595–613.
[14] Rawls, J., *A Theory of Justice*, revised edition, Cambridge, MA: Belknap Press, 1999.
[15] Hufschmidt, M.M., 'Benefit-cost Analysis: 1933–1985', *Water Resources Update*, 2000, vol. 116: 42–49.
[16] Kahn, M.E., *Green Cities: Urban Growth and the Environment*, Washington, DC: Brookings Institution Press, 2006.
[17] Newman, P.W.G. and Kenworthy, J.R., 'Gasoline Consumption and Cities: A Comparison of US Cities with a Global Survey', *Journal of the American Planning Association*, 1989, vol. 55 (1): 24–37.
[18] Deffeyes, K.S., *Hubbert's Peak: The Impending World Oil Shortage*, Princeton, NJ: Princeton University Press, 2003.
[19] Chen, D. and Jakowitsch, N., 'Transportation Reform and Smart Growth: A Nation at the Tipping Point', *Funders' Network Newsletter*, August 2001.
[20] Gordon, P. and Richardson, H., 'Are Compact Cities a Desirable Goal?' *Journal of the American Planning Association*, 1997, vol. 63 (1): 95–106.
[21] Ewing, R., 'Counterpoint: Is Los Angeles-style Sprawl Desirable?' *Journal of the American Planning Association*, 1997, vol. 63 (1): 107–126.
[22] Kunstler, J.H., *The Geography of Nowhere*, New York: Simon and Schuster, 1993.
[23] Lockwood, S., 'The Current State of Non-motorized Transportation Planning and Research' in *Proceedings of the James L. Oberstar Forum on Transportation Policy and Technology*, Minneapolis, MN: Center for Transportation Studies, 2006.
[24] Giuliano, G. and Hanson, S., 'Managing the Auto', in S. Hanson and G. Giuliano (eds.), *The Geography of Urban Transportation*, New York: The Guilford Press, 2004, p. 417.
[25] Levine, J., 'Access to Choice', *Access*, 1999, vol. 14 (Spring): 16–19.
[26] Handy, S.L., 'Urban Form and Pedestrian Choices: Study of Austin Neighborhoods', *Transportation Research Record*, 1996, vol. 1552: 135–144.
[27] Levis, A., 'Architectures', in A.P. Sage and W.B. Rouse (eds.), *Handbook of Systems Engineering and Management*, New York: John Wiley & Sons, 1999, pp. 427–454.
[28] Crawley, E., de Weck, O., Eppinger, S., Magee, C., Moses, J., Seering, W., Schindall, J., Wallace, D., and Whitney, D., *The Influence of Architecture in Engineering Systems*, Engineering Systems Division Monograph, Cambridge, MA: MIT Press, 2002.
[29] Söderbaum, P., *Ecological Economics: A Political Economics Approach to Environment and Development*, London: Earthscan Publications, 2000.
[30] Levinson, D., 'Identifying Winners and Losers in Transportation', *Transportation Research Record: Journal of the Transportation Research Board*, 2002, vol. 1812: 179–185.

Chapter 11

Designing

"The compact neighborhood is the true architecture of nature."

Andreas Duany

Fresno, California was the thirty-seventh largest city in the United States at the turn of the twenty-first century, and one of the fastest growing. With more than 482,000 residents as of the 2000 Census, it anticipates a population of 790,000 by 2025. With that size comes the problems that face many US cities: what to do with downtown, how to bring people downtown who do not work there, how to keep jobs downtown, and so on. Fresno has an advantage in that most of its suburban-type development has occurred within its city limits, allowing the city to capture a substantial proportion of the growing tax base. Historically, growth in Fresno has made a steady march to the north of downtown, leaving downtown today near the southern edge of the city. While the city hopes to annex land and direct growth to the southeast in order to re-center itself around its historic downtown, that area is not what the city government or many residents hope it to be; it hasn't been what they want for nearly four decades.

Urban planner Victor Gruen wanted to change that. He designed Southdale—often identified as the first fully enclosed, climate-controlled shopping mall—in the Minneapolis suburb of Edina. Subsequently, Gruen turned his attention from creating new pedestrian realms to reconfiguring old ones, and was a major proponent of auto-free streets and districts. [1] One place Gruen was able to at least partially implement his ideas was the heart of Fresno. Following his 1958 plan, the six blocks of Fulton Street, the main street through downtown, was converted in 1964 to Fulton Mall, a pedestrian mall with no motor vehicle traffic. Cars were diverted to parking garages one block above or below the street, and the cross-streets remained with pedestrian signals at the intersections.

The mall, as a symbol of government interest and investment in the area, was initially successful in attracting some private development. But the macro-trend of suburbanization toward the north overpowered the micro-investment focused on downtown. The mall soon resumed its course of decline, driven in part by the closing of department (and other) stores that chose to save their investments for greener (financially at least) pastures in the suburban areas of North Fresno. Office buildings, too, went unoccupied. Although never completely abandoned, downtown was not operating at full capacity either. Efforts to attract residents to the downtown area have included construction of a stadium for the Fresno Grizzlies, the minor-league baseball team affiliated with the San Francisco Giants, and the opening of several small museums. The Fulton Mall, shown in Figure 11.1, is home to a number of businesses serving the Hispanic community, but not all storefronts are full, and are certainly not getting the rents that owners would like to see.

Proposed solutions have included opening up the mall to vehicle traffic. A study from Eugene, Oregon, is cited by advocates of de-malling, including the City of Fresno government. In 1989, the City of Eugene Planning and Development Department surveyed 35 cities that had built pedestrian malls; 18 of the malls had already been removed. Fresno's mall was described at that time as "doing poorly," with "downgraded retail." These conditions remained largely unchanged throughout the 1990s. [2]

There are opponents to reopening Fulton Mall to motor vehicles, including the Fresno arts community, which on February 28, 2006 staged a "March on the Mall" to attract media attention to the possibilities of improving the mall as a pedestrian space. [3] Ideas included providing free wireless Internet access to attract lunch-goers to dine outside, building additional housing, and improving public transit. Posters displayed as part of the March (which was really more of an assembly, or even Theatre of the Absurd) advocated "Fresno, Clean Air Leader" and "Fresno, an Entrepreneurial Giant." Fresno has some of the worst air quality in the United States, primarily because of its automobile orientation, centrality in an agricultural region consisting of artificially irrigated desert, and location in a basin (the Central Valley). In fact, it has the nation's third highest mortality rate from asthma, and in 2001 beat Los Angeles in terms of days in violation of the Environmental Protection Agency's ozone standard. [4]

Clearly, a great deal of effort has been put into making Fulton Mall work. The site has benches, shaded areas (important in the summer sun), play areas for children, aesthetically pleasing (and working) water features, clear signage, ample parking, a brand-new stadium, and farmers' markets. These elements illustrate the evolution of urban design ideas from the 1960s to the present. One certainly feels safe walking in daylight hours. The government has tried to assist by occupying a number of buildings that had been vacated by private firms.

Figure 11.1 The Fulton Mall

Figure 11.1 (continued) The Fulton Mall

What places are appropriate to be exclusively automotive? What places should be exclusively pedestrian? Where might these travel modes co-exist peacefully? In the corner of auto-exclusivity we have freeways; in the corner of pedestrian-exclusivity, we have skyway networks, underground cities, shopping malls, airport terminals, houses, and office buildings. Are there streets where cars do not belong?

These questions all evoke concerns about the overall design of land use and transportation systems. The term *design* generally refers to notions of patterns, arrangements, or even blueprints. Although most people agree that design has a spatial or geographical component, it is commonly associated in land use-transportation contexts with the scale of buildings or the layout of roads. Limiting the meaning of the term to the smallest or most local attributes (i.e., the building), however, fails to recognize larger matters of how multiple buildings coalesce with streets in the design of cities.

For instance, one could refer to instances where key characters in film aspired to design things larger than the individual building. Take George Bailey in *It's a Wonderful Life*, when asked by his father what he wants to do when he gets out of college. His response: "Oh, well, you know what I've always talked about ... build things ... design new buildings—plan modern cities!" Or consider the episode of the hit comedy show, *Seinfeld*, in which a teenager seeking a scholarship initially says he wants to be an architect and later realizes, "I think I'd really like to be a city planner. Why limit myself to just one building, when I can design a whole city?"[a] After all, the prospect of designing a single building pales in comparison to the grandiose plans of many of the greatest planners. Take, for instance, Daniel Burnham's plan for the Commercial Club of Chicago or the Regional Planning Association's *Regional Plan of New York and its Environs*.

In this book, we therefore employ the term *design* to describe how elements of place (land use) and plexus (particularly transportation) arrange their parts into a whole on a variety of scales, from the neighborhood to the metropolis.[b] After reading literary accounts, witnessing presentations on new land use-transportation paradigms, and personally observing countless communities and cities, we conclude that there are four key design tenets—hierarchy, morphology, layers, and architectural content—that play out differently for place versus plexus, forming the Diamond of Design (Figure 11.2). In the discussion that follows, we trace two distinct threads: one for place, another for plexus. For each tenet, we introduce and discuss predominant concepts, describe their genesis (where appropriate), highlight current trends, and warn of possible pitfalls.

Hierarchy

Embedded within all forms of place and plexus is a sense of hierarchy. Not all locations or roadways are equal in composition or stature. Both infrastructure networks and government agencies are hierarchically organized. In the US, it is typical to find homeowner associations at the lowest level, followed

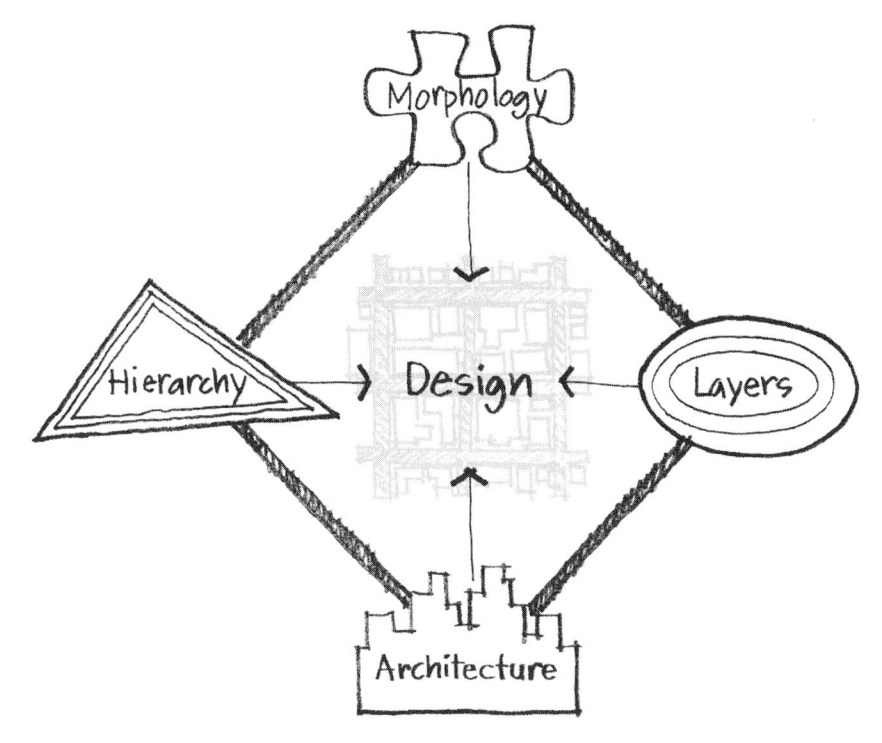

Figure 11.2 The Diamond of Design

The Diamond of Design suggests there are four different tenets important to consider when designing any land use and transportation system—hierarchy, morphology, layers, and architecture. Hierarchy refers to a system of levels that is embedded within all forms of land development or transportation facilities (e.g., a hierarchy of roads comprises freeways at the top, which are designed primarily to move people, and residential cul-de-sacs at the bottom, which aim foremost to allow individuals to access property). Morphology addresses the character or configuration of a community and its overall road network (e.g., the New Urbanist movement suggests certain protocol for street patterns, building shapes and sizes, and patterns and composition of different types of land uses). Layering is a structural property that is endemic to the design of a system (e.g., traffic markings are placed on the road surface, which falls within a right-of-way constrained by existing buildings). Finally, architecture addresses the physical nature of any development and the networks connecting them (e.g., employing traffic calming on select streets).

by towns and counties, then states, and finally the federal government at the highest level. However, the slope of the hierarchy (the number of levels it possesses) varies from case to case. Management by a government layer that is geographically too small or too large brings about costs which can be avoided by associating the infrastructure with the most appropriate level of government.

Hierarchy of place

Hierarchy is an intrinsic characteristic of any location. Why does New York City exist? Why is the next largest city, Los Angeles, a continent away from

New York? Answers to these questions have to do with the scales and spatial patterns of human settlement and commerce.

In 1933, after making a series of assumptions about available land—mainly that it is a flat, featureless plain—Walter Christaller devised a groundbreaking theory to predict the number, size and distribution of a system of cities (rather than concentrating on a single city/location). Christaller identified three principles: marketing, transportation, and administration. We describe the marketing principle here; the others add complexity to the model by relocating settlements so as to minimize transportation costs and simplify administrative boundaries. The marketing principle says that because of economies of scale, organizations are most efficient when they serve markets of a certain size. The threshold is the minimum market area required to support a particular service. This principle suggests that a place of rank A in Figure 11.3 serves itself and one-third of the six surrounding places of the next rank, B. For that reason, the place of rank A is designated K=3 (where 3 refers to 1 +1/3*6). Similarly, a place at rank B serves itself and one-third of the six surrounding areas of rank C. Each A functions as an area of rank B and C as well, and each B also functions as an area of rank C. In Figure 11.3, each central place A contains nine hexagons of market area, each B contains three hexagons of area, and each C contains one hexagon of area.

The results are simple: there are fewer big places and more smaller ones, and big places have larger hinterlands than smaller ones. This general observation comports with the distribution of retail activities in urban and suburban America, though reality is far messier than the hexagons suggest. Brian Berry

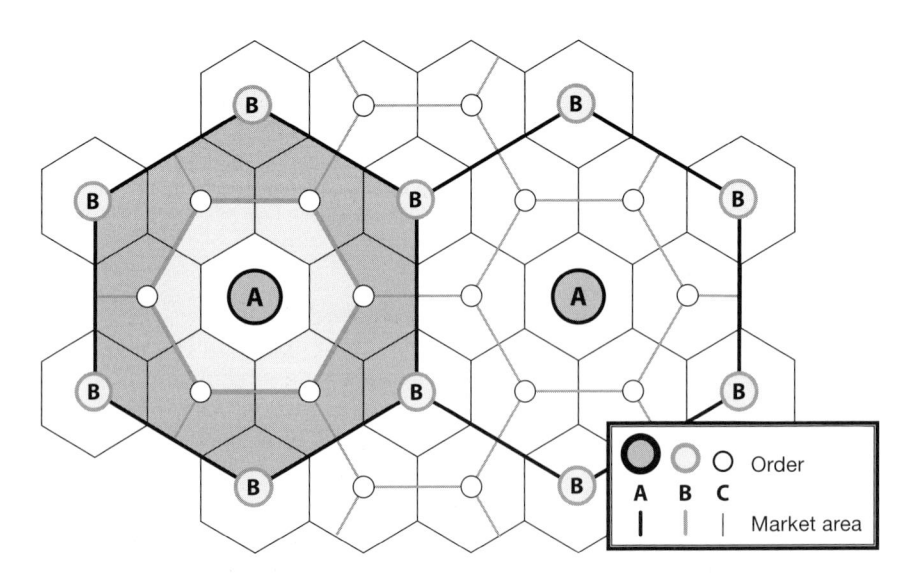

Figure 11.3 Christaller's central place theory

and Bill Garrison [5] validated the presence of a hierarchy of central places with data from Snohomish County, Washington, in one of the first applications of computers to quantitative, empirical geography. They showed that places are clustered in various levels of a hierarchy, rather than lying on a continuum.

James Vance criticized the Christaller model, describing the theory as "essentially bureaucratic in conception and unfriendly to innovation. Its very units of measurement are those of a rigid Germanic political order, and its warmest reception is to be found among the architects of planned or command economies." Vance asserted that the model is "not an explanation of economic enterprise but rather a Geography of Imposed Economic Order." [6] Vance, along with Allen Pred, argued that cities form for a variety of reasons related to mercantilism and trade with the outside world. Further, cumulative causation would be at work; once a place became settled, it would gain advantages over other potential settlements. In other words, the temporal sequence of development would matter more than some arbitrary timeless geography as suggested by Christaller. August Lösch, another German economist, extended Christaller's model from the service sector to manufacturing.

Central place theory, as it soon became known, suggested centralization is a natural principle of both hierarchy and order that human settlements follow. Furthermore, several rules governed what types of places sprung up and their character. Such "rules" suggested that: (a) the larger the size of cities, the fewer in number they will be, (b) the larger cities grow in size, the greater will be the distance between them, (c) the larger the size of a city, the greater the range and number of functions it will serve, and (d) the larger the city, the more higher-order services it will provide.

For example, one of the transportation principles embedded within central place theory is that the market area of a higher-order place includes one-half of the market area of each of the six neighboring lower-order places. This generates a hierarchy of central places that produces what is considered to be the most efficient transportation network. There are maximum central places possible located on the main transportation routes connecting the higher-order centers. Box 11.1 considers central places within a system of cities.

Hierarchy of plexus

Hierarchies of transportation networks, on the other hand, are usually delineated in terms of capacity or speed. Capacity is defined as the number of vehicles or persons per hour that can move past a point. At the two extremes, there are freeways and local streets; in between are collectors, minor arterials, principal arterials, and regional or county highways. Each has its own advantages and disadvantages. For example, the virtue of freely flowing roadways such as the United States' Interstate Highway system—uninterrupted travel—is also their liability. Major highways and freeways serve long-distance or pass-through movement, and have no local street connections, much less access to land and actual destinations.

Box 11.1 Systems of cities

"Cities are systems within systems of cities."

Brian Berry [7]

In *Place and Plexus,* we have focused on the relationship between cities and networks within the metropolitan area. As Hohenberg and Lees [8] point out, there are really two interacting systems. Cities are not isolated entities, but have trading relationships with rural areas (which, following Christaller, they call *Central Place Systems*) and with other metropolitan regions (which they call the *Network System*). They also have political relationships with the countries that govern them.

As discussed in Box 7.1, Eric Raymond wrote about *The Cathedral and the Bazaar,* two models for the development of computer software. This work also describes two models for the development of cities and regions: the cathedral is a top-down centrally organized system, while the bazaar is a bottom-up, decentralized system that exhibits "order from chaos" as complexity theorists might put it. Central Place Theory describes a hierarchy of places, and various researchers (e.g., Krugman) have tried to develop mechanisms to understand the formation of central places. Central place theory views systems of cities as cathedral-like, with a central city fed in a tree-like manner from subsidiary cities.

Network cities do not have a strictly tree-like hierarchical relationship with higher and lower places. Peer cities trade with each other. The modern epitomes of network cities are Hong Kong and Singapore, which are literally city-states that developed with essentially no economic hinterland and today are two of the world's largest ports. When networked, towns that appear to be in the central place hierarchy of one city interact and trade with those in another hierarchy without passing through the primary city. The nature of the city (whether it is predominantly a network city or part of a central place hierarchy), affects its character in several ways.

Whether a city is a central place and/or a node on a network affects not only its trade relations, but also its internal structure. The term "Bohemian" is commonly used to refer to artists and others with unconventional standards of behavior, and is derived from the region of Bohemia in the modern Czech Republic; Bohemia was once wrongly believed to be the origin of the Roma people, popularly known as "Gypsies" and frequently celebrated in art and literature for their supposedly unconventional and romantic lifestyle. Bohemia is a landlocked place, far upstream on the Elbe River. The term, however, best applies to places that are trading cities, with influxes of large numbers of individuals from different cultures. The mixing of cultures is associated with network places. In the United States, such places historically include certain neighborhoods of New York City, New Orleans, and San Francisco, which are (or were) the dominant ports on the East, Gulf, and West Coasts, respectively.

Richard Florida's work [9] on the "Creative Class," although suspect, [10] ranks the top US cities with more than one million residents by a "Creativity Index": Austin,

San Francisco, Seattle, Boston, Raleigh-Durham, Portland, and Minneapolis come out on top. Notably, each of these cities is a port or the seat of a major university, or both. Ports historically were the gateways from which travelers from far off would enter, bringing with them exotic goods and novel ideas. Following the transformation of international transportation away from the steamship to the jet airplane, many more cities now have international (air)ports, the legacy of the port-culture remains. University towns also have intellectual interaction, in addition to many international students and faculty, and being the homes of ideas, they are also characterized by big turnover as many members of the community (its students) change every four or so years. In a sense, universities serve as ports to adulthood.

The mixing of cultures that are conventional in one place but unconventional in another, and local tolerance for the strange behavior of new immigrants and visitors in the name of trade, allows a culture of tolerance to emerge. In many cases, foreign traders would locate within a selected neighborhood, self-selecting their residences in ghettos or outside city walls (and in other cities, the residence in the ghetto would be required). This was true not only of Jews in European cities, but in many cases where a diaspora of one culture was a trading minority in the city of another. [11] Trade diasporas were established phenomena from at least as early as 3500 BCE, enabling the cross-cultural flow not only of merchandise, but of ideas as well. Although many of the communities created by trade diasporas were eventually erased by wars or changes in economic methods (such as the use of native agents rather than locally based compatriots), others remain, such as the Chinese minorities through southeast Asia and Indians of East Africa. The Hanseatic League of independent cities began with traders from Cologne establishing trading colonies in many northern European port cities, and the founding of Lübeck, Germany in 1158. Despite the relatively modest ethnic differences between Hanseatic traders and locals, even they tended to be kept in isolated neighborhoods.

Markets (sometimes called transit markets) often formed at natural points such as the most upstream (or downstream) navigable point on a river, or at a junction between two navigable rivers. Markets at natural ports linked land-based and water-based trading systems. In Asia, Africa, and the Americas, trade networks were overlaid on each other. The first indigenous (or at least non-European) networks were largely land-based. Traders from China and India developed some sea-based networks. On top of these, European colonizers established, negotiated for, or conquered port cities that served as interfaces between local trading networks and the global trade economy of the European powers. Several of these systems (British, French, Dutch, Portuguese) operated in parallel.

Many studies of systems of cities focus on questions about city-size distributions and the determination of primary and secondary cities, how cities of different sizes differ, and whether the same patterns of systems of cities at the national level is replicated at the urban level in systems of edge cities. [12] George Zipf's "Law" (or the rank-order rule) governing city size distribution is a remarkable observation; the second-largest city in a nation tends to have half the population of the first, and the third largest city has one-third the population, and the ninety-seventh largest city has about one-ninety-seventh the population of the largest city. A number of researchers have attempted to explain this phenomenon, while others have disputed whether this fact is, in fact, so. [13] Several issues arise with elevating this observation to the status of "Law." First, it implies that cities will grow in lock-step, and does not allow for dynamics. Second, it breaks down in many countries (in the United States the cities ranked first and second in population—New York City and Los Angeles—are similarly sized, with Los Angeles being far larger than one-half the size of New York). Third, definitions matter greatly, and what comprises a city (or metropolitan area) will dictate how good a fit (and how close the fit of the model, especially the estimate of a is to the ideal: $P_n = P_1(1/n^a)$) where: P_n is the size of the nth ranked city, n= rank and $a = 1$) is observed. Fourth, the rank-order rule is inconsistent with central place theory, which posits a set of equally sized places feeding into larger places, and implies that the relationship between rank and size is stepped rather than continuous and linear. Systems of cities are much more complex than size metrics alone can explain.

Cities for which the central place system dominates are part of the world trading system, but they are not enmeshed in it to the same extent as cities for which the network system dominates. Network cities, although nominally driven by the exchange of goods, have in many cases found the exchange of ideas to be more valuable, as new techniques permitted innovation and created wealth.

The specific location of a facility, and its relationship with other links (both in terms of what traffic it carries, how much, and how fast) define its position within the hierarchy of roads. Building a new 100 km/h freeway across a grid of 50 km/h streets may have several effects. If the new highway is laid diagonally across the grid, it allows movement in a new direction. It attracts traffic from local streets onto the new limited access roadway. It may disconnect existing links, thereby further channeling traffic onto the new roadway because there are fewer alternative paths than before. In contrast, access to land is primarily served by slow, and low-flow, local streets, distinguished by their many curb cuts and turning opportunities. Collectors connect local streets to each other and to higher-level roads. These concepts are represented in Figure 11.4 where planners and engineers seek to design roads approximately along the diagonal represented in the conceptual diagram. Functionally separating movement from network access is important.

There are several rationales for network hierarchy, including:

- aggregating traffic (economies of scale);
- separating access and movement functions reduces conflict (makes both safer);

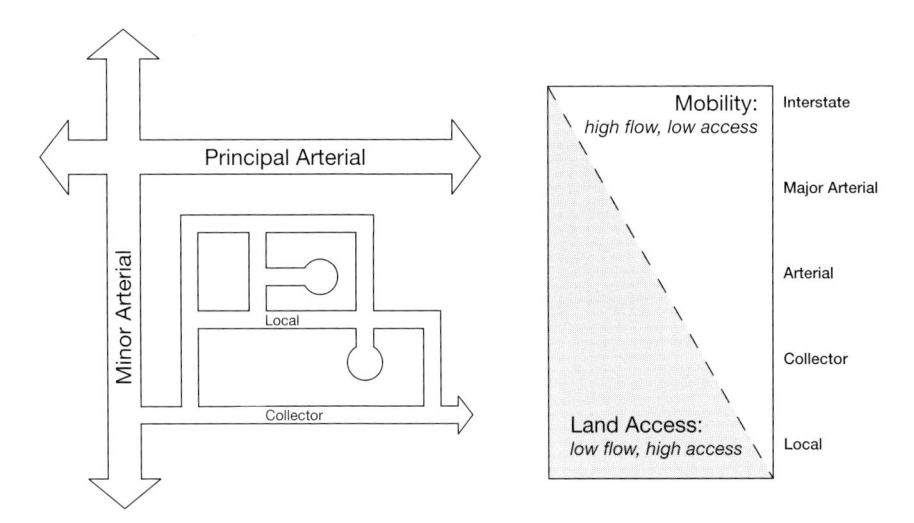

Figure 11.4 Hierarchy of roads

- keeping residential neighborhoods quiet;
- reducing redundancy in the transportation system;
- excluding higher levels and separating layers (making financing by different agencies easier).

At the same time, there may be a range over which the additional use engendered by a higher quality facility outweighs its cost. On the road network, capacity is not only measured by links. For example, transforming an at-grade intersection to a grade-separated interchange also adds capacity. Similar logic applies to transit systems; there are local buses, streetcars, light rail, commuter buses, subways, and commuter trains.

Aggregating traffic creates additional users who create the consumption economies of scale (the more traffic, the more consumers over which to spread the cost) necessary to justify the extra expense in building the faster, grade separated road. Therefore, although it is more expensive per lane-km, it is often cheaper per passenger-km traveled to build larger facilities rather than smaller ones.

If people had driveways entering onto freeways, those freeways would not be free-flowing. However, by limiting the number of entrances and exits, the roads become more vulnerable to delays: while it is easy to route traffic around a collision on a signalized arterial, the same collision on a freeway keeps a lot of traffic bottled up. There is a risk-seeking strategy at work here—planners trade higher travel time variability on freeways for lower average travel time.

A consequence of a hierarchy of roads is that some links serve more local traffic and others serve more through traffic. A rational way to govern this kind of hierarchy is to assign roads to the most local unit of government from which a large share of traffic is coming or going. [14] It is misguided to have a township manage an Interstate highway, or the federal government manage an alley. When the management of the road is mismatched with the type of road, problems emerge and are hard to deal with. These problems cut both ways, local traffic congesting roads designed for long distance travel (e.g., interstates), and cut-through traffic speeding down what should be local streets.

The bulk of the transportation attention in a metropolitan area is devoted to freeways or other large-capacity roadways; these are the routes carrying the bulk of the traffic. The irony is that, in the United States, while the Interstate system represents a mere 1 percent of all roadways by length, it carries on the order of 30 times that share in terms of daily distanced traveled in metropolitan areas.[c]

Other disadvantages of a hierarchy of plexus include:

- increased travel distance (backtrack costs);
- increased criticality of specific points (less redundancy means greater vulnerability);
- increased difficulty in navigation compared to flat networks.

Hierarchy is not limited to roads, though that is where we now see it most often. Railway development in Japanese cities and suburbs followed the Streetcar Suburb model [17] promulgated in the United States. Lines were constructed from the city center, usually with a terminus on the Central Loop of the city's heavy rail line, and another terminus in an existing village that was to become a suburb of the city, with perhaps an extension to an amusement park. A number of the nodes where the suburban rail lines connected with the city lines (such as Shinjuko and Shibuya in Tokyo) became enormous shopping and office districts in their own right. Retail development was significant around most stations. The rail companies developed land around stations, but in many cases held on to the land (rather than selling it, as had been the case in the US), giving a stronger base of financial support for the railroad. Also, because automotive mobility arose much later in Japan than in the US, the railroads became more deeply entrenched in transportation life and were more difficult to dethrone (along with higher densities making them much more valuable and practical).

Morphology

The term "morphology" is often employed by design-oriented professionals within the discipline of city planning to address the physical form of the city—focusing on, among other things, the street patterns, building shapes and sizes, and patterns and composition of different types of land uses. We employ it here to address the character or configuration of a community and its overall road network.

Morphology of place

Ever since the early 1990s, there has remained considerable enthusiasm in the urban planning profession for a movement known as the New Urbanism (also referred to as Transit Oriented Development or Smart Growth, or Traditional Neighborhood Design). The movement crystalized at a conference at a resort hotel in Yosemite Park in 1991, where the "Ahwahnee Principles" were crafted. The doctrine largely promotes compact development, mixing land uses, relatively narrow storefronts, plus urban design changes (e.g., predominantly gridded streets and small blocks).

Although the application of such principles to various communities is prone to misinterpretation, several conferences, societies, and other publications have been formed around the idea that there is a preferred morphology of human settlement. Subsequently, there has been an exorbitant amount of study examining the degree to which New Urbanist styles of development advance goals related to, for example, travel behavior generally, [18] pedestrian access, [19] sense of community, [20, 21] watersheds, [22] or aesthetic qualities. [23]

The New Urbanism movement has style—by showcasing select communities with architectural splendor. And it has history as a precedent—yearning for the tried and true communities of yesteryear. Politicians love it. Advocates stand behind it. Planners embrace it. Even a smattering of economists, traffic engineers, and housing specialists endorse it. Legions of students seemingly return to graduate school to further their careers as city planners and perpetuate doctrines based on New Urbanist principles. Subsequently, hundreds of developments worldwide now lay claim to being New Urbanist in nature, the most renowned being Playa Vista and Laguna West in California, Seaside in Florida, and Kentlands and King Farm in Maryland. Many New Urbanists themselves often refer to Portland, Oregon, with its light-rail transit orientation and urban growth boundary, as their model city.

Interestingly enough, however, many of the principles upon which New Urbanism is based are themes that have existed within city planning since the birth of the profession:

- As part of the 1920 New York Regional Plan, Clarence Perry sketched out a "neighborhood unit" as an essential component of a town and defined its size based upon a five-minute walking radius;
- In 1929, Clarence Stein and Henry Wright designed Radburn, a planned community, based on the concept of the "new town" to further the work of Ebenezer Howard and Patrick Geddes;
- In 1968, Konstantinos Doxiadis [24] proposed several solutions for rapidly growing cities, one of which was for city planners to leave room for expanding the city core along a predetermined axis so that most urban expansion would be channeled in a single direction. New, self-contained urban centers, he contended, would encourage better communication and transportation links between them. Doxiadis envisioned *ekistics*, a name that derives from the ancient Greek term *oikizo* meaning "creating a

settlement," as an interdisciplinary effort to "arrive at a proper conception and implementation of the facts, concepts, and ideas related to human settlement;"
- Of course, there is always the late Jane Jacobs, espousing hybridization: mixed-use planning, buildings of a variety of ages, the pedestrianization of the streets, and organized complexity. Such qualities produce plurality and visual discord—the ultimate goals in cities and neighborhoods.

Working to develop a bottom-up rather than top-down conception of urban form, architect Christopher Alexander's 1977 book *A Pattern Language* [25] develops a set of 253 patterns (which we might call the atomic elements of morphology) that can be used to build pleasing urban forms and buildings. Alexander's book is more normative than positive, but after 30 years on the market, it remains one of the most popular books on urban design, and perhaps the best-selling of all time in that genre. Although Alexander has been given a lifetime achievement award by the New Urbanists (Congress for the New Urbanism XIV), his philosophy of unfolding development starkly contrasts with the New Urbanist master-planned approach. Whereas the Pattern-Language and New Urbanist movements share some desired "ends," Alexander believes the path to the ends (the means) matters.

Morphology of plexus

Before *A Pattern Language*, Christopher Alexander authored "A City is Not A Tree" [26] in which he argued that, unlike a strictly hierarchical tree, a living city has interconnections between all its elements, not just connections up to higher elements and down to lower elements.

People often think of the overall configuration of a road network as one of two forms: tree-like or web-like. However, in his recent work, *Streets and Patterns*, Stephen Marshall [27] cautions against this conception. Marshall discusses what he refers to as the pattern or configuration of streets; an appendix of the book identifies over 94 different terms describing patterns from various authors. Clearly there is no consistency across each of the authors, though many of the patterns are similar. Marshall therefore reduces the list to 20 distinct types of streets and patterns.

The morphology of a city's plexus (sometimes called topology) is important because the street network is nearly immutable once it is laid down. As development occurs, new networks are constructed (certainly this happens locally in any new subdivision, where new roads fill the interstices within the existing coarser network). The examples of cities trying to rebuild themselves after disasters demonstrate this point. Cities rebuild themselves using the same street patterns that existed before the disaster, despite the fact that rebuilding represents the best opportunity in hundreds or thousands of years to reorient their transportation networks. Streets in London today are more or less where they were in the mid-1600s despite the Great Fire of London in 1666 and Christopher Wren's plan, because it was difficult to change so much legally defined property. On occasion, new roads are built through and across existing networks—the US Interstate Highway system, the parkways of Robert Moses in New York City, and the Boulevards of Baron Haussmann in Paris—are classic examples that come to mind. Technology reshapes the dominant morphology of communities: many older cities evolved from transit-based, radially oriented, dominant-node cities to automobile-based, dispersed, grid-like, multi-nucleated cities.

The morphology and queuing properties of the plexus (its supply and demand) ultimately determine both the efficiency of the network in moving people and the efficiency of the land use. Radial (hub-and-spoke) networks allow easy access to the center but create inconvenient sharply angled parcels. In contrast, 90-degree grids maximize travel times (for anyone traveling in a diagonal direction) but create efficient parcels. A major issue with network topology is the interconnectedness of the network. Interconnected networks, be they grid or radial in nature, enable and even encourage through traffic, while a tree-like network discourages that problem. The topology of the network, grid, radial, organic (curvilinear) or otherwise, affects its performance.

The regular grid (with occasional interruptions) is arguably the most common topology for cities. It has been employed in cities for millennia. In the United States, the most influential legislation affecting the morphology of roads was the Land Ordinance of 1785. In many respects, it laid the foundation for future land use-transportation policy by adopting the Public Land Survey System, creating townships and subdividing them into 36 sections of one square mile (259 hectares) and 144 quarter-sections of 0.25 square mile (65 hectares) each (Figure 11.5). Roads delineating each of the sections were referred to as "section roads." Subsequently, many urbanizing areas continued to use the centerlines of those roads as the location of present day arterials; the arterial networks are often further broken down into a finer grid of blocks.

A key point that has not been generally considered is the flexibility that the uniform and undifferentiated mesh networks (termed "grids" here) provide to changes in land use. A uniform grid allows alternative spacing between activities, spacing that can change with economies of scale. For instance, consider retailing. As described in Chapter 9, many stores—especially grocery stores—have been getting larger, while their numbers have dropped. Many New Urbanists, who advocate small-scale neighborhood retail, bemoan this phenomenon. Suppose that economies of scale indicate that it is efficient for the average retail store of a certain kind to increase in size from 1,000 to 2,000 ft^2 (93 to 186 m^2). Previously there may have been one such store every 10 blocks (one for every 100 square blocks); now there can be one every 14 blocks (one for every 200 square blocks). A grid allows the flexibility for re-spacing while keeping nearly optimal size stores. Box 11.2 considers the notion of flexibility.

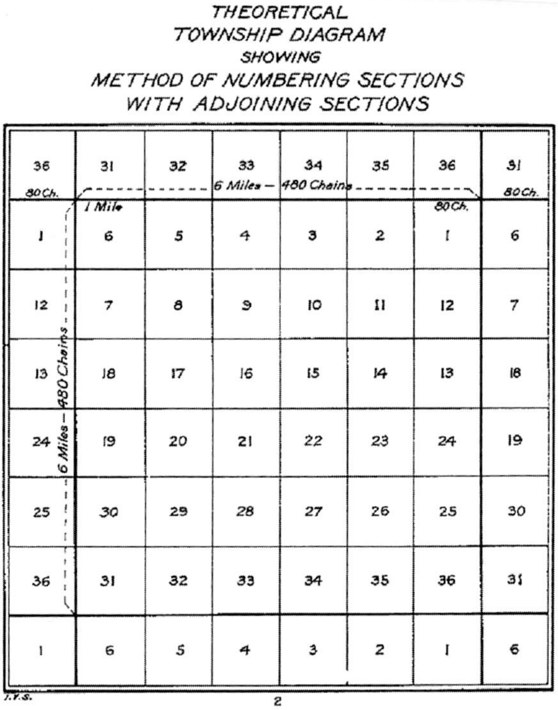

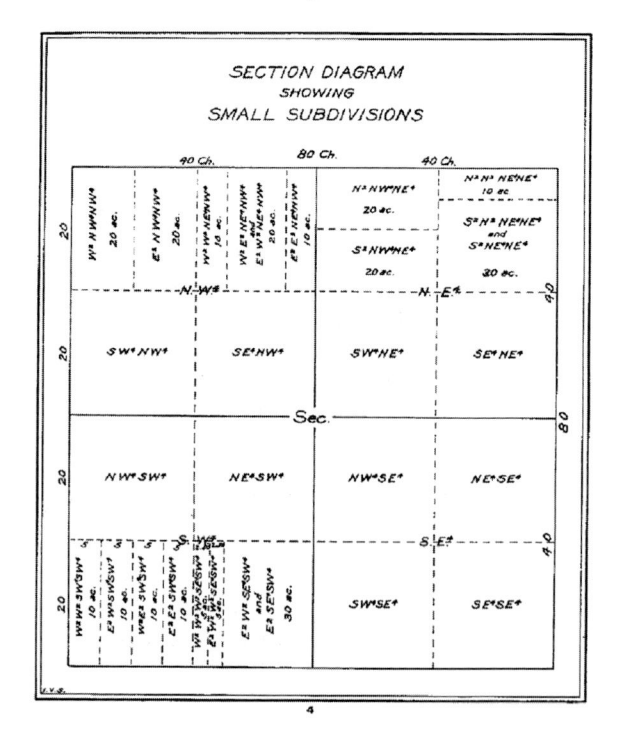

Figure 11.5 Grid-based morphology shaped by the Land Ordinance of 1785 (taken from original General Land Office drawings)

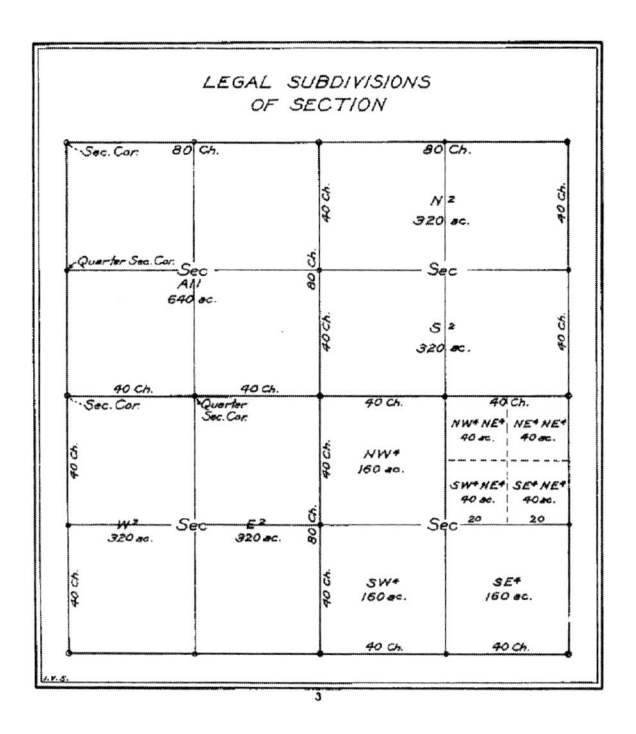

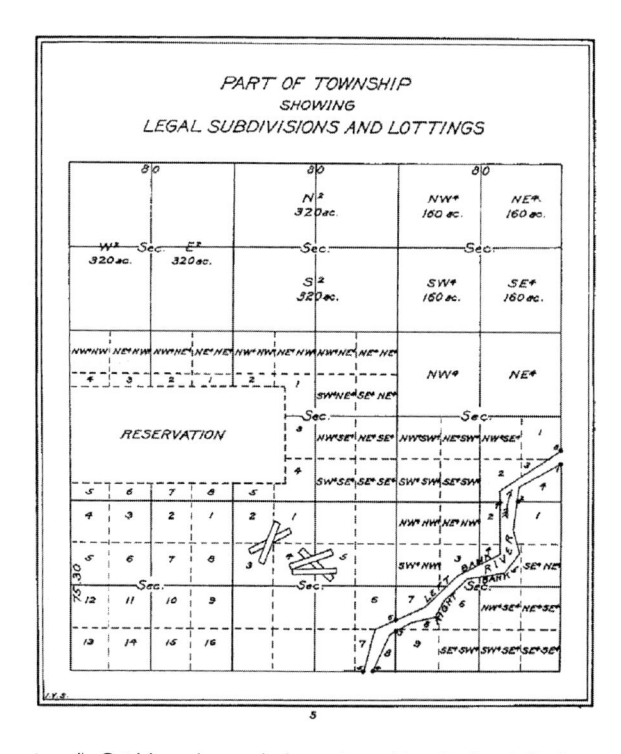

Figure 11.5 (continued) Grid-based morphology shaped by the Land Ordinance of 1785 (taken from original General Land Office drawings)

A tree network, in contrast, fails to provide such flexibility; a store can locate either at the neighborhood center, at the community center, or at the regional center; it can serve perhaps 5,000 people, 15,000 people, or 60,000 people. A store optimally sized to serve 10,000 people cannot be located at a consistent node level—or, if it is, it cannot be efficient. A firm may need to locate stores in some neighborhood centers and not others, causing people to go into other neighborhoods in some places.

Recognizing that grid-based road networks might not lend themselves to locations that were not situated on flat, featureless plains, designers introduced several variations. To conform to the contours of the land, Frederick Law Olmstead employed curving streets in many of his designs (e.g., Roland Park, Maryland). Permutations continued to evolve over the years, and the "loop" and "lollipop" designs became the standard in suburban settings.

Some new towns were designed to prohibit or discourage through traffic, like Radburn, New Jersey, or Columbia, Maryland, and a well-designed hierarchical road network of particular topologies accomplishes this goal—though with other costs. However, the fully connected grid network that appears in many US cities results in many streets carrying a large proportion of non-local traffic.

Box 11.2 Flexible design

In further considering the role of certain design tenets—particularly morphology, layers and architectural content—one needs to think long-term in creating adaptive communities. A primary example is schools. When a community of single-family homes is first developed, it is typically occupied by young families whose children attend elementary schools. Over time, some of those families move out, others move in, but both the housing stock and many of the residents age in place. Thus the elementary school, which was once teeming—if not bursting—with children from the surrounding neighborhood changes. As those children mature and graduate from high school, the neighborhood elementary school begins to resemble a ghost town, and the school board considers merging the school with that of another neighborhood.

On a well-defined tree network, some children continue to go to school in their neighborhood, others are bused in from longer distances to compensate for declining enrollment. A flexible grid allows some children to go to school ten blocks away while others come from 20 blocks away. However, the neighborhoods in such a regime are more fluid in their ability to adapt over time and are less firmly defined by their structures. The identity of the neighborhood is lost to gain adaptablity and identification with a larger whole. In more extreme cases—and in the spirit of form-based codes—the one-time school may even be converted to lofts in order to adapt to changing market conditions.

Layers

In his 1994 work *How Buildings Learn*, Stewart Brand endeavored to change the practice of building and the use of buildings. [28] Consistent with the philosophy of Christopher Alexander, Brand described the evolution of the built environment in terms of people changing the structures in which they live and work to adapt to new circumstances. Using a model of a house, he describes six nested systems: Site, Structure, Skin, Services, Space, and Stuff. The site is the property on which the building sits; the structure consists of the physical building skeleton, including the foundation and load-bearing components; skin encases the structure and forms the exterior of the building; services are the utilities that people use within buildings, including plumbing, heating, and wiring; space describes the layout of the building, its floor plan; and stuff includes the objects and people the building contains. Adding to Brand's six Ss, Scheer (cited in [29]) suggests that a seventh layer—the Street—needs to be considered.

Interestingly, Brand has participated in the computer industry since the 1960s.[d] It is therefore not surprising that he has developed a layer-based system to describe buildings, resembling the systems developed by computer networking professionals for the electronic world; the layers he offers in his 1994 work bear a close resemblance to computer structures. In particular, the Open Systems Interconnection (OSI) Reference Model is a widely accepted protocol that defines a networking framework in which protocols are implemented in seven layers. Control is passed from one layer to the next, starting at the application layer at one station, proceeding to the bottom layer, over the channel to the next station, and back up the hierarchy.

We cannot claim there is an exact counterpart to the OSI model in place and plexus, but some interesting connections emerge. There is a role for considering layers to a land use-transportation system (Table 11.1), where differences that exist include the inherent contribution of the user to the production of transportation. Layering is a structural property of the architecture of any system, and may almost be thought of as a natural law, in contrast with the design choices of hierarchy, morphology, and architectural content. However, layering is not amenable to policy or decisions in the same way.

Layers of plexus

Each of the layers in the elements in transportation is specified, organized, operated, and financed separately. Each layer has rules of behavior, some of which are physical (either deterministic or probabilistic), and some of which are legal or customary rules that are occasionally violated.

There are other points of comparison to consider. The structure of a building can be considered analogous to signs and markings on roads; signs and markings maintain traffic flow on the road, while the structure of the building transmits loads from above to the ground. The skin maps better to the vehicle, although services and traffic signals may be related, as they both deal with flows from utilities that control the operation of the system. Finally, drivers

Table 11.1 Layers of Place, telecommunications, and plexus

Layer #	Buildings (termed Place) (from Brand and extended by Scheer)	OSI Model (from Open Systems Interconnection Reference Model)	Plexus (from the authors)
1	Street	Physical	Site: Right of Way and Alignment (Horizontal and Vertical)
2	Site	Data link	Street: Road Structure (including Pavement)
3	Structure	Network	Signs and markings
4	Skin	Transport	Signals
5	Services	Session	Vehicle
6	Space	Presentation	Driver
7	Stuff	Application	Purpose of travel

(and their passengers and freight) are the principal items being transported on roads, while stuff is what buildings contain. On roads, drivers are going to places that have activities, while in buildings, space contains the stuff, which is at the highest layer.

Within the physical layer, there are elements that can be aggregated. In transportation, we model the world as if we have nodes or junctions, links, turns, paths, routes, and networks, as shown in Table 11.2. The node is the basic building block, and the other elements are collections of nodes. A node is just a point in space represented by coordinates. A link is represented as two nodes, and may have associated with it specific attributes (e.g., number of lanes, free-flow speed, permitted modes). A turn describes a movement at an intersection, which requires identifying the intersection (the *at*-node), as well as the start and end of the turn. A four-way intersection can have up to 16 turns (allowing for U-turns). A path or route is just a sequence of nodes, and can be used to portray the route a bus may take, or a traveler may pursue from origin to destination. A network can be represented in a model as a table of nodes and links.

Furthermore, networks are useful as a mechanism for differentiating space. Space without networks requires people to travel in a straight, but unimproved, line from origin to destination. Networks can provide faster, if more circuitous, routes between two points, but they cannot connect every pair of points directly because of costs.

Layers of place

Comparing the buildings and roads at layers 1 and 2 of Table 11.1, we see that buildings are constrained by streets and then by sites, while roads are

Table 11.2 Data structures representing the transportation network

Element	Representation
Network	node 1
	node 2
	node 3
	...
	node N
	link 1
	link 2
	link 3
	...
	link L
Path or Route	start-node, node 1, node 2, node 3, end-node
Turn	at-node from-node to-node
Link	i-node j-node
Junction or Intersection	node x-coordinate, y-coordinate

constrained by right-of-way (which can be thought of as a type of site) and then by road structure (or streets). In other words, the roads are the principal constraint on the buildings, and buildings are the principal constraint on roads. Thus, we have a *mutually reinforcing constraint structure* between place and plexus.

Buildings change most quickly at the lowest level (stuff) and most slowly at the highest level (site). Even when buildings are destroyed in great numbers (as in the Great Fire of London in 1666, the Great Chicago Fire of 1871, or the San Francisco Earthquake of 1906), the sites remain. Buildings are reconstructed on the same sites as their predecessors; there is no reallocation of land. Despite plans to reorient the structure of cities— Christopher Wren in London and Daniel Burnham in San Francisco immediately come to mind —in the aftermath of disasters, the cities' buildings remained similar. New buildings capitalized on the existing foundations (which often were not destroyed), and certainly the property lines, to accelerate construction. Another example of this phenomenon is discussed in Box 11.3.

As described above, the street changes very little even when the streets are surrounded by suddenly empty lots. Scheer argues that urban designers should focus on slow-moving elements such as streets and sites, not faster-moving elements such as space and stuff. If the streets are in the right locations, it is much easier to adjust the other things to fall into place. The lack of radical restructuring of clearly inefficient cities at the point of maximum opportunity (the time immediately after a disaster) is evidence for the mutually reinforcing constraint structure that buildings (and property) impose on networks and networks, in turn, enforce on buildings and property.

Box 11.3 Japan's building-line system: attaching plexus to place

Japan offers a different model of planning and development than the one usually presented in the English-speaking world. From the outsider's perspective, the country appears highly regulated, and its rail network appears very strong, yet its planning controls were relatively weak. Transportation-land use planning in a systematic way began in the twentieth century. The land readjustment (land reform) system in Japan was originally targeted at rural areas and aimed to consolidate small parcels into larger and more useful ones, which would be improved with features such as irrigation and drainage. After the 1918 City Planning Law, the system was applied in urban fringe areas; governments were allowed to impose a "betterment levy" on landowners who gained from public improvements, and also to expropriate land for those projects. Under this system, landowners would pool their land, designate 30 percent for public uses like roads and parks, and benefit after the road or park was built. If a two-thirds majority of owners in an area supported the land readjustment, all owners in the area could be forced to cooperate. The most notable feature of land expropriation is the idea, modeled on Haussmann's Paris, of "excess condemnation" of an area far larger than needed for a project (such as a road). The excess lands could then be sold for a profit by the government after the new road was constructed.

The 1919 Urban Buildings Law in Japan adapted Germany's "building-line system." At the foundation of this system were three rules: any public right of way greater than 2.7 m (9 feet) was declared a public road; the edges of public roads were deemed "building-lines"; and new building could only occur on building lines (so every building would have frontage on a road and none would be "land-locked"). The aim of the system was to bring order to development—to establish controls on what was perceived to be an anarchic system.

The width of 2.7 m is quite narrow for roads in a modern sense, less than a single standard road lane in the United States (which is 3.65 m on a freeway). This designation had the effect of "grandfathering" interior service lanes in Tokyo's slums, and may have been too inclusive, enabling building on roads that were too narrow.

After roads were designated, they needed to be turned into streets. Many roads remained unpaved. The planning laws allowed nearby land owners to be charged for up to one-third of the cost of street improvements. For example, in Kyoto, one-sixth was charged to adjacent land owners and one-sixth to nearby but non-adjacent land owners.

Although the building-line system may not have brought as much order to urban Japans as its proponents desired, in undeveloped areas it provided a way to regulate new development, ensuring both that development was carried out in an orderly manner along designated roads, and that right-of-way was preserved for those roads. The system also made landowners hoping to develop an undeveloped parcel responsible for the road in front of their building.

> Interestingly, if a city designated a new road on a map but was not ready to build it, landowners could still develop across the right-of-way so long as the building was two stories or less in height and could be easily removed (which was easily accomplished with the wooden structures of the era). When the city was ready to build the road, the owner would be compensated at fair-market prices.
>
> The building-line system was abolished in 1950 with adoption of the Building Standards Law and the Road Location Designation System, which attempted to ensure that roads in new residential developments would be four meters wide (two meters from building edge to center line). However, even that standard remained unenforced because many roads were unsurveyed, and roads and construction remained more organic than planned. Thus, planners lost their leverage over new development, since there was no way to ensure it followed planned roads, instead the roads could follow the development, and planners could not control the shape, location, or design of development. After 1950, only arterials were designated, and smaller roads were haphazardly routed.

Architectural content

A final pillar of design addresses the physical nature of what is contained within development. Architecture is often considered the science or method of designing and constructing buildings; computer scientists, however, broaden our notion of architecture to span the overall design, content, and structure of a computer system, including the hardware and the software required to run it, especially the internal structure of the microprocessor. We have interest in both when it comes to place and plexus—the physical nature of such and places and the networks connecting them.

Architectural content of place

Altering the physical nature of existing communities is particularly difficult, however, because of the inconsistency between many contemporary urban planning initiatives and many of the current policies on the books. For example, the bulk of current building zoning law prescribes a separation of land uses and focuses on regulating the types of activities permitted on land parcels. Current building codes mandate setbacks from the road. Current codes mandate minimum parking requirements.[e] Changing the physical architecture of a place, many argue, requires changing current land use policy. Towards this end, several leading urban planning practitioners have urged the abandonment of current zoning practice, [31] claiming it is time to do away with codes regulating use, height, and bulk.

Many New Urbanists suggest replacing "old-school" land use codes with prescriptions that classify building type and bulk, but are broadly permissive as to use. Labeled "form-based codes," [32] they include zoning ordinances and maps that would specify different types of building forms in varying configurations of bulk and height. The form of the building would dominate over the management of uses, through parking regulations, sign controls, and other specifications. Because form-based codes unlink use from building type, their proponents argue that such codes are particularly relevant to central cities experiencing infill development that must be fitted into the context of existing neighborhoods (but also that the codes be applicable to suburban development). Form-based zoning codes emphasize the aesthetic qualities of developments in order to shape the growth of cities. Rather then regulating the use of buildings, form-based codes control building styles and characteristics, while allowing developers significant latitude in determining the use of land parcels. Embraced by the New Urbanist movement, form-based codes are seen as a method to rehabilitate depressed downtowns. The supposed advantages of this approach are that it reduces the risk of building abandonment as markets fluctuate, encourages greater public participation in the planning process, and streamlines the regulatory permitting process.

But are cities putting themselves at risk by allowing greater free-market control? How will land prices, which have historically been tied to land speculation, be affected? Despite considerable enthusiasm, it is doubtful that communities will be able to quickly craft form-based codes to replace conventional zoning regulations. To date, conventional zoning has been principally used as a means of implementing a physical plan for a community, and often serves as a placeholder until development arrives at a site.

Architectural content of plexus

Finally, we arrive at consideration of the specific role local streets can play in influencing the design of a community. The science and architecture of roadways include a variety of perspectives, ranging from domains of engineering (e.g., geometric design) to the burgeoning field of road ecology. This section focuses on how the streets are designed—particularly on the ways street design capitalizes on architectural features to affect user experience both on and off the street.

The bulk of such initiatives fall under the banner of "traffic calming," where central objectives are threefold: (1) to influence traffic speed, by changing the character of the road and drivers' perception of the character of the road; (2) to affect traffic volume, by relying on physical diversion; and (3) to mitigate the severity of crashes (which is largely a function of speed). Many of these techniques are illustrated in Figures 11.6 and 11.7.

Delft, in the Netherlands, is acclaimed as the birthplace of traffic calming, where in the late 1960s neighborhoods proposed the *woonerf* (the Dutch root words *woon werf* are translated as "living yard," but the term is commonly taken to mean "street for living") in order to reduce cut-through traffic. The *woonerf* puts the needs of automobile drivers secondary to the needs of users of the street as a whole. Thus, although they remain connected, many neighborhood streets appear such as driveways, with features including a realignment of the travel path, the institution of double parallel parking so that travel lanes would go around two parked cars, the use of brick pavers to inform and slow drivers, and the placement of planters and other furniture in what had been the roadway. The same transportation lane is used by pedestrians, bicycles, and motor vehicles. The local success of the *woonerf* (it was endorsed by the Dutch national government in 1976) encouraged its spread to other European cities, primarily in the Netherlands, Denmark, and Germany initially. *Woonerf* techniques, especially the use of diverters, were adopted in mainstream planning in the United States in the 1970s in cities such as Berkeley (California), Seattle (Washington), and Eugene (Oregon); they have since spread to countless others. Moving out from the city center, several efforts have been made to integrate these design principles into mainstream suburban environments. [33]

Traffic calming is most often applied on residential streets that otherwise receive a great deal of through traffic. But it may also be appropriate for shopping streets where a more pedestrian-oriented realm is desired, without completely excluding vehicles. There are a variety of techniques for traffic calming [34, 35] as well as strategies for funding such improvements. [36] The techniques include:

Figure 11.6 A woonerf in Utrecht, the Netherlands (photo by the authors)

- vertically altering the terrain via speed bumps, speed humps, speed tables, raised crossings, undulations, and variations in road texture or paving material;
- horizontally altering the terrain via traffic circles and roundabouts, curb extensions (bulb-outs, neckdowns, chokers, chicanes/lateral shifts), median or pedestrian refuge islands, or the use of edgelines to narrow wide roadway and create bicycle lanes, parking lanes, or shoulders;
- linearly altering the terrain via full closures or cul-de-sac conversion, half-closures (closing one direction), the installation of diverters (barriers at intersections to prohibit or require certain movements) or intersection realignment.

Most traffic calming environments have simple signage, markings, and signals; they construct traffic diverters and channels and may even change the type of pavement surface (see signs indicating the beginning and end of a *woonerf* in Figure 11.7). All are intended to indicate to drivers that the nature of the street is different, and to make drivers uncomfortable traveling at high speeds.

Figure 11.7 Signs indicating the start and end of a woonerf in Utrecht, the Netherlands (photos by the authors)

An even more radical notion has been proposed by Hans Monderman, a Dutch traffic engineer. At busy intersections in communities where traffic speed is a problem, he suggests that the signs and signals be removed so drivers focus on each other and on pedestrians (and vice versa). Without explicit instructions from signs and signals, drivers are more cautious and enter the intersection more slowly, but do not necessarily need to come to a stop, depending on conditions. This strategy has been dubbed "designing for negotiation" where the negotiation takes place between users of the roadway, which is now a "shared space" rather than a facility for automobiles alone.

An alternative notion in the United States is labeled "complete streets;"[f] a complete street is defined as a street that works for motorists, bus riders, bicyclists, and pedestrians, including people with disabilities. A "complete streets" policy is aimed at producing roads that are safe for users of all modes of transportation. This policy is not so radical, however, that all signs and markings are eliminated. Rather, it suggests that rights-of-way be allocated so that pedestrians have sidewalks, bicyclists have bike lanes, transit buses have bus bays, and automobiles have adequate space for movement, including turn lanes, in an environment that is appropriately landscaped to make it attractive for travelers using all modes.

Wrap up for designing

In the introduction to this chapter, examining Fresno's Fulton Mall, we asked "Are there streets where cars do not belong?" We argue that the answer is yes. There are plenty of examples of automobile-free streets and zones, and considering the fact that many streets in the heart of our cities were created before the automobile was widespread, autos should not have an automatic claim to free rein. The expectation of being able to drive to the front door of every building, rather than walking a block, may not be uniquely American, but it certainly plays to the stereotype of the American. If the street were enclosed, the belief that driving would somehow make it thrive would be dismissed out of hand. That the automobile-free zone in Fresno did not save downtown is not due to the zone's lack of automobile traffic; it is due to more fundamental structural factors that Fresno has yet to deal with. Greenfield development (so to speak; because the area around Fresno is largely covered by desert or irrigated agricultural land, the term "green" is something of a misnomer) is cheaper than brownfield development and redevelopment. As described in Chapter 3, there is a preference for the new. Not only does Fresno not seriously regulate development, the city encourages new development through significant subsidies. Race is an issue, as is the frequently observed desire of people to self-segregate according to income into districts with the best schools that can be afforded. Keeping automobiles off six blocks downtown will not change those factors. Whether people should dine, play, or surf the Internet outdoors in an area violating national ozone standards an average of 101 days a year is another question—one that we leave to the discretion of the participants.

The example of Fresno illustrates the interplay of the four tenets of the Diamond of Design. Hierarchy of place is apparent, as this is downtown, and the appropriate placement of Fulton Mall on the hierarchy of roads also emerges. The morphology is closely related to hierarchy. The massing of buildings is largely given in downtown Fresno. Though there has been some new construction, it is modest compared to the existing stock. Creating an auto-free street disrupted the regularity of the local grid, but did not eliminate it entirely. The mutually reinforcing constraint structure dictated by the layers of buildings and layers of networks greatly limits what can be done. The roads limit where buildings can go; the buildings limit where new transportation axes can be laid. The question of architectural content emerges: what should the road network look like, how should the buildings and road interrelate? In contrast to the hopes of physical determinists of all stripes, no amount of urban design effort alone can reshape the Mall. Street furniture without solid economics behind it fails to induce use.

In contrast to strategies for building and operating systems of place and plexus (which tend to be more science-oriented in their applications and are described in the following two chapters), the design of these systems weaves together art and science. Design involves injecting a sense of art into the field of planning, and deciding what types of designs are superior to others. Rating forms of art, however, is difficult. Interestingly enough (and unknown to many), town planning was once an Olympic event, in which gold, silver, and bronze medals were awarded in the 1928, 1932, 1936, and 1948 games (Table 11.3).[g] Granted, each of the "entries" (i.e., plans) had to be inspired by sport and also be original (that is, not published before the competition), but the determination of which plans were superior was based on a set of criteria.

Proper system designs, whether they be they Olympic medal contenders or not, set the stage for everything else. The overall design of a community's transportation and land use patterns occurs over four different dimensions: the network possesses layers, architecture, hierarchy, and morphology or topology. The hierarchy results in some links serving greater flows at higher speeds than others. It is necessary to think about how many links operate at each level in the hierarchy of roads. The topology is also important: roads such as Manhattan's Broadway break the grid—opening up movement in a new direction (e.g., diagonal), and thereby removing traffic from local streets. There is often a risk-seeking strategy at work. In terms of design, a steep hierarchy helps separate the mobility and the access functions of roads, allowing local streets (which serve the access function), to accommodate uses beyond the personal vehicle. Streets existed before motor vehicles, and they will exist after motor vehicles, so why should they be the exclusive province of motor vehicles? Designs that incorporate traffic calming can ensure safe use of local streets for both the automobile and the pedestrian.

Table 11.3: Medalists in the town planning event of the international Olympic Games [37]

Year	Gold	Silver	Bronze
1928	Adolf Hansel (GER) Stadium at Nuremberg	Jacques Lambert (FRA) Stadium at Versailles	Max Lauger (GER) Municipal park at Hamburg
1932	John Hughes (GBR) Design for sports & recreation center with Stadium for Liverpool	Jens Houmøller Klemmensen (DEN) Design for a stadium and public park	André Verbeke (BEL) Design for a marathon park
1936	Werner March (GER) National sports field	Charles Downing Lay Marine park, Brooklyn (USA)	Theodor Nussbaum (GER) Cologne city plan for sports facilities
1948	Yrjö Lindegren (FIN) The centre of athletics in Varkaus	Werner Schindler and Eduard Knupfer (SUI) Swiss federal sports and gymnastics training center	Ilmari Niemelainen (FIN) Athletic centre

Many overlapping plexuses have been addressed so far, first among them surface transportation, including highways, transitways, and facilities for non-motorized transportation. But cities are connected by a variety of networks. In other chapters, we have also considered in detail social networks, communication networks, information networks, media networks, and economic networks (supply chains). There are other networks that we have mentioned only in passing: water and sewer networks (which are pipeline networks that transport fluids), electrical networks, financial networks, wildlife corridor networks, ecological networks (the food chain) and so on. Literature on these topics refers to a variety of networks; this literature, however, fails to analyze networks as such, and uses the idea and language of networks in an amorphous manner.[h] This book suggests that an important network lies in concrete structures, which have flows on them—flows that change dynamically on structures that also change. But not all networks are physical and concrete; even social and information networks require the exchange of information (bits and memes) through some type of physical interaction.

Returning to the five Es from the Diamond of Evaluation, cities can design place and plexus for different combinations of attributes: Efficiency (Does the hierarchy result in a more reliable or faster network?); Equity (How does the design affect different users?); Environment (Does the configuration of land uses or roads increase or decrease impacts on the environment?); and Experience (How does the morphology affect users' interaction with the facility?). Expediency leads one to balance design objectives.

Notes

a   Other instances of city planners in film include the TV comedy series, *The New Gidget*, where Gidget's husband, Moondoggie, was a city planner.

b   Many may consider our use of *design* synonymous with *urban form*, a term which fails to have a universally agreed upon definition. Most consider the latter to consider the land use pattern (the location) *in concert with* the transportation system (the network) (or vice versa).

c   Authors estimate based on surveying a variety of sources from metropolitan areas including vehicle miles of travel along specific facilities. Although this may strike the reader as a bit excessive (why is such a small portion of the roadway system carrying so much of its travel?), such a phenomena is not out of line with other behaviors. Consider, for example, the bulk of church donations come from roughly a quarter of the congregations, that the top 1 percent of the population owned 38.1 percent of the wealth of the United States (whereas the bottom 40 percent of the population owned 0.2 percent) [15], or that the bulk of the writing and editing for Wikipedia's 1.2 million entries (as of this writing) on the English-language site is done by a geographically diffuse group of 1,000 or so regulars. [16]

d   According to Brand's biography, in 1969 he participated in Doug Engelbart's demonstration of "Augmented Human Intellect" at the Fall Joint Computer Conference in San Francisco. It has been described as "the mother of all demos," since it inspired much of what was to come in personal computers. Brand was the progenitor of the Whole Earth Catalog, and founded the Well online community.

e   For example, Duany and Plater-Zyberk [30] suggest that so-called suburban sprawl is not the product of natural urban evolution. Rather, it is the direct result of current zoning codes that dictate wide streets, huge lots, attached two-car garages, and the absolute separation of houses from shops and workplaces.

f   See "Let's Complete America's Streets" by Complete the Streets www.completestreets.org/ accessed July 4, 2006 and "Complete the Streets for safer bicycling and walking" by America Bikes: www.americabikes.org/bicycleaccomodation_factsheet_completestreets.asp accessed July 4, 2006.

g   The event was known as the art competitions and included architecture, literature, music, painting, and sculpture. Two categories existed for architecture. In the general architecture category, prizes were awarded from 1912 to 1948; the town planning category was added in 1928. The division between the two was not always clear, and some designs were awarded prizes in both categories.

h   Perhaps the most widely read of the amorphous users of the term "network" are Castells, Manuel (1996) *The Rise of the Network Society*, 2nd edn, 2000, Oxford: Blackwell Publishers; and Lefebvre, Henri (1974, translation 1991) *The Production of Space*, Oxford: Blackwell Publishing, which in the end, are ambiguous in the meaning of network. For a review of the literature see Schaick, Jeroen van (2005) 'Integrating Social and Spatial Aspects of the City', in *Shifting Sense*, (eds) Edward Hulsbergen, Ina Klaasen and Iwan Kriens, Amsterdam: Techne Press.

References

[1]   Gruen, V., *The Heart of Our Cities*, New York: Simon and Schuster, 1964.

[2]   ELS Architecture and U.D. SWA, *Historic Fresno: Fulton Street and the Downtown*, 2002. Available at: www.fresno.gov/downtown/fulton/legacy.html (accessed March 12, 2006).

[3]   Clough, B., 'Marchers see Fulton Full of Promise: Show of Support for Fresno's Downtown Mall Accompanies a Report to City Hall on Revitalization', *The Fresno Bee*, March 1, 2006: A1.

[4] Grossi, M., 'Valley Tops L.A. as Bad-air King: Region Violated a Key Ozone Standard More Times than South Coast in 2001', *The Fresno Bee*, January 6, 2002: A1.
[5] Berry, B.J.L. and Garrison, W.L., 'The Functional Bases of the Central Place Hierarchy', *Economic Geography*, 1958, vol. 34 (2): 145–154.
[6] Morris, P., *Lecture Notes for Urban Geography*, 2006. Available at: http://homepage.smc.edu/morris_pete/urban/ugnotes05industrial.html.
[7] Berry, B., 'Cities as Systems within Systems of Cities', *Papers and Proceedings of the Regional Science Association*, 1964, vol. 13: 147–163.
[8] Hohenberg, P.M. and Lees, L.H., *The Making of Urban Europe: 1000–1950*, Cambridge, MA: Harvard University Press, 1985.
[9] Florida, R., *Rise of the Creative Class*, New York: Basic Books, 2002.
[10] Markusen, A., 'Urban Development and the Politics of a Creative Class: Evidence from the Study of Artists', in *Regional Science Association Conference on Regional Growth Agendas*, 2005, Aalborg, Denmark.
[11] Curtin, P.D., *Cross-cultural Trade in World History*, Cambridge: Cambridge University Press, 1984.
[12] Bourne, L.S. and Simmons, J.W. (eds.), Systems of Cities: Readings on Structure, Growth, and Policy, New York: Oxford University Press, 1978.
[13] Soo, K.T., *Zipf's Law for Cities: A Cross Country Investigation*, 2002. Available at: www.few.eur.nl/few/people/vanmarrewijk/geography/zipf/kwoktongsoo.pdf (accessed January 21, 2007).
[14] Levinson, D., *Financing Transportation Networks*, Cambridge: Edward Elgar, 2002.
[15] Levy Economics Institute, *The Distribution of Income and Wealth*. Available at: www.levy.org/default.asp?view=research_distro (accessed April 14, 2006).
[16] Hafnew, K., 'Growing Wikipedia Revises Its "Anyone Can Edit" Policy', *The New York Times*, June 17, 2006: A1.
[17] Warner, S.B., Jr., *Streetcar Suburbs: The Process of Growth in Boston, 1870–1900*, Cambridge, MA: Harvard University Press, 1962.
[18] Ewing, R. and Cervero, R., 'Travel and the Built Environment: A Synthesis', *Transportation Research Record*, 2001, vol. 1780: 87–112.
[19] Lund, H., 'Testing the Claims of New Urbanism: Local Access, Pedestrian Travel, and Neighboring Behaviors', *Journal of the American Planning Association*, 2003, vol. 69 (4): 414–430.
[20] Talen, E., 'Sense of Community and Neighbourhood Form: An Assessment of the Social Doctrine of New Urbanism', *Urban Studies*, 1999, vol. 36 (8): 1361–1379.
[21] Talen, E., 'The Social Goals of New Urbanism', *Housing Policy Debate*, 2002, vol. 13 (1): 165–188.
[22] Berke, P.R., MacDonald, J., White, N., Holmes, M., Line, D., Oury, K., and Ryznar, R., 'Greening Development to Protect Watersheds: Does New Urbanism Make a Difference?' *Journal of the American Planning Association*, 2003, vol. 69 (4): 397–414.
[23] Ellis, C., 'The New Urbanism: Critiques and Rebuttals', *Journal of Urban Design*, 2002, vol. 7 (3): 261–291.
[24] Doxiadis, C., *Ekistics: An Introduction to the Science of Human Settlements*, New York: Oxford University Press, 1968.

[25] Alexander, C., *A Pattern Language: Towns, Buildings, Construction*, New York: Oxford University Press, 1977.
[26] Alexander, C., 'A City is not a Tree', *Architectural Forum*, 1966, vol. 122 (1, 2): 58–62.
[27] Marshall, S., *Streets and Patterns*, New York: Routledge, 2005.
[28] Brand, S., *How Buildings Learn: What Happens After They're Built*, New York: Viking-Penguin, 1994.
[29] Postrel, V., *The Substance of Style*, New York: Perennial Books, 2004.
[30] Duany, A. and Plater-Zyberk, E., *Towns and Town-making Principles*, Cambridge, MA: Harvard University Graduate School of Design, 1991.
[31] Duany, A., Plater-Zyberk, E., and Speck, J., *Suburban Nation: The Rise of Sprawl and the Decline of the American Dream*, New York: North Point Press, 2000.
[32] Peirce, N., *Zoning: Ready to be Reformed?*, 2003. Available at: www.postwritersgroup.com/archives/peir0127.htm (accessed June 2, 2006).
[33] Ben-Joseph, E., 'Changing the Residential Street Scene: Adapting the Shared Street (Woonerf) Concept to the Suburban Environment', *Journal of the American Planning Association*, 1995, vol. 61: 504–515.
[34] Federal Highway Administration, *Descriptions and Pictures of Traffic Calming Devices and Techniques*, 2001. Available at: www.fhwa.dot.gov/environment/tcalm/part2.htm (accessed April 29, 2006).
[35] Montgomery County Maryland Department of Public Works and Transportation, *Appropriate Traffic Calming Measures*, 1996. Available at: www.dpwt.com/TraffPkgDiv/artvspri.htm (accessed April 29, 2006).
[36] Weinstein, A. and Deakin, E., 'How Local Governments Finance Traffic Calming', *Transportation Quarterly*, 1999, vol. 53 (3): 75–87.
[37] Wikipedia contributors, *Olympic Medalists in Art Competitions*. Available at: http://en.wikipedia.org/wiki/Olympic_medallists_in_art_competitions (accessed May 14, 2006).

Chapter 12

Assembling

> "I give you the Springfield Monorail! I've sold monorails to Brockway, Ogdenville and North Haverbrook, and by gum, it put them on the map!"
>
> Lyle Lanley, The Simpsons

To coincide with the 1962 World's Fair in Seattle, organizers commissioned the city's iconic Monorail. Not only was it America's first full-scale commercial monorail system, serving a stretch of approximately 1.5 kilometers (one mile) between the Central Business District and the Fairgrounds, it was a new form of transit featured in the Elvis Presley movie *It Happened at the World's Fair*. In 1968 and again in 1970, ballot initiatives were put forward in Seattle to develop a slightly different transit technology: light rail. Both ballots were rejected. In subsequent years, traffic on highways and arterial streets escalated.

Approximately 30 years after the construction of the original monorail, the region began (again) considering investments in mass transit. The Washington State legislature subsequently created Sound Transit, an agency charged with developing and managing a mass transit system to serve the three-county Puget Sound region.

In November 1996, one of the authors, having recently moved to Seattle, voted for a referendum to fund regional bus and rail systems, including a 40 km (25-mile) electric light rail line. Despite voter approval of such a major light rail building initiative for the region, Seattle-centric transportation activists were intent on developing a transportation alternative that would serve the City of Seattle specifically, and would use elevated monorail technology, as had been employed in the original 1962 Seattle Monorail.

The public quickly became versed in the complexities of different forms of mass transit. For the next eight years the media, residents, council members, and community organizations would discuss the relative costs and benefits of alternative technologies, routes, funding strategies, and service areas. In 1997, Seattle voters approved a ballot initiative to create a private organization to plan and seek funding for a 64 km (40-mile) monorail system that would criss-cross the city. In June of 2000, the City of Seattle considered repealing the initiative, but when faced with public and legal pressure, did not.

In November of 2000, Seattle voters continued to endorse an initiative to plan a monorail system, and to commit funds to develop it. Two years later, in November 2002, Seattle voters again approved a ballot initiative to use public funds (from a motor vehicle excise tax) to build the planned Monorail system. The Elevated Transportation Company received legislative authorization from Washington State and approval from the voters to begin building the first phase of a monorail system, the 22.5 km (14-mile) Green Line. After the tax funds fell short and costs escalated, and it became apparent that interest payments would be required (which were presented in the media as a total value to be paid over 50 years rather than a present value, thereby making project costs seem much larger than in the earlier proposal) Seattle's Mayor, Greg Nickels, and City Council withdrew support. Finally, a fourth ballot initiative, in November 2005, asked Seattle voters to approve funding arrangements for the Monorail and the creation of an organization that would supervise its construction. This time voters said enough was enough; the monorail effort was subsequently terminated.

Why were there so many ballot initiatives? How many times must a community agree on a proposal before infrastructure is built? All of the ballot initiatives asked voters to support a monorail in Seattle, but each of them entailed differing organizations, funding sources, and routes (the proposed routes got smaller each time, just as the proposed construction costs grew and grew). Simultaneously, an alternative technology, light rail, was proceeding apace with a similar round of route revisions, reductions in the length of the route, and pitched battles with neighborhoods over siting of the stations. Interestingly enough, the Monorail and the Light Rail were not even coordinated with each other, much less integrated; though they served different communities (the Monorail connecting northwestern and southwestern areas with downtown, LRT connecting northeastern and southeastern areas), it sometimes seemed as if they were in direct competition. The plans for any Monorail-type project subsequently grew defunct, perhaps because of the complicated process of trying to build something that, depending on one's view, either connects or divides the city. Meanwhile, Sound Transit construction hobbles along with usual planning delays.

Debates like those in Seattle stem from where and how to invest in infrastructure to most effectively move people from one point to another, given the consequences for people who live along the corridor. The need to connect points is the central reason for governments to provide transportation services of any kind. The issue is that building major investments—be they transportation systems such as monorails, supertrains, light rail or freeways (or bridges as described in the introduction to Chapter 10)—together with new sports facilities, are highly contentious issues in civic debates (Figure 12.1).

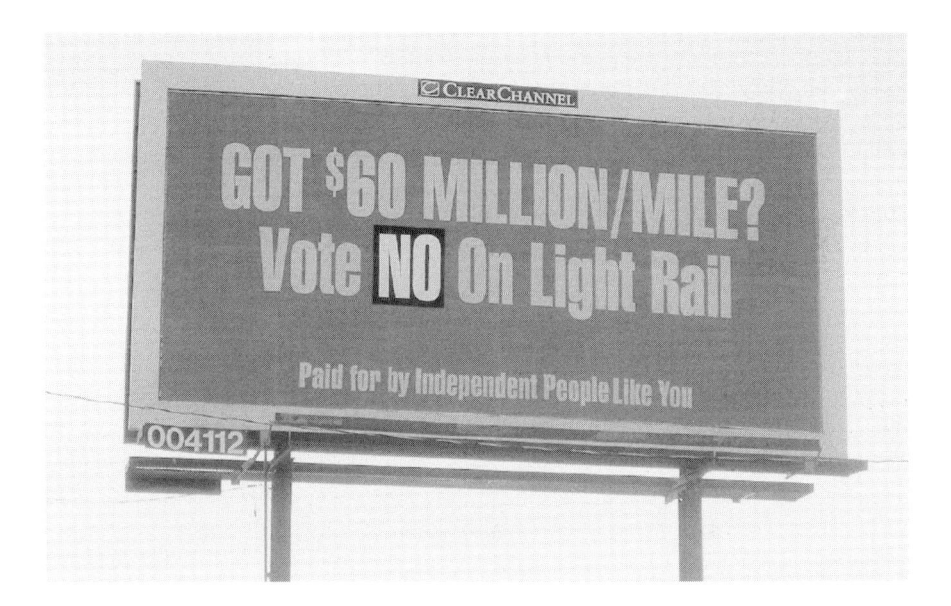

Figure 12.1 Examples of political rhetoric surrounding the assembly of infrastructure investments (photos by the authors)

Motivations for transportation infrastructure

The stakes in major infrastructure developments are high; the identity of a city often hinges on them. The more distinctive the infrastructure, the stronger the identity. It is difficult to visit New York without encountering the subway. One rarely sees a movie filmed in San Francisco without an iconic image of cable cars on that city's steep streets. Residents and visitors alike in London lament that, as of December 2005, the famous Routemaster double-decker red buses have been withdrawn from regular service, leaving only two heritage routes for nostalgia's sake. [1] Transportation systems are living monuments to their cities, and contribute to the image the city wishes to project (Figure 12.2). There are

Figure 12.2 Modes of distinction: the Routemaster double-decker red bus in London and examples of identifiable modes of transportation in other cities

City	Mode
Amsterdam and Davis, California	Bicycle
Calgary	Plus 15 Skyway
Chicago	The El (elevated rail)
London	Underground, Black Cabs, Double Decker Bus
Los Angeles	Freeways
New York	Subway
Paris	Boulevards
San Francisco	Cable cars
Seattle	Monorail
Vancouver	Skytrain
Venice, Italy	Canals and Gondolas
Venice Beach, California	Roller skates
Tokyo	Metro, Shinkansen Bullet Trains

(probably) more cities than distinct types of transportation infrastructure. Cable cars can be found outside San Francisco, but the "City by the Bay" owns the image of cable cars in most people's minds.

If unable to rely on a unique mode of transportation, a city may seek to assemble transportation infrastructure without regard to its distinctiveness. Witnessing the success of world-class cities, inferior cities try to emulate them, in order to present themselves as world-class cities as well. For example, some perceive that light rail, like large convention centers and domed or retro athletic stadia, qualifies a city as "world class." The number of United States cities adding light rail in the 1990s is, in part, a component of these cities' efforts to boost their image to one successful enough (dense enough, large enough, "big-league" enough, permanent enough) to warrant a rail system. Such motivation harkens back to days of early Britain where, until the sixteenth century, a town was bestowed "City" status by the Crown if it had a diocesan cathedral within its limits. Thus, just as "City" status requires a cathedral, a world-class city (one that cannot support a metro heavy rail system) needs light rail.[a]

The motivations for infrastructure construction may be symbolic and practical, while simultaneously destroying older symbols. Elevated expressways are dominant features of the Tokyo landscape. Opening in anticipation of the 1964 Olympics (much like the first Shinkansen (bullet train) line), the expressways succeeded in connecting disparate parts of the city but, in the process, covered much of the city's architectural and landscape glory with concrete. Japan was driving for economic growth at any cost, and sacrificing some aesthetics seemed a bearable price at the time. Elevated roads were also built in Osaka and Nagoya, though the highways in Nagoya faced local opposition as they covered (and destroyed) at-grade boulevards that had provided an important form of open space. A few freeways were blocked (in the suburbs of Tokyo) though the freeway revolt in Japan was nowhere as successful as that in the United States.

Phoenix, Arizona, is apparently now the largest US city (and metropolitan area) without rail, since Houston opened their light rail line in 2004, though Phoenix has a system under construction at the time of this writing. Phoenix, meanwhile, is also the largest US city not served by intercity passenger rail. If the people of Phoenix identify the same factors as others about what makes a world-class city, they may be developing an inferiority complex.

Supply and demand

For analysts seeking to rationally explain why it is important to build transportation infrastructure, it is worrisome that the justifications discussed so far have little to do with demand—a concept which, under rational planning regimes and together with supply, typically underscores major transportation investments. This section, therefore, describes supply and demand in the context of assembling new infrastructure investments.

The logical outcome of any additional transportation infrastructure (e.g., roads, light rail, or bicycle paths) is increased use. Although increased use of a facility may not constitute a "good" in and of itself, consuming more or better activities at the ends of the trips is a good for those consumers (otherwise they would not be traveling to the places to which they travel). This phenomenon is illustrated using classic supply and demand curves of microeconomics in Figure 12.3. The downward-sloping demand curve, representing the willingness to use a facility given its cost, intersects an upward-sloping supply curve representing costs to travelers using a particular facility. The demand curve slopes downward because demand drops as the price increases.

To bring home this important concept, imagine consuming your favorite beverage. At $1/liter you will consume a certain quantity; at $10/liter you will consume considerably less; at $0.10/liter you will consume somewhat more. The exact shape of the curve depends on individual preferences and incomes. There may be a price at which you will not drink the beverage and seek less expensive substitutes; there may be a price so low that you are satiated, so that a lower price will not entice more consumption. The curve may be convex, or linear, or concave, or even more complex. Although the shape of the demand curve depends on how thirsty you are, it is basically downward-sloping. When we are dealing with transportation and the relatively low price of gasoline in the United States, time spent in travel is often considered the most important cost. Analysts therefore typically use the vertical axis to measure time and the horizontal axis to measure consumption (e.g., the number of trips using a facility).

The supply curve, in contrast, slopes upward. This represents the effects of congestion; the more travelers use a given facility, the higher the travel time.[b] Chapter 5 discussed network externalities and how additional travelers help improve one's travel by reducing travel time (by increasing the frequency of service or decreasing the spacing of roads). This phenomenon is true to a point (that is, it represents the left side of a U-shaped curve). Beyond a certain point, however, and in the relatively short term, additional travelers cost users more time in congestion than they save in additional service or by supporting more-direct and faster connections. This is an inherent characteristic of mature transportation systems.

Induced demand

Conventional theories of how much people travel and at what cost serve as the foundation for debates about future transportation investments. "Induced" is the label transportation-land use professionals use to signify that a particular condition (such as the amount and nature of travel demand) might be indirectly caused by another condition (such as improvements in transportation infrastructure).[c]

Imagine a set of origins and destinations, and high levels of travel demand between the two—so high that congestion is a rampant problem. To provide

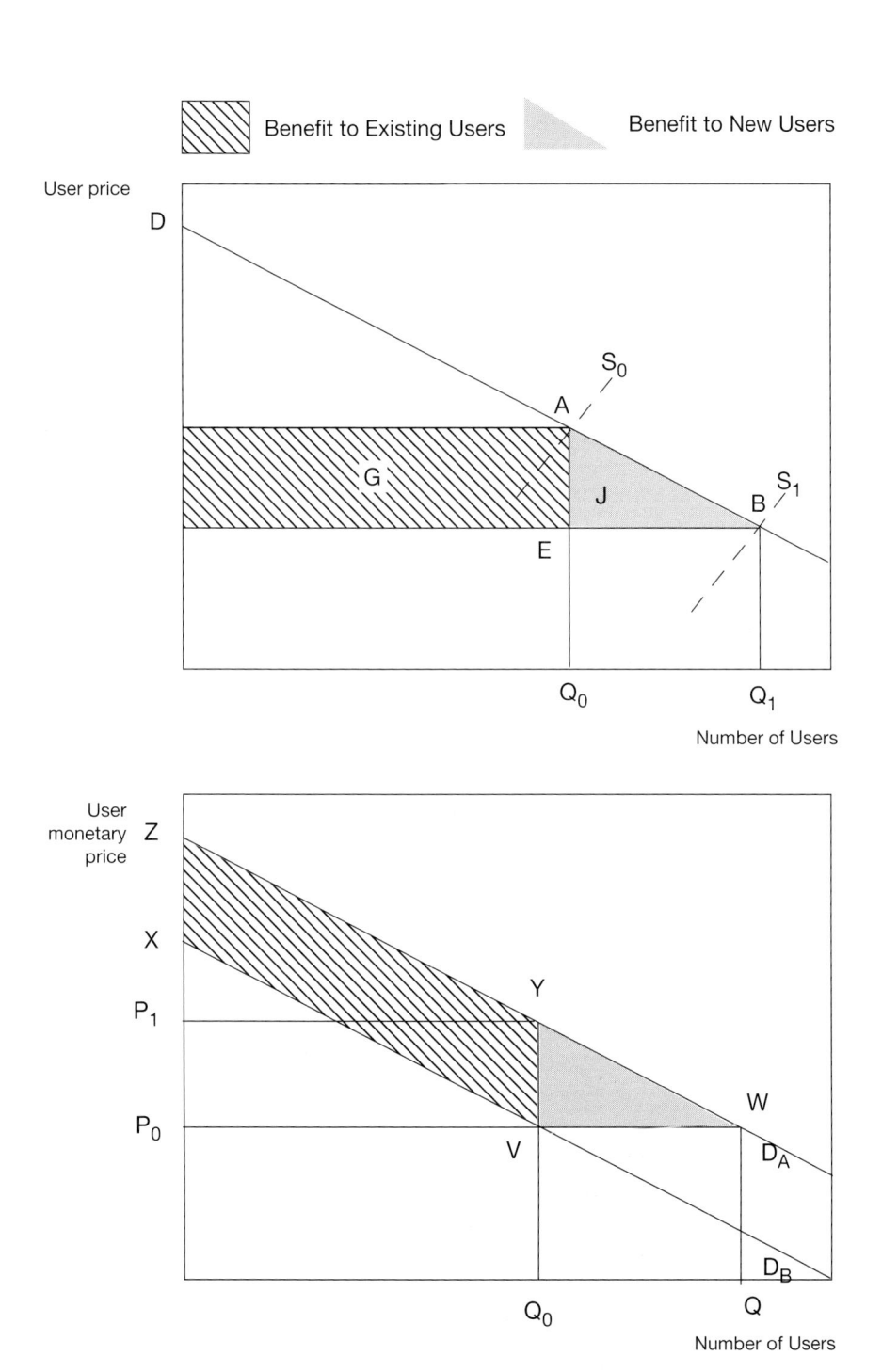

Figure 12.3 Change in consumers' surplus due to a shift in (a) the supply curve, (b) the demand curve

a real-world example, consider travel to and from Bloomington, Minnesota's beloved Mall of America (MOA) and downtown Minneapolis (Figure 12.4). Any discussion of "inducement" usually involves temporal considerations that result from improving the transportation environment. We typically assume demand between origins and destinations is fixed—such that any changes in travel that people make are changes in the short term. This is often referred to as induced traffic. [2] Any improvements, however, may have long-term consequences that affect the overall attractiveness of associated locations, thereby affecting development patterns and the overall demand. This is commonly referred to as induced demand. Box 12.1 describes the estimation of induced demand from a microeconomic perspective.

Induced traffic

Primary automobile routes for the pair of origins and destinations are Interstate Highway 35W (I-35W) and Cedar Avenue, a major arterial street. If these

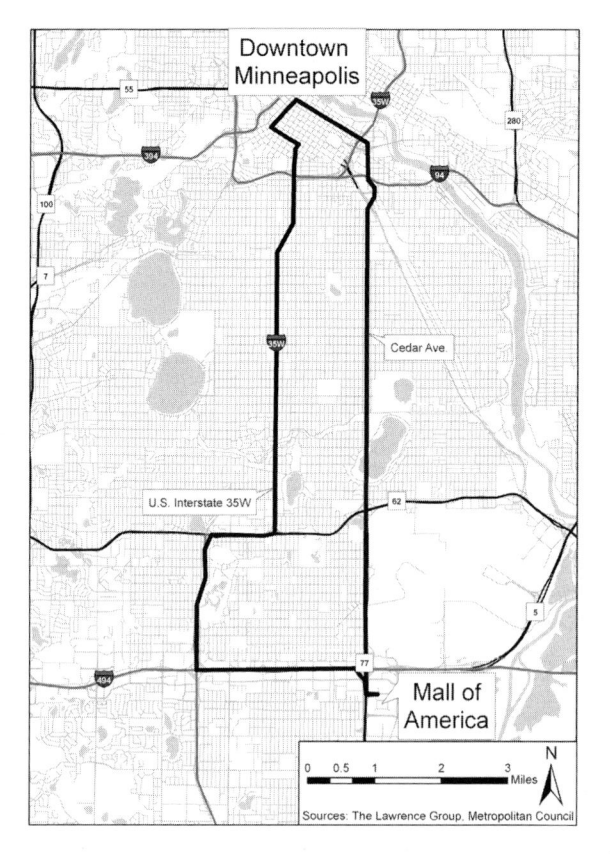

Figure 12.4 Primary travel routes between the Mall of America and Downtown Minneapolis

Box 12.1 Consumers' surplus

Expanding capacity is often represented in economics by moving the supply curve outward, from S_0 to S_1. Doing so triggers two events. First, the cost of traveling for existing travelers (travelers 0 to Q_0) drops from P_0 to P_1. In economic jargon this is known as increasing the consumer's surplus, which is nothing more than the difference between what consumers are willing to pay (referred to as the reservation price) and what they actually pay (the actual price). Change in consumer welfare can be measured by the rectangle defined the number of travelers (Q_0) multiplied by the change in price ($P_0 - P_1$) (Area G in Figure 12.3a).

Second, the number of travelers increases from Q_0 to Q_1. Those new travelers are said to be "induced" by the construction of the road. They would not travel if the cost were above P_0, but if the cost falls below P_0, they will. This is alternatively referred to as "latent" or "induced" demand, another term with which transportationists should be familiar. These new travelers also gain welfare (they are better off traveling than not traveling, otherwise they wouldn't travel). Their consumers' surplus is measured by the triangle defined by the change in travelers ($Q_1 - Q_0$) and the change in price ($P_0 - P_1$) (Area J in Figure 12.3a). The area of the triangle is one-half the base times the height, so it is $0.5 * (Q_1 - Q_0) * (P_0 - P_1)$.

Combining the user benefits (B) associated with the change for the existing and the new travelers, we have a trapezoidal region (G + J), whose area is measured using the so-called Rule of Half[a]:

$$B = Q_0(P_0 - P_1) + 0.5* (Q_1 - Q_0) * (P_0 - P_1) = 0.5 (Q_1 + Q_0) (P_0 - P_1) \quad (1)$$

where:

Q = Quantity of Trips
P = Price of travel

The same logic also holds in reverse when measuring a decrease in user benefits resulting from a decrease in capacity or an increase in cost. This decrease results from the loss of travelers and from diminished welfare for those who continue to travel. [3]

Movement along the demand curve is the first element of change in consumers' surplus. The second element is a shift of the demand curve. Improved trip quality will lead individuals to pay more (in money or time) for that trip. This quality shift can be achieved through real savings in the quality of trip (making it faster, for instance) or in certainty about the trip (the traveler is sure a trip will take 20 minutes, rather than have some probability that it will take 40 minutes).

Figure 12.3b illustrates a change in the quality of the trip which results in no change in the monetary price users pay. Again, there are two groups for which benefits must be measured, the old users and the new users. Here the demand curve shifts from D_B to D_A.

The change in benefits to old users is defined by the area ZXVY. This benefit comes about because old users would be willing to pay more money to receive a

> higher quality of service (a traveler would pay more to travel at Level of Service A, at which she would spend 25 minutes on the road, than at Level of Service B, at which she would spend 30 minutes on the road, all else equal). In practice, this is extremely hard to measure because doing so requires knowledge of the shape of the demand curves at all price levels, and generally the demand curve is really only understood in the area around the actual price (P_0). The change in benefits to new users is defined by the triangle VYW.
>
> If the price level increases, the benefit increase will be smaller. If price increases to P_1, then the net benefit will be zero and there will be no new users. The difference between P_1 and P_0 is the value of the additional quality of service.
>
> [a] Strictly speaking, the rule only holds if the demand curve is linear, which is in all likelihood an approximation, but even if it is non-linear, the rule of 1/2 gives a good estimate of the benefits.

routes are heavily used, then the relief provided by any type of improvement will reduce travel time and lower overall costs. One effect of improving a link (I-35W) is to attract traffic from other competing links (Cedar Avenue). The travel time and traffic levels on those other links decrease. A second effect is increasing traffic on feeding links (arterial connectors), which complement the link (I-35W) that was expanded. Those unimproved feeder links will thus suffer potentially worsened traffic conditions.

Capacity expansion also allows travelers to shift departure times. When congestion diminishes, travelers who had avoided a route at a particular time because of congestion can now shift to their preferred time (e.g., traveling at 8.00 a.m. rather than waiting until 9.00 a.m. to avoid traffic). Travelers who had reorganized their daily schedules to accommodate the inconvenience of congestion are able to return to a preferred schedule. Similarly, some travelers may have previously used the Light Rail Route to avoid congestion. These former transit travelers may now turn to the auto-based route take advantage of faster travel.

This phenomenon is referred to as the Law of Triple Convergence (also known as the Iron Law of Congestion), a term coined by Anthony Downs in a seminal 1962 work. [4] Because of additional "assembling," travel along a main route becomes cheaper. In response, three types of convergence—spatial, temporal, and modal—occur on the improved facility. In the case of roadways, many drivers who formerly had used alternative routes during peak hours switch to the improved expressway (spatial convergence); drivers who had traveled just before or after the peak hours start traveling during those hours (temporal convergence); and travelers who had used public transportation during peak hours switch to driving, because it has become faster (modal convergence). Among these distribution factors, only mode shifts add new auto trips.

Induced development

Changes can also occur over a longer timescale. Travelers may choose a better but more distant location for activities, taking advantage of reduced travel times to seek out better values or higher-quality opportunities at a lower cost. Or, travelers may now pursue new activities that had previously been too costly (in terms of time). All of this *may be* attributed to expanding the capacity of the transportation facility. Although an improvement will increase the demand for that infrastructure facility, there are many roads that a traveler must use; improving a single link may not dramatically shift the demand between an origin and a destination. On the other hand, each link serves thousands of different origin-destination pairs; even a small change in each market, when added together across many markets, may have a substantial effect. By reducing the costs of travel, not only are the people and firms already better off, but the increase in accessibility will likely prompt more people and firms to move to that place. Reducing cost may increase the number of buildings (density) and the occupancy of offices in Minneapolis and stores at the MOA, thereby increasing development. Assuming minimal net increases in demand, increased new development in downtown Minneapolis decreases new development elsewhere.

Furthermore, transportation markets are coupled; that is, the demand in one affects the supply characteristics of another. Reducing cost in one market will increase the demand in that market. That demand will use links shared by other markets, where the supply was not expanded. Thus, an increase in demand in some markets will increase costs and decrease trips in others. A direct benefit accrues to a market where the improved link is used or the improved link is at least a partial substitute for a link that is used. In this framework, with variable demand, many markets that do not receive the benefit directly will receive a net loss of transportation welfare.

Alternatively, there are only so many people demanding so many houses. Those houses have to go somewhere. All things equal, they will go to places with lower travel costs, so lowering travel costs will attract housing development to that place, but as a consequence there will be less housing development in other places. Transportation infrastructure organizes and reorganizes the location of activities and buildings, but improving infrastructure is much less likely to increase total development. Building a road is not going to measurably increase society's need for a greater number of housing units (though it may increase the size of houses), but will increase the relative advantage of a place. On the commercial side, transportation infrastructure may increase the total amount of development, as lowering the costs of production will increase the amount of production and may lead to economic growth. Building a road increases relative accessibility for a place, but also increases absolute accessibility for the economy as a whole.

Diamond of Assembly

Ultimately, building infrastructure sets in motion several processes [5] that lead to a cycle of effects presented in the Diamond of Assembly (Figure 12.5). The behaviors of individuals, developers or governments are shown on each axis of the diamond; effects are depicted at the corners. Suppose a community builds or expands infrastructure to further long-term goals such as expanding the capacity between origins and destinations, reducing congestion, or reducing travel times. These infrastructure investments affect one dimension of accessibility by (generally) reducing travel times. The changes in travel time

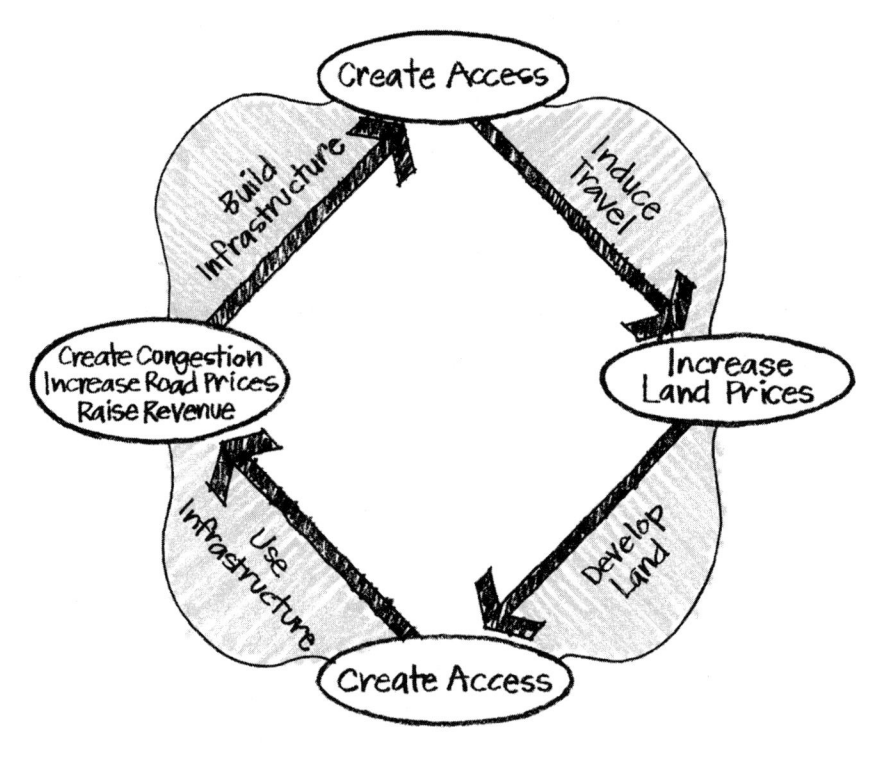

Figure 12.5 The Diamond of Assembly

Any effort to build infrastructure sets in motion several processes that lead to a cycle of effects. The behaviors of individuals, developers, or governments are shown on each axis of the Diamond of Assembly; effects are depicted at the corners. Suppose a community builds or expands infrastructure to further long-term goals such as expanding the capacity between origins and destinations, reducing congestion, or reducing travel times. These infrastructure investments affect one dimension of accessibility by (generally) reducing travel times. The changes in travel time prompt shorter-term affects: diverted trips, new trips, or longer trips. Travel that is now facilitated to the locations served by these infrastructure components increases the attractiveness of the locations, often leading to increases in land prices. These effects, in turn, initiate longer-term effects such as shifts in land use (i.e., induced development), decreases in automobile ownership, or even increases in the rate of transit use. When communities develop land, they influence a second dimension of accessibility by changing the availability of opportunities. Again, increased development produces short-term effects such as use of new infrastructure, resulting in increased capacity utilization, higher levels of congestion, or even shorter trips.

prompt shorter-term affects (described above and below) under the label of induced traffic such as diverted trips, new latent trips or longer trips. Travel that is now facilitated to the locations served by these infrastructure components increases the attractiveness of the locations, often leading to increases in land prices [6, 7] or, possibly, to increased economic productivity. [8, 9] These effects, in turn, initiate longer-term effects such as shifts in land use (i.e., induced development), [10] decreases in automobile ownership [11] or even increases in the rate of transit use. When communities develop land, they influence a second dimension of accessibility by changing the availability of opportunities. Again, increased development produces short-term effects such as use of new infrastructure, resulting in increased capacity utilization, higher levels of congestion or even shorter trips.

Debates swirling around the concepts presented in the Diamond of Assembly center on the question of whether the stated effects are large enough in scale and scope to "matter." [12, 13] The magnitude of the induced demand is referred to as the *elasticity of demand*—an index often used to describe the sensitivity of the relationship between two phenomena. Elasticity of demand is formally defined as the percentage change in quantity divided by percentage change in price, a topic more comprehensively explained in Box 12.2. The question is, given a certain increase in price (including travel time, operating costs, and accidents, as well as user charges), what would be the corresponding decrease in demand?

Roadways are the mode of transportation infrastructure receiving the most study, and subsequently, the most debate. Notwithstanding a variety of methodological challenges to such analysis, most studies confirm a relationship between capacity and demand, [14] though it is relatively inelastic. Noland reviewed several studies [15] and estimated that a 10 percent increase in lane-miles per capita was associated with between an increase of between 3 percent and 11 percent in miles traveled. In a thorough review of the literature, Cervero found that average elasticities varied by geographic level and by time frame. For example, elasticities of vehicle travel as a function of capacity ranged from 0.15 to 0.30; over a 10-year horizon they increased to 0.30 to 0.40, and from 0.40 to 0.60 across a 16-year horizon. Looking at matters at a larger unit of geography (i.e., the county), elasticities ranged from 0.32 to 0.50; and at an even larger unit of analysis (i.e., metropolitan scale), short-term elasticities were 0.54 to 0.61. Similar studies of removal of roads (increasing the time cost instead of the money cost) have shown just as increased capacity results in increased demand, capacity reduction reduces demand. [16]

When it comes to estimating elasticities for other modes, unfortunately, there has been less robust, empirical research. The main question for these other modes is: If you build it (i.e., additional transit lines, bike paths) will they (users) come? Studies examining transit, walking or cycling, however, are even more fraught with the problems afflicting roadway studies (lack of good use data, cross-sectional data). [17] There is a dearth of research focused

> **Box 12.2 Elasticity**
>
> When referring to the elasticity of various goods, it is common to differentiate between elastic and inelastic goods, with a value of −1 used to distinguish them. An elasticity of the demand curve nearer to 0 than to −1 means the percentage change in quantity is less than the percentage change in price, and the curve will be steep and inelastic; under these conditions, it will take a large change in price to affect demand (Figure 12.6). Goods characterized by such demand curves tend to be things that are necessities for consumers in their daily lives. The price of gasoline is perhaps the most often talked about transportation elasticity and, over the years, we have found that its price is largely inelastic.
>
> Alternatively, if the elasticity of demand takes on a value nearer to −1 than to 0, that suggests the percentage change in quantity is greater than the percentage change in price, and the curve will be relatively flat and elastic; small price changes will have large effects on demand. Elastic products may be readily available in the market, but a person may not necessarily need them in his or her daily life; dining out in a restaurant is an example.
>
> In the context of induced travel and demand, elasticity measures are usually presented as the percentage change in travel demand given a 1 percent increase in roadway capacity (or some other measure of supply-side improvement). (So the signs will be different than price elasticities.) For example, an elasticity of 0.5 signifies that for every 1 percent increase in roadway capacity, there is a 0.5 percent increase in traffic—that is, roughly half of the added capacity gets absorbed by additional traffic. Discussions of elasticity are not limited to induced demand, however. Similar findings can be inferred from any analysis looking at, for example, price premiums on homes closer to the central business district (see the example presented in Chapter 3) or changes in transit demand prompted by increases in fares (Chapter 13).

on "before and after" analysis of induced use that results from the construction or improvement of transportation facilities. [18, 19] Transit studies looking at elasticity are seemingly obsessed with looking only at direct price (monetary) elasticities, and tend to neglect the question of how changing levels of service may decrease other costs. Where the two factors have been differentiated, ridership has been found to be more sensitive to changes in the level of service attributes than to changes in fares.

Forecasting and the principal-agent problem

The accuracy of forecasts is an important consideration in any decision to expand (or contract) infrastructure. Forecasts of demand (along with estimates of consumer's surplus) help inform the benefits side of a benefits-cost calculation, while forecasts of costs determine the costs side. There are opportunities

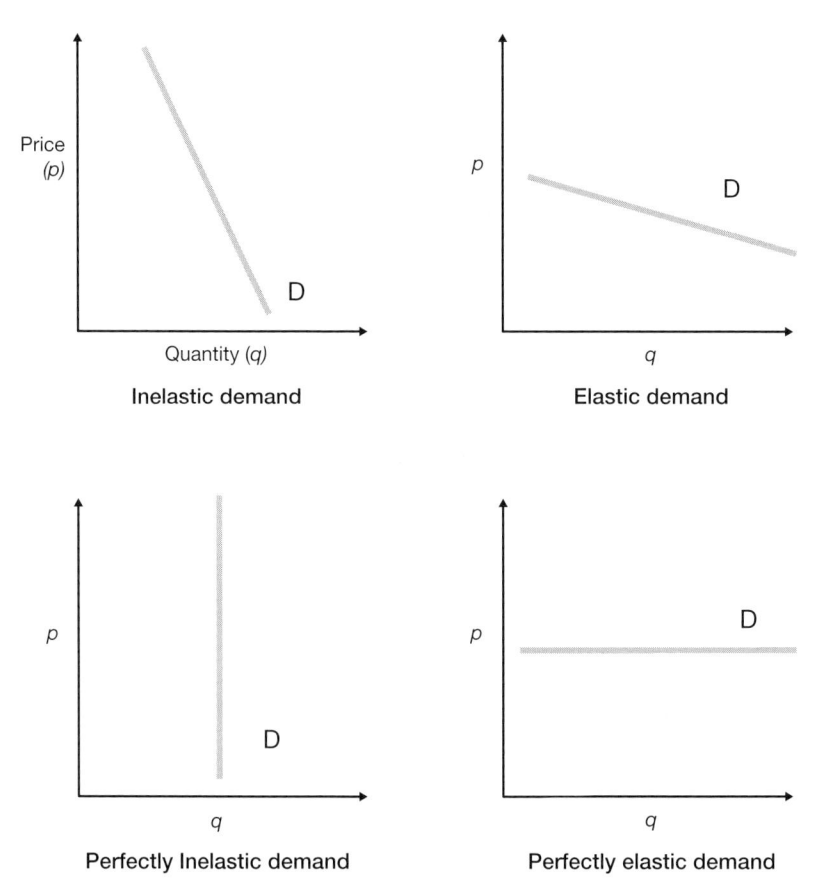

Figure 12.6 Examples of elasticity curves

for bias in any forecast, and some researchers—including Bent Flyvbjerg, [20–28] John Kain, [29, 30] and Martin Wachs [31]—have taken issue with planners and their ethics, discussing situations that lead to clearly unethical outcomes in forecasting and infrastructure planning.

Flyvbjerg, in a series of papers and books, has examined the mis-forecasts and misuse of forecasts in public works, including both road and rail projects. He shows that rail forecasts have been systematically biased (corroborating controversial work by Donald Pickrell [32] and Jonathan Richmond [33]) while road forecasts (mostly pertaining to non-tolled roads) have not. Even the late anti-highway urbanist Jane Jacobs [34] acknowledged the problem of light and heavy rail lines missing their goals and soaking up resources that could have supported buses running on city streets. She was especially disdainful of metropolitan governments running bus systems that provide excess routes in low density suburban areas while starving the central city of service, a phenomenon noted by others as well. [35] Toll road forecasts have faced the same problem of excessive optimism as rail forecasts. [36]

One need only consider the incentives for forecasters (or their employers) to understand why this happens. A high forecast is likely to lead to project funding and implementation; a low forecast is unlikely to do that. When agencies compete with one another for scarce funds (either from federal transit funding programs or, in the case of toll roads, from capital markets), there is an incentive to lean the forecast one way. For the forecaster, there is no retribution for poor forecasts. A high forecast implies losses greater than expected or profits lower than expected, but that burden falls on the funder. First, the forecaster can always claim conditions have changed between the time the forecast was made and the project opened—and this is likely to be the truth. Second, the long delays between forecast preparation and the period being forecasted will result in most forecasters having moved on to new projects or employers. The lack of documentation of the numerous detailed procedures and assumptions in most forecasting exercises makes them candidates for the *Annals of Irreproducible Results*.[d] There are few requirements for detailed peer review of forecasts or for the transparency required in scientific analysis.

Economists often talk about the principal-agent problem. This arises when information is incomplete and asymmetric, and the interests of the two parties (the principal and the agent) are not aligned. The information and incentives available to the agent (a forecaster) differ from those affecting his employer (a transportation department or other public agency) and from those of the public they serve.

The problem with forecasting is a matter that will not easily be resolved by the introduction of better forecasting methods. Rather, perhaps forecasting should be downplayed in importance, and facts "on the ground" be given more significance. Forecasting is not necessary to identify today's heavily used routes (transit or highway), and there are enough needs apparent today that it is not really necessary to go looking for problems. Perhaps the best suggestion comes from Jacobs, who endorsed the Toronto Transit Commission's "Subways Second Strategy" [37] of building subways where the demand already exists, rather than where planners might want it to be. She posited that, if a rail line is being proposed, a high-quality bus service should be run on that route; if it does well, the bus service should be expanded. Only when the bus ridership is high enough that rail will be more economical and provide better service should a new rail line be constructed. Rather than looking at available rights-of-way and then seeing where a line could be run, the service should follow the demand.

Network design and growth

Ideally, from a microeconomic perspective, networks are designed to maximize the (positive) difference between benefits and costs. This is often assessed on a project-by-project basis using benefit-cost analysis. However, designing the combination of links that should be added is much more difficult. This is largely because of the combinatorics involved. There are many possible links; there are many more possible link combinations. For instance, if there are four possible links (A, B, C, D), the 15 possible combinations are (A), (B), (C), (D), (A,B), (A,C), (A,D), (B,C), (B,D), (C,D), (A,B,C), (A,B,D), (A,C,D), (B,C,D), (A,B,C,D). Evaluating 15 combinations would be feasible, but assessing each of the millions or billions or trillions of combinations possible on a real urban network is, for practical purposes, impossible. There are a variety of algorithms that simplify the analysis, [38] and don't require testing every combination. In general, the network design problem takes a large part of the network as given, and considers which of a set of links, or expansions to existing links, to add to that network, subject to a budget constraint. Some additions complement both the existing network and other proposed links while competing with other existing or proposed links. Competing with existing congested links is good for users, as it reduces travel time; competing with uncongested links, however, draws away resources that could be spent else-where. Given a system-wide objective, adding new links that compete with each other is unlikely to be a good strategy. Solutions to the network design problem (NDP) may determine not just which set of links to build, but the optimal order in which to build them.

Observing the existing transportation network leads one to believe that the network is not truly optimal from any perspective, and that in practice, the network design problem has not been solved. Instead, think of a network growth process occurring. Analysis of network growth is concerned with understanding the rules that describe what links and nodes in the network are added/deleted or expanded/contracted. These rules are not optimal in the sense of maximizing system efficiency, but have been judged by the jurisdictions that use them to be expedient, and perhaps to satisfy the need for a simple decision process. Rules such as expanding links when the traffic per lane exceeds some threshold are typical. [39] More developed jurisdictions typically have more sophisticated and complex procedures for assessing alternatives—though in some senses, the relative maturity of networks in those areas renders such sophistication less necessary. Younger and faster-growing areas, in contrast, have simpler rules, even though the decisions made in these areas will have greater impacts over the long term.

There are several things to note about network growth processes. First, the hierarchy of roads (discussed in depth in Chapter 11) is designed with and embedded in many engineering policies (such as the AASHTO Green Book [40] in the United States); however, such hierarchy would be present even if were not adopted in policy guidebooks. [41] This is one example illustrating the challenge of assessing causality (as discussed in Box 12.3).

Managing transportation systems

Strategies to gain additional throughput without building additional infrastructure involve making better use of the infrastructure that has already been laid down. Technological means to this end range from the simple (using "highway helper" tow trucks to clear incidents more quickly) to the complex

Box 12.3 Causality

Prominent in any study of induced travel demand is the issue of causality. Might growth in traffic or other types of behavior induce infrastructure investments every bit as much as infrastructure investments induce growth in traffic? The land use-transportation community is learning, not surprisingly, that findings from cross-sectional research do not tell the whole story of travel demand. Analyzing a single policy or environmental change without fully capturing other important influences may lead to erroneous conclusions; in some cases such an analysis may even overstate outcomes of a particular policy or environmental change. Trying to unravel this decision-making web by isolating the specific role of infrastructure facilities is a complex endeavor.

Put another way—as any reliable textbook on statistics points out—correlation does not mean causation. It is important to distinguish between observed correlations of use and new investments from the claim that new investment induces use. The majority of previous work on the subject has not adequately differentiated the two positions. For example, as we learned in Chapter 3, residents (or families) often select locations to match their desires to engage in certain behaviors, such as cycling. This is an option that home buyers prioritize in choosing their home location. This finding suggests, for example, that differences in rates of cycling between households in different areas of the city with different levels of access to cycling facilities should *not* be credited solely to the existence or proximity of the facilities; the differences should be attributed to self-selection. In other words, people who are likely to cycle choose to locate in a given neighborhood where they have a better chance of cycling.

These considerations are particularly vexing for researchers attempting to shed light on the debates and discussions taking place around causality. Proving statistical association is not the same as proving causality; in fact, one can never prove causality. Two phenomena can move together due to chance, or there could be bi-directional causality. There is no satisfactory statistical test for causality. What is the researcher of travel behavior left to do? How can one reliably say that new transportation investments induce use?

Although one can never prove causality, social scientists have provided several guidelines that help move us closer to reliably inferring causality. John Stuart Mills, who suggested that at least three conditions need to be met, reportedly first provided some of the most relied-on guidelines:

1. *Concomitant variation* is the extent to which a cause X and an effect Y occur together or vary together in the way predicted by the hypothesis under consideration (e.g., rates of cycling and the presence of a cycling facility).
2. The *time order of occurrence condition* states that the causing event must occur either before or simultaneously with the effect; it cannot occur afterwards (e.g., the cycling facility was constructed before heightened levels of cycling were observed).
3. The *absence of other possible causal factors* means that the factor or variable being investigated should be the only possible causal explanation. This condition is the most difficult to satisfy. For example, an additional causal explanation leading to heightened levels of cycling may be a result of what is commonly referred to as the "Lance Armstrong factor." This refers to the overall cycling boom in the United States in 1999 and 2000—and hence increased rates of bicycle commuting—attributed to the increased popularity American Lance Armstrong brought to cycling after winning his first of seven Tour de France titles.

(using ramp meters to smooth the inflow of traffic and thereby diminish the amount of braking at freeway merges, maximizing capacity utilization). A common feature is the effective use of information. Knowing the state of the transportation system in real time is crucial for many applications. These techniques were initially called Transportation System Management (TSM) or Transportation Supply Management, but many technology applications have been collected under the Intelligent Transportation Systems (ITS) rubric. Regardless of their name, they aim to make better use of infrastructure, and can be thought of as increasing either vehicle-carrying or person-carrying capacity. A related set of tools, dubbed Transportation (or Travel) Demand Management (TDM), aims to reduce vehicle demand and is discussed in the next chapter. Some tools (e.g., High Occupancy Vehicle Lanes) aim to do both.

Major Transportation System Management Tools:

- Access Control (driveway coordination/removal);
- High Occupancy Vehicle Lanes;
- Park and Ride Lots;
- Bus Bays;
- Reversible Lanes;
- Traffic Signal Coordination;
- Ramp Metering (See Box 13.1);
- Variable Message Signs;
- Freeway Service Patrols/Incidence Detection and Clearance;
- Events Management;
- Freight Management;
- Electronic Toll Collection.

One advantage of TSM tools is that they are usually uncontroversial (ramp metering notably excepted). A disadvantage is that, with their relatively modest effects on traffic, TSM has been dubbed "too small to matter." They are a rational means to better manage mature transportation systems, but are unlikely to create radical improvements in people's daily travel. Perhaps the most clearly effective is the replacement of manual tollbooths with electronic toll collection systems, and over the past ten years, these have been widely deployed on toll roads and bridges. Fears about privacy have largely been overcome (unidentifiable smart cards can be used), though some toll authorities are lagging in deployment because of union/patronage issues and other forms of bureaucratic inertia. Freeway service patrols and other incident detection and clearance programs are also clearly effective with little opposition (aside from private tow truck companies who may feel usurped).

Electric plexus

> We used to tell people that if you wanted to know where we operated you only had to look at the roads. If the road was paved that was private power territory. If it was a dirt road, that was our own territory. [42]

Physical transportation systems are important networks on which everyone relies, but they are not the only networks that affect place and plexus. The idea of anything other than private power may seem strange to residents who pay the utility bill monthly to a private corporation, but many places are not profitable to serve with electricity, and would otherwise be bypassed on the paved road to modernization. The New Deal solution to this dilemma was to support cooperatives owned by customers of the utility in providing power. In 1935, the Rural Electrification Administration (REA) was created to help cooperatives electrify the American farm. In 1930, before the REA was created, only one in ten rural homes in the United States had electricity, while by 1940 nearly a quarter did, and presently the number is near 100 percent.[5]

The lack of power in rural communities was clearly seen, along with unpaved roads, as an impediment to development. The missing elements of connectivity in rural areas handicapped them compared with cities, which had been electrifying since the late nineteenth century. The underdeveloped rural plexus is one factor that led to rural depopulation and increased urbanization throughout the twentieth century. Pictures of urban life in the early twentieth century are dominated by the image of the streetcar—a transportation technology permitted by the electric plexus.

Streetcars were introduced in the United States in the 1880s and 1890s, at the same time as their enabling technology, electricity. Steam trains had been running on railroads since the 1820s, and at grade and on elevated lines in cities, but their operating characteristics made them inferior to electric trains for on-street service. For one, electric trains did not require a separate engine car or coal storage, since power generation was accomplished at a central facility rather than onboard the train itself. In addition, because electricity was generated off-site, pollution was restricted to a single source rather than being spread up and down every street. Electric street lighting came to prominence in the same period, replacing gas lighting because of its lower cost and higher quality illumination. The lights allowed city businesses to stay open longer, creating even more intense opportunities for interaction. Electricity changed activity patterns when installed. People adapted, a few traditional communities such as the Amish aside. A century of highly reliable electricity led to a century of adaptations and systems built upon systems that assumed the presence of electricity.

Stephen Doheny-Farina [43] describes from first-hand experience what happens when electricity is suddenly removed. Reading his account of the 1998 ice storms that disrupted electrical service in the northeastern United States and Canada for weeks, one realizes our society's dependence on the various grids that support urban life—besides the transportation systems on which we have been concentrating this book: electricity, gas, water, sewers, telecommunications.

The story of the northeast ice storms suggests that because of the dependency society has created by relying on electricity, society is perhaps worse off without it than if electricity had never existed in the first place. The high-quality service provided by electricity has led us to disconnect the precursor technologies, which could have served as a fall-back, such as wood- or gas-burning stoves, candles, or oil lamps.

The City of New Orleans, Louisiana was completely evacuated in the wake of Hurricane Katrina—not because all the homes were uninhabitable (many were not habitable, of course, but some were on higher ground and not destroyed by the hurricanes or floods)—rather, the city was evacuated because the networked infrastructure supporting those houses had become unreliable if not outright hazardous. Perhaps any one network could have been substituted (if the water system fails, bottled water can be trucked in; if power is out, back-up generators can be used; if bridges collapse, ferries can be pressed into service), but with all of the supporting networks down, the physical plexus on which New Orleans depended had collapsed, and much of the city became a ghost town.

All plans and policies are shaped by the experiences of those who create them. [44] The collapse of infrastructure is a catastrophic experience for those who live through it, or merely observe it from afar. Lessons can be learned that will minimize the effects of the catastrophe the next time one occurs. However, sometimes the wrong lessons are learned: after 9/11, the United States responded with a focus on terrorism as the primary threat facing the nation. Natural disasters such as blizzards and ice storms, earthquakes, and most notably hurricanes, seemingly received second billing.

Coruscant, accessibility and development

Coruscant is the city-planet that serves as capital of the Galactic Republic in George Lucas's Star Wars saga; nearly the entire planet is covered by structures (streets, squares, or buildings).[f] Coruscant faces some congestion, but it also has flying cars, so traffic is not confined to the two dimensions of the planet's surface. The city has towers that rise far above the surface, seen through the window as a backdrop in the Jedi Temple, but in lower rent districts close to and below the surface are many of the traditional signs of a red-light district such as whiskey and wagering.

Is Coruscant a model of the future, or merely a vividly drawn backdrop for a story about the ambiguity of the sources of good and evil? Applying what we have learned about accessibility, we propose relatively straightforward tenets when considering such matters, comprising networks, accessibility, and development. First, by definition, network expansion that reduces travel time creates accessibility. Second, accessibility creates demand for development. Third, land development creates travel demand, creating resources for building more networks. Fourth, development increases accessibility of its own accord (refer to Figure 12.5, the Diamond of Assembly, for an illustration of the logic).

So, if the model is correct, why don't we live in Coruscant? One answer is we just don't live there yet; the effects of feedback loops may take time to materialize. A second answer is that some of us do; a visit to Hong Kong,

Tokyo, Sao Paulo, Calcutta, London, or Manhattan suggests a crowded world where the accessibility feedback loop has flourished. Third, our formulation of the model has not introduced the limiting factors (at least in the short run) of congestion and perhaps pollution. Fourth, there are other factors that limit development, including population size, food supplies, energy availability, etc. Although the process still works under limits, it works more and more slowly until development seems to have stopped, waiting for the constraints to be relaxed. Finally, people have preferences for attributes of lower density, which checks the process.

Perhaps a better explanation for the self-limiting phenomenon can be discovered if we think in microeconomic terms. The demand curve for a typical good is downward-sloping; the more something costs, the less will be consumed. However, the demand for a network good rises with the number of members of the network, as user of the network creates a positive externality for other users.

Figure 12.7 constructs a revealed demand curve for positive network externalities. The network is more valuable the more members it has, the more units are sold, the greater the accessibility, etc. With only one consumer,

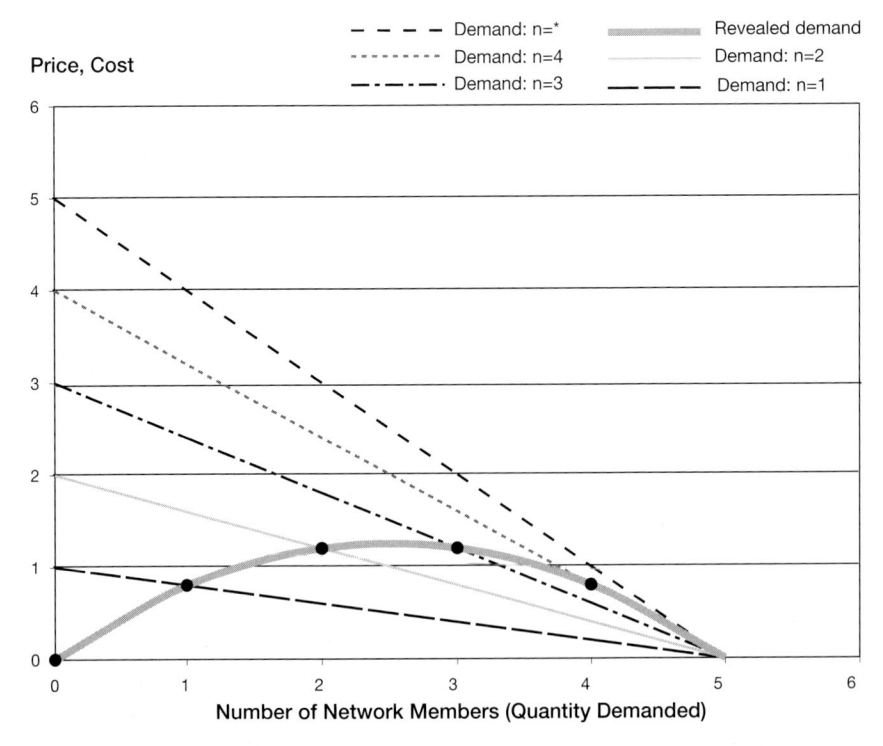

Figure 12.7 Construction of revealed demand (fulfilled expectations) curve with positive network externalities

($n = 1$), the network is not particularly valuable, so the implicit demand at $n = 1$ is low; demand is higher at $n = 2$, and higher still at $n = 3$, etc. Drawing a line between the number of consumers (n) and the implicit demand curve at that number (D_n) traces out an approximately parabolic shape: the revealed demand curve, or the set of prices that the nth consumer would actually pay to join a network that would sustain n consumers. This model shows that willingness to pay (the revealed demand curve) to join a network rises with the size of the network, up to a point. At that point, the network is ubiquitous, and the value of additional members is negligible.

In 1962, Richard Meier published *A Communications Theory of Urban Growth*, [45] which posited that it is transactions of both information and goods that create the need for cities. In short, we can rewrite the Sun Microsystems slogan "The Network is the Computer" to "The Network is the City." Our world contains many local networks that we call cities, as well as one global economy represented by many megalopolises, with even more primary business districts. The cinematic vision of a worldwide Coruscant has not yet been manifested on Earth, but the world's economy has become integrated to the point where we already inhabit a global city that operates 24 hours a day and never sleeps.

Wrap up for assembling

> But I still love my car.
> Environmentalist Linda Powell, played by Kyra Sedgwick, counters the proposal by transportation planner Steve Dunne (played by Campbell Scott) for a Supertrain in Seattle in the movie *Singles*

The question of what type of transportation infrastructure a community such as Seattle should assemble to address the travel behavior needs of its residents is a complex one. Sometimes, history can inform—if not guide—such deliberations, as in the parable of mountains. The nominal objective of most government planning agencies is to maximize welfare of their respective citizens and pursuing various means to achieve that objective (e.g., deploy transportation infrastructure to maximize welfare). Analogous to mountain climbing, governments aim to get as high in the welfare category as possible, as quickly as possible.

From a modal perspective, assume there are two mountains, Mount Auto and Mount Transit, both of which are very tall and separated by a deep valley. The peaks of both mountains are obscured by clouds. By climbing one of the mountains, we travel further away from the valley that separates them, and thus further from the peak of the other (not only do we have its height to overcome, we must retrace our steps through the valley below). We have no assurances about the true height of either mountain, only forecasts that are little better than astrology. Some prophets who preach "Pedestrian Friendly

Design" or "Transit Oriented Development" warn that atop Mount Auto are dragons and monsters such as Global Warming, Environmental Destruction, Alienation From Community. Prophets of Mount Auto, for their part, warn us against unwise use of government resources and the dangers to individual rights posed by socialist societies.

Beginning in the 1920s, western society had already made good strides up Mount Transit (though our progress was in large part because we had not yet fully learned of Mount Auto's alluring presence). In the era of a single downtown and high-density cities, the United States reached the summit of Mount Transit, and from that vantage point determined Mount Auto was higher (Figure 12.8). The United States, and cities in many other countries for that matter, has seemingly retraced its steps and for the past three-quarters of a century has been steadily climbing Mount Auto ever since. The US is now fully committed to that mountain and has long surpassed the peak of Mount Transit.

There are those who now claim, however, that because of complex tectonic activity, advances in technology, or changing societal preferences Mount Transit may now be taller than it once was. These factions are starting to retrace their steps down Mount Auto by building light rail systems (or advocating monorails) in cities of all shapes and sizes (some in relatively dense downtowns such as Seattle, some in places where Mount Auto still stands tall!). Perhaps the clouds will eventually part, revealing a bridge between the two peaks—or, as seems more likely, we will see a third and taller mountain in the distance, and start building infrastructure to climb its peak.

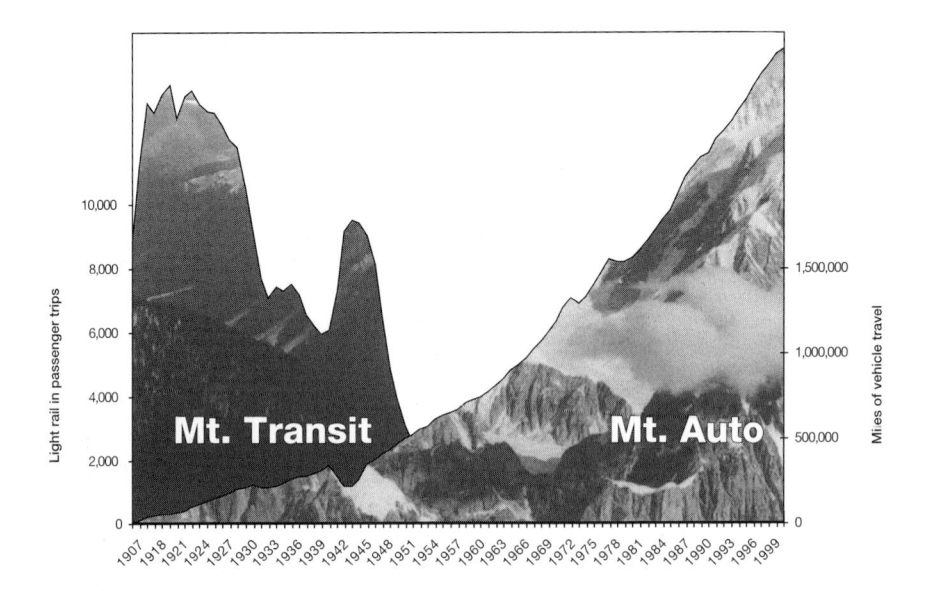

Figure 12.8 Mount Transit and Mount Auto

Notes

a   There is no shortage of other cavalierly mentioned justifications. Some liken the pursuit of light rail construction and associated technology with macho images, phallic imaging, or a sort of a twentieth-century bureaucratic idolatry. The federal government has made billions of dollars available for light rail, for example, and cities have great incentives to "bring home the bacon"; hence, these individual cities would wrestle as quickly to build monoliths, were federal funds available. A pointed difference, however, is that pyramids, unlike light rail, don't have operating costs and may be a cost effective use of public funds.
b   For further discussion of this phenomenon, see section on queuing in Chapter 13.
c   For our purposes here, the term "induced" is synonymous with the term "latent" introduced in Chapter 3.
d   The *Annals of Irreproducible Results* should not be confused with the *Journal of Spurious Correlations* (www.jspurc.org/), which represents a more legitimate venue for publishing peer-reviewed research findings.
e   See cite on Rural Electrification Administration, available at: http://en.wikipedia.org/wiki/Rural_Electrification_Administration (accessed November 29, 2006).
f   Presumably on Coruscant food and other natural resources are imported from other planets or are grown in buildings in agricultural factories.

References

[1] Jack, D., 'London Routemasters Approach End of the Line', *Buses International*, March 2004: 1–3.
[2] Lee, D.B.J., Klein, L., and Camus, G., 'Induced Demand and Induced Traffic', *Transportation Research Record*, 1999, vol. 1659: 68–75.
[3] Neuberger, H., 'User Benefit in the Evaluation of Transport and Land Use Plans', *Journal of Transportation Economics and Policy*, 1971, vol. 5 (1): 52–77.
[4] Downs, A., 'The Law of Peak-hour Express-way Congestion', *Traffic Quarterly*, 1962, vol. XVI (3 July): 393–409.
[5] Button, K., 'What Can Meta-analysis Tell us about the Implications of Transport?', *Regional Studies*, 2005, vol. 29 (6): 507–517.
[6] Mohring, H., 'Land Values and the Measurement of Highway Benefits', *The Journal of Political Economy*, 1961, vol. 69 (3): 236–249.
[7] Ryan, S., 'Property Values and Transportation Facilities: Finding the Transportation–Land Use Connection', *Journal of Planning Literature*, 1999, vol. 13 (4): 412.
[8] Boarnet, M., 'Highways and Economic Productivity: Interpreting Recent Evidence', *Journal of Planning Literature*, 1997, vol. 11 (4): 476–486.
[9] Boarnet, M., 'Spillovers and the Locational Effects of Public Infrastructure', *Journal of Regional Science*, 1998, vol. 38 (3): 381–400.
[10] Wheaton, W.C., 'Residential Decentralization, Land Rents, and the Benefits of Urban Transportation Investment', *The American Economic Review*, 1977, vol. 67 (2): 138–143.
[11] Holtzclaw, J., *Using Residential Patterns and Transit to Decrease Auto Dependence and Costs*, San Francisco, CA: Natural Resources Defense Council, 1994.
[12] Cervero, R., 'Induced Travel Demand: Research Design, Empirical Evidence, and Normative Policies', *Journal of Planning Literature*, 2002, vol. 17 (1): 3–20.

[13] Ruiter, E.R., Loudon, W.R., Kern, C.R., Bell, D.A., Rothenberg, M.J., and Austin, T.W., *The Vehicle-miles of Travel-urban Highway Supply Relationship*, Washington, DC: Transportation Research Board, National Cooperative Highway Research Program, 1980.
[14] Kiefer, M. and Mehndiratta, S.R., *If We Build it, Will They Really Keep Coming? A Critical Analysis of the Induced Demand Hypothesis*, Preprint paper, Transportation Research Board Annual Meeting, 1998.
[15] Noland, R., 'Relationships Between Highway Capacity and Induced Vehicle Travel', *Transportation Research A*, 2001, vol. 35 (1): 47–72.
[16] Cairns, S., Hass-Klau, C., and Goodwin, P.B., *Traffic Impact of Highway Capacity Reductions: Assessment of the Evidence*, London: Landor Publishing, 1998.
[17] Krizek, K.J., Handy, S.L., Forsyth, A., and Clifton, K., 'Explaining Changes in Walking and Bicycling Behavior: The Transportation Researcher's Full Employment Act', *Active Communities/Transportation Research Group, Working Paper 07–02*, 2007.
[18] Barnes, G. and Krizek, K.J., 'Estimating Bicycling Demand', *Transportation Research Record*, 2006.
[19] Birk, M. and Geller, R., 'Bridging the Gaps: How the Quality and Quantity of a Connected Bikeway Network Correlates with Increasing Bicycle Use', *Proceedings of the 85th Annual Meeting of the Transportation Research Board*, 2005.
[20] Flyvbjerg, B., 'The Dark Side of Planning: Rationality and Realrationalitat', in S. Mandelbaum, L. Mazza, and R. Burchell (eds.), *Explorations in Planning Theory*, New Brunswick, NJ: Center for Urban Policy Research Press, 1996, pp. 383–394.
[21] Flyvbjerg, B., *Rationality and Power: Democracy in Practice*, Chicago, IL: University of Chicago Press, 1998.
[22] Flyvbjerg, B., 'Delusions of Success: Comment on Dan Lovallo and Daniel Kahneman', *Harvard Business Review*, 2003: 121–122.
[23] Flyvbjerg, B., *On Measuring the Inaccuracy of Travel Forecasts: Methodological Considerations*. Manuscript submitted for publication, 2004.
[24] Flyvbjerg, B., Bruzelius, N., and Rothengatter, W., *Megaprojects and Risk: An Anatomy of Ambition*, Cambridge: Cambridge University Press, 2003.
[25] Flyvbjerg, B., Holm, M.K.S., and Buhl, S.L., 'Cost Underestimation in Public Works Projects: Error or Lie?' *Journal of the American Planning Association*, 2002, vol. 68 (3): 279–295.
[26] Flyvbjerg, B., Holm, M.K.S., and Buhl, S.L., 'How Common and How Large are Cost Overruns in Transport Projects?' *Transport Reviews*, 2003, vol. 23 (1): 71–88.
[27] Flyvbjerg, B., Holm, M.K.S., and Buhl, S.L., 'How (In)accurate are Demand Forecasts in Public Works Projects? The Case of Transportation', *Journal of the American Planning Association*, 2004, vol. 71 (2): 141–146.
[28] Flyvbjerg, B., Holm, M.K.S., and Buhl, S.L., 'What Causes Cost Overrun in Transport Infrastructure Projects?' *Transport Reviews*, 2004, vol. 24 (1): 3–18.
[29] Kain, J.F., 'Deception in Dallas: Strategic Misrepresentations in Rail Transit Promotion and Evaluation', *Journal of the American Planning Association*, 1990, vol. 56: 185–196.
[30] Kain, J.F., 'The Use of Straw Men in the Economic Evaluation of Rail Transport Economics', *American Economic Review*, 1992, vol. 82 (2): 487–493.

[31] Wachs, M., 'Ethical Dilemmas in Forecasting for Public Policy'. *Public Administrative Review*, 1982, vol. 42 (2): 562–567.
[32] Pickrell, D.H., *Urban Rail Transit Projects: Forecasts Versus Actual Ridership and Cost*, Washington, DC: US Department of Transportation, 1990.
[33] Richmond, J.E.D., *New Rail Transit Investments: A Review*, Cambridge, MA: John F. Kennedy School of Government, 1998.
[34] Jacobs, J., *Dark Age Ahead*, New York: Vintage Press, 2004.
[35] Taylor, B., 'Unjust Equity: An Examination of California's Transportation Development Act', *Transportation Research Record*, 1991, vol. 1297: 85–92.
[36] Bain, R., *Traffic Forecasting Risk: Study Update 2004*, Infrastructure Division of Standard and Poors, 2004. Available at: www.people.hbs.edu/besty/projfinportal/S&P_Traffic_Risk_2004.pdf (accessed May 10, 2007).
[37] Hall, J., *Subways Owe Big Debt to Streetcars*, 2002. Available at: http://transit.toronto.on.ca/archives/data/200206102338.shtml (accessed June 10, 2002).
[38] Magnanti, T.L. and Wong, R.T., 'Network Design and Transportation Planning: Models and Algorithms', *Transportation Science*, 1984, vol. 18 (1): 1–55.
[39] Montes de Oca, N. and Levinson, D., 'Network Expansion Decision-making in the Twin Cities', *Transportation Research Record*, 2006, vol. 1981: 1–11.
[40] American Association of State Highway and Transportation Officials (AASHTO), *Green Book: A Policy on the Geometric Design of Highways and Streets*, 2001.
[41] Levinson, D. and Yerra, B., 'Self Organization in Surface Transportation Networks', *Transportation Science*, 2006, vol. 40 (2): 179–188.
[42] Interview with Bill Sager from Jay County, Indiana, Rural Electric Membership Corporation, 1990.
[43] Doheny-Farina, S., *The Grid and the Village*, New Haven, CT: Yale University Press, 2001.
[44] Garrison, W.L. and Levinson, D.M., *The Transportation Experience: Policy, Planning, and Deployment*, New York: Oxford University Press, 2005.
[45] Meier, R.L., *A Communications Theory of Urban Growth*, Cambridge, MA: MIT Press, 1962.

Chapter 13

Operating

"In no other major area are pricing practices so irrational, so out of date, and so conducive to waste as in urban transportation."

William S. Vickrey (1963)

Ken Livingstone (nicknamed 'Red' Ken because of his leftist leanings) became London's first directly elected Mayor in 2000. He had made transport a central issue in his campaign. As Mayor, although he may have lost some political capital in transportation circles by decommissioning the beloved double-decker Routemaster buses discussed in Chapter 12, he gained significant political capital (and some monetary capital) through his persistence, and subsequent success, in leading one of the great transportation experiments in one of the great world cities.

Livingstone's campaign stressed the need to ease traffic congestion in central London by persuading people to switch from private cars to public transport. He promised to accomplish this through pricing—specifically, by introducing a congestion charge while at the same time dramatically increasing the number of buses on London roads. Under the scheme, private car drivers entering central London initially paid a daily fee of five pounds (about eight US dollars); later, the fee increased to eight pounds. The congestion-charging scheme was at the heart of a larger transport strategy designed from the outset to tackle four key transport priorities for London: reducing congestion; improving bus services; improving journey time reliability for the remaining road-users; and making the distribution of goods and services more reliable, sustainable and efficient. In addition, it was also designed to raise substantial funds for London's transport system.

Since the congestion charge was commissioned in February 2003, it has been met with generally positive reviews. Livingstone was re-elected in May of 2004, at least suggesting the populace tolerates it. The charge succeeded in reducing street traffic an estimated 20–30 percent. From a land use-transportation perspective, Livingstone's program is revolutionary in its intent, scope, and success in allocating a scarce resource—roadway space—by charging for its use.

This chapter describes how government agencies operate by balancing matters of supply and demand. Governments both ration the use of transportation infrastructure (a relatively short-term matter) and regulate the use of land (a relatively longer term matter) using two strategies, The two time scales, coupled with the two strategies, form the basis for this book's final diamond, the Diamond of Operation (Figure 13.1). The theoretical underpinnings of long- and short-term strategies are similar but not identical. We begin this chapter by examining short-term allocation of scarce road space through queuing, the congestion many travelers face each day. We consider other short-term strategies, such as moderating automobile use through pricing. The second part of this chapter examines government actions over longer-term horizons; they can employ a variety of growth controls by prescribing when and where development will occur (queuing) or requiring developers to pay for infrastructure directly in the form of, for example, taxes or impact fees, or indirectly through mechanisms such as exacting proffers in exchange for approval (charging). The chapter is organized into four sections (see Figure 13.1), each describing primary tenets of different strategies.

Queuing (short term)

Congestion

Many books about transportation and land use gloss over traffic congestion and queuing as "engineering problems." Queuing theory, however, is a fundamental aspect of land use-transportation dynamics. Unbeknown to many, congestion operates as an extremely effective default policy strategy. Most communities balance infrastructure supply and travel demand in the short run by using congestion as a rationing mechanism. Although this may not often be considered outright or explicit strategy, it is the modus operandi in many cities; it is what occurs when nothing else is done. Although the "do-nothing" alternative may lead to negative experiences for some (e.g., sitting in congestion), there are some positive outcomes which often go unacknowledged.

Among them is the fact that, like sports and the weather (and unlike politics and religion), traffic congestion is an accepted topic of discussion at cocktail parties—so higher congestion levels mean more to talk about. Second, congestion is often the result of success and prosperity. As Brian Taylor notes, "it is a drag on otherwise high levels of accessibility, not a *cause* of economic decline and urban decay. So while we can view congestion as imposing costs on metropolitan areas, the costs of inaccessibility in uncongested places are almost certainly greater ." [1]

Queues (a more formal term for congestion) occur when the amount of traffic (e.g., cars, pedestrians) exceeds, for some period of time, the infrastructure's capacity to serve them. Queues can most simply be represented by

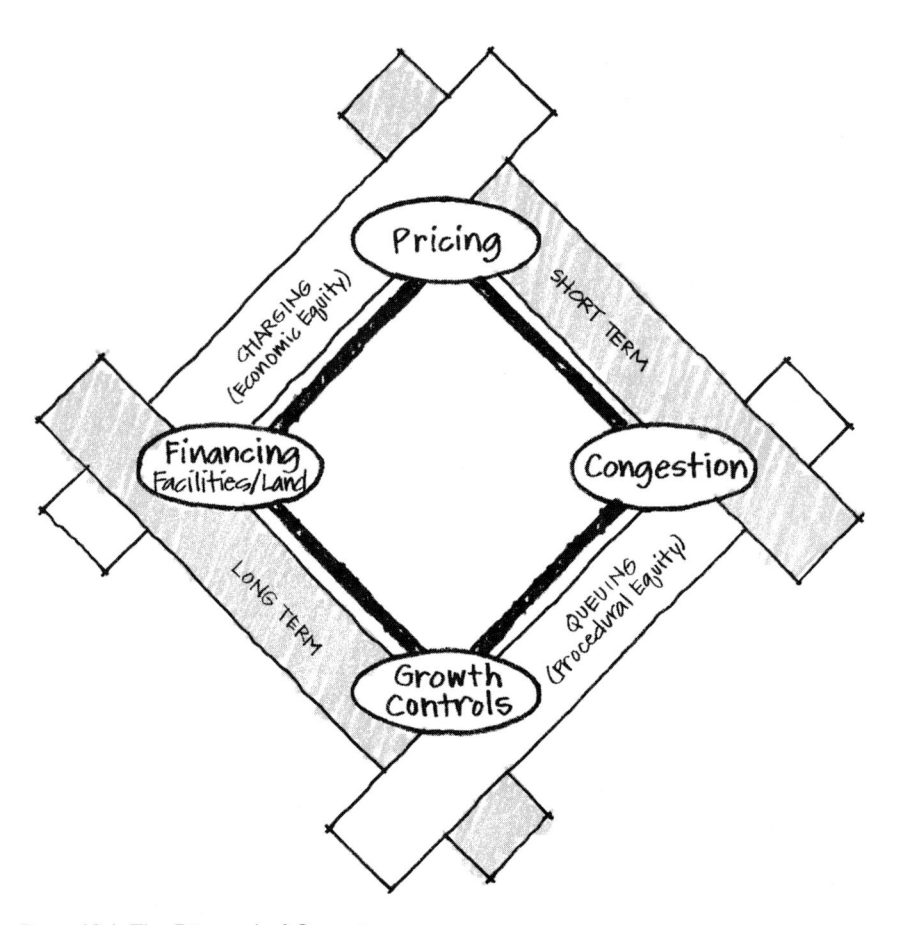

Figure 13.1 The Diamond of Operation

Governments both ration the use of transportation infrastructure (a relatively short-term matter) and regulate the use of land (a relatively longer-term matter) using two strategies: queuing and charging. The former strategy (queuing) is widely considered to be more equitable; the latter strategy (charging) generally leads to greater economic efficiency but is widely criticized for being inequitable. The two time scales, coupled with the two strategies, form the Diamond of Operation (Figure 13.1), which organizes ways of allocating scarce resources related to land use and transportation. The theory behind long and short term strategies is similar but not identical. Looking at the short term, governments can allocate use along transportation infrastructure by relying on congestion (queuing) or pricing mechanisms (charging). Over longer-term horizons, they can employ a variety of growth controls by prescribing when and where development will occur (queuing) or requiring developers to pay for infrastructure directly in the form of, for example, taxes or impact fees, or indirectly through mechanisms such as exacting proffers in exchange for approval (charging).

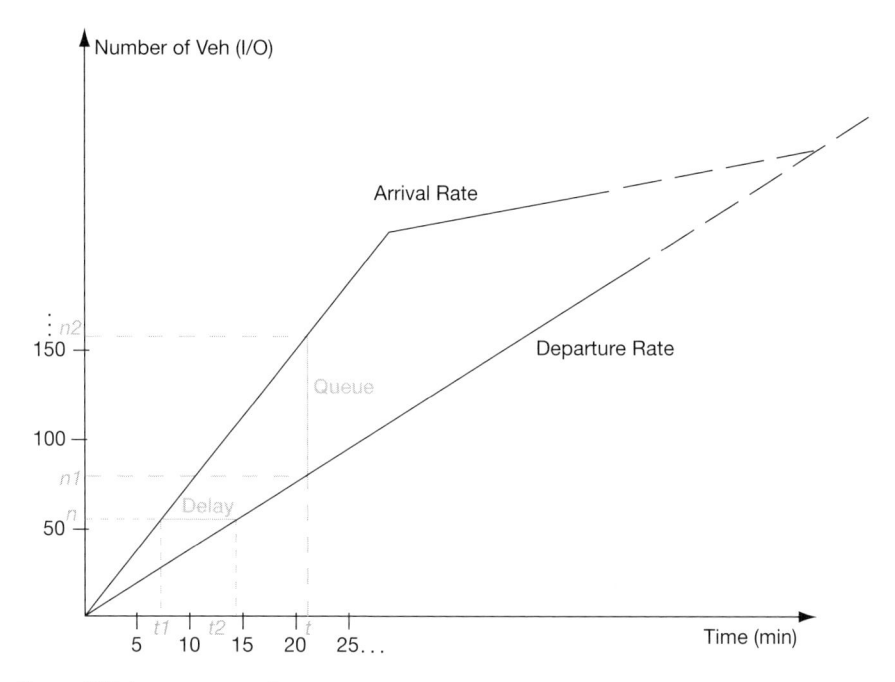

Figure 13.2 Input–output diagram

cumulative input-output (IO) diagrams (Figure 13.2). Ultimately, what enters must eventually exit; otherwise the queue would grow to infinity. Although frustrated commuters often see long backups on expressways (in which delays may seem infinitely long), these backups do eventually clear. In fact, queuing, and traffic flow in general, along with "free-flow" travel times, access times, and schedules, shapes the accessibility contours that have been discussed in previous chapters. Queues are fundamental to understanding accessibility, which unlike our simple models, is not constant across the day, but varies with congested and uncongested travel times.

The first point to note about the IO diagram is that delay varies for each driver.[a] The average delay can be measured easily (the total area in the triangle is the total delay, the average delay is just the area of that triangle divided by the number of vehicles). The variation (or standard deviation) can also be measured statistically. As the total number of vehicles increases, the average delay increases.

The second point to note about the IO diagram is that the total number of queued vehicles (the length of the queue, or the spatial extent of congestion), can also be easily measured. This, too, changes continuously; the back of the queue gets longer or shorter with changes in the arrival rate. Only if the arrival rate exactly equals the departure rate would we expect to see a fixed queue length. If the queue results from a management practice such as ramp metering, we can control the departure rate to approximately match the arrival rate and ensure that the queue remains on the ramp and does not spill over to neighboring arterials.

We have so far discussed queuing rather than congestion, because it is often a clearer way to think about the issues in play. Queues occur at bottlenecks; were it not for bottlenecks, travelers could move at free-flow speed to their destinations. The bottleneck is where maximum flow possible drops. Consider an hourglass, or a funnel, or the neck of a bottle for physical analogies. If there were no bottlenecks (which can be physical and permanent, such as lane drops or steep grades; or variable and by design, like traffic control devices; or temporary, due to a crash or a slow-moving funeral procession), there would be very little congestion.[b]

Maximum flow

What defines maximum flow possible? Traditional queues have "servers." For instance the check-out clerk at the grocery store can serve, say, one customer every 150 seconds (or 24 customers per hour), or one purchasable good every 10 seconds (360 goods per hour). In the context of road capacity, however, the term is a misnomer. Capacity is determined by the driver, more precisely by the driver's willingness and ability to follow the driver ahead. Drivers willing and able to follow behind one another with very small gaps (spacing between vehicles) and drive at high speeds significantly increase the number of vehicles per hour that can use the road. However, although some compression of vehicles occurs in heavy traffic, this situation is unstable because drivers tap their brakes, or even release the accelerator, for any number of reasons: to change lanes, to respond to someone else trying to change lanes, to avoid an object in the road or to limit the centripetal forces experienced when rounding a corner. These actions lower the speed of the vehicle, which in turn lowers the flow. Risk-averse drivers will slow down even more than others, in order to establish an even larger gap to accommodate the behavior of unpredictable drivers. In this way, maximum possible flow (i.e., capacity), is a function of the driver. Of course, the road shapes a driver's willingness to take risks. Drivers will slow down around curves, vehicles may have difficulty accelerating uphill, and merge zones require drivers to take time in order to avoid a collision.

Bottlenecks

There is a maximum capacity assuming drivers are controlling vehicles, which may vary with road conditions. In a series of segments, the binding section is the section with the lowest capacity, shown in Figure 13.3 (also known as a bottleneck).

Although flow at the bottleneck usually remains at or near capacity when there is a queue (there may be a small drop associated with inefficient vehicle behavior because of acceleration and deceleration), the flow per lane (but not

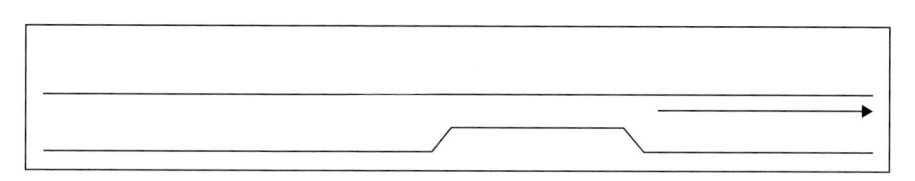

Figure 13.3 Bottleneck

the flow per link) on the upstream links necessarily drops, as traffic queues behind the bottleneck. This is often perceived as congestion, but this is just the symptom. The problem is the travel demand through the bottleneck at a given time exceeds available capacity.

Systems are often too complicated to be quickly analyzed with graphs. Therefore, tools such as computer micro-simulation are used; in micro-simulation, the movements of virtual "vehicles" are governed by rules about following other cars and changing lanes, and are subject to physical constraints of the road (number of lanes), the desire to avoid collisions (no two vehicles can share the same space at the same time), accounting for complex topologies, and negotiating entrance and exit ramps. However, the real-world observation that traffic generally moves at free-flow speed until capacity is approached or exceeded remains true. Only bottleneck links operate at capacity, while most other links operate below capacity—either when they are under free-flow conditions (demand is low), or when queuing is in place (demand is high). This leads to the observation that the same flow can be achieved on many links at two different speeds.

Some refer to this as the "backward bending" phenomenon, referencing a figure with a backward bending curve. [2] This discussion tends to confuse input flow (or demand) with output flow (which in congested conditions is constrained by capacity). The flow is realized flow, not demanded flow. Under any given demand pattern, flow and speed are a unique pair. When demand is below the active downstream bottleneck's capacity, a flow on an upstream link can be achieved at high speed. When demand is above the active downstream bottleneck's capacity, the same flow on the upstream link can only be achieved at a low speed because of queuing. But the ability to have high flow at high speed on an upstream link depends on the absence of a downstream bottleneck. A flow upstream at a level higher than the capacity of the downstream bottleneck can only be temporary. Looking at traffic upstream of the bottleneck is interesting, but does not get to the root of the problem: the bottleneck. Shock waves, produced by vehicles arriving at the back of queues, coupled with the queue clearance rate, indicate where the back of the queue is. This is where the traveler first suffers delay, but it is not the source of the delay.

Every vehicle eventually clears the bottleneck. If all travelers could arrive at the congestion exactly at the time they would have cleared, there would be no delay. This requires either that vehicles be controlled upstream somewhere (by technologies such as ramp meters, discussed later in this chapter), or that demand be manipulated to arrange this pattern. Matching supply and demand can be accomplished in a number of ways.

Restaurants, for instance, have reservation policies. Most markets use prices to match supply and demand. These two ideas can be combined in the form of reservation pricing. It would not be best to price the road at the time a driver reaches the bottleneck—by then it is too late. Instead, it is best to sell reservations (an allocated number of slots to arrive between 5.00 p.m. and 5.05 p.m.). Drivers attempting to use the road without a reservation are either turned away or required to pay a premium. The technology exists for such control, though it has yet to be properly packaged or deployed. In a related strategy, engineers in many transportation management agencies have started implementing ramp metering programs (see Box 13.1).

Charging (short term)

Determining the appropriate size of infrastructure to move people and vehicles in the long term depends on what is done in the short term. This issue often comes to a head in the form of road pricing, a topic long on rhetoric in the transportation industry, but until recently, short on action. As evidenced by this chapter's epigraph, calls for the pricing of roadways date to the 1960s [3] and probably earlier. Little has been done to execute such strategies, however, largely because of a lack of political will. As Martin Wachs has poetically observed, "the prospects for widespread adoption of congestion pricing are limited, because the only constituents in favor are academics and environmental zealots, hardly influential groups." [4] This is why Livingstone's initiative, described earlier in the chapter, is so noteworthy. The topic of congestion pricing has received a good amount of attention over the years, focusing on equity aspects, [5, 6] financial aspects, [7] political aspects, [8, 9] planning aspects, [10] and mobility versus accessibility aspects. [11]

The central issue is planning for the "right" amount of infrastructure, an amount which may differ considerably in cases with and without road pricing. Road pricing (sometimes called congestion charging, congestion pricing, or value pricing) assumes a number of forms. Purposes of pricing include raising revenue and punishing driving, but in an ideal case it balances the benefits of use of infrastructure with the costs incurred by that use. There are several variables that can be manipulated: where (what facilities are covered), when (what time periods are covered), and how much (at what level the toll is set).

The conventional explanation of road pricing uses a variation of Figure 13.4. On the vertical axis is the cost of travel. On the horizontal axis is traffic flow in vehicles per hour. The short-run average cost curve is the delay drivers face—the more drivers, the higher the average delay (because of congestion). In the absence of tolls, equilibrium occurs at (Q_o, P_o), where the short-run average cost curve intersects demand. Travelers who value their trip at higher

Box 13.1 Ramp meters

In the autumn of 2000, the ramp meters in the Twin Cities were turned off for eight weeks to assess their effectiveness. Although this assessment focused on the efficiency of the system, considering mobility and safety particularly, a transportation equity analysis of the delay distribution across space was also conducted. This section estimates the relationship between mobility and equity for O-D pairs on Route 169, a suburb-to-suburb limited access highway, connecting the North and South legs of the region's beltway, with and without ramp meters. In order to make the results comparable, the data used for the analyses (ramp metering on and off) were collected on Tuesdays: March 21 and November 7. November 7 is the third Tuesday after ramp signals were shut down. The calculation methodology is described more fully elsewhere. [12] The analysis assumes that traffic was approximately in equilibrium so that day-to-day traffic patterns were stable.

What ramp meters bring in mobility and equity can be shown by the comparison of the two cases. Previous research indicates that ramp meters can increase the mobility of freeway networks, which is confirmed by our findings. With ramp metering, the average travel speed (taking ramp delay into account) of the highway increases from 37 km/h to 62 km/h; travel delay per mile decreases from 136 seconds to 112.5 seconds, and the average travel time for one trip decreased from 610 seconds to 330 seconds.

No previous results can be relied upon to guide our analysis of equity. In contrast with our consideration of efficiency, when looking at trips we find a drop in the Gini coefficient (a measure of equity, where 0 is perfect equity and 1 is perfect inequity) in the absence of metering. This suggests the system becomes fairer when meters are removed. This drop is observed for three primary measures: travel time per kilometer, travel speed, and travel delay per kilometer. Figure 13.4 shows the trends in the change of the mobility and equity with and without metering for trips. Note that in the figure, the shortest trips (those on the right side of the graph) actually are penalized in mobility terms by ramp metering, while the longest trips, (those on the left side) benefit the most.

than P_o will travel. However, travelers impose a cost on other drivers—my presence costs you time. This is reflected in the short-run marginal cost curve, which shows how the total cost rises with additional traffic. It would be economically efficient if travelers faced the short-run marginal cost rather than the short-run average cost. This can be achieved by a toll of the amount of the difference between the marginal and average costs. The shaded area on the graph is the benefit lost when tolls are not imposed. Imposing a marginal cost toll moves the equilibrium to (Q^*, P^*) and eliminates the welfare loss.

Congestion charging

Facility-specific tolls are commonplace in many metropolitan areas, charges for bridges are an example. The rationale for such charges is to pay for the facility, especially expensive facilities. By varying the tolls according to the time of day (or level of demand), such facility-specific tolls can achieve, at least locally, some of the benefits of congestion pricing. Such strategies include substituting money for time, by charging users more for travel during peak times (thereby encouraging them to travel in the off-peak, when they will impose little or no delay on their fellow travelers, or to take less congested routes, or change mode, or not make the trip at all). The revenue collected by those who do still choose to travel in the peak can be used for any number of purposes, among them, expanding choices for travelers.

A variant on the facility-specific toll is the HOT (high occupancy toll) lane. HOT lanes are lanes designated for use by high-occupancy vehicles that may be used by single-occupant vehicles if a toll is paid. In general, they are parallel to "free" lanes, and so the solo drivers choose to pay the toll in order to avoid congestion occurring on non-HOT lanes.

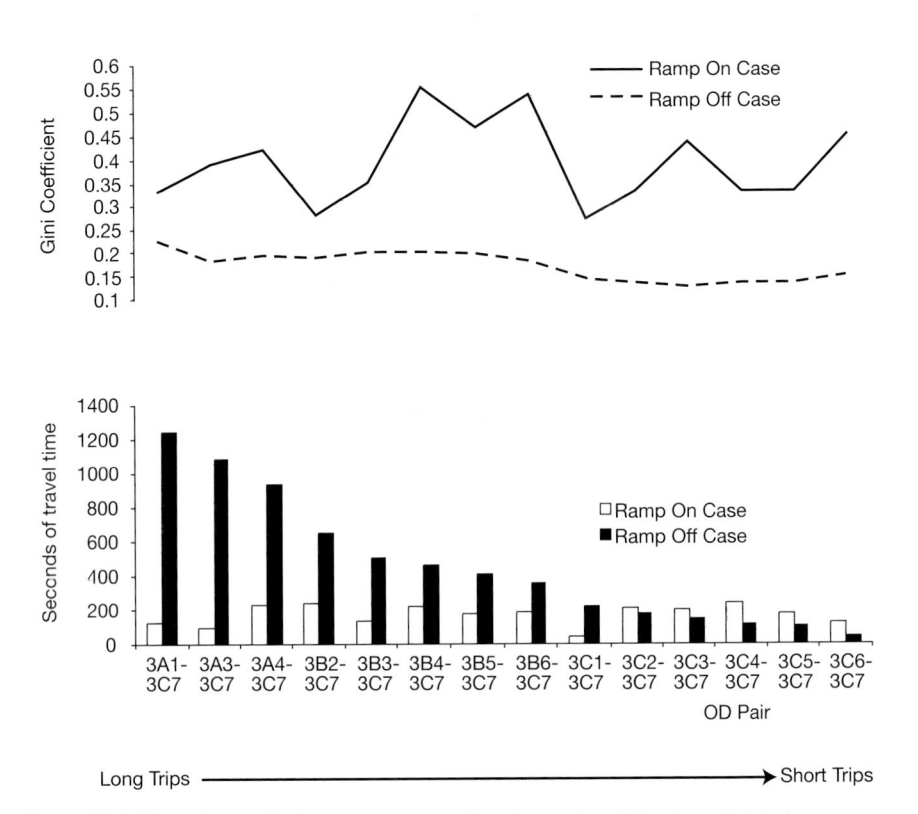

Figure 13.4 The relationship between temporal equity and mobility (travel delay)

Another alternative is cordon tolls, which are simply charges to pass a boundary, or cordon. These are used in a number of cities, including Singapore, Oslo, Bergen, and Trondheim, as well as in a more limited form in the United States in Manhattan, Long Island (near New York City), and San Francisco, where one has to pay tolls to enter. Still another alternative is area-based pricing, a variant on the "pay to enter" approach which could be termed the "pay to be in" approach. The London-Based Congestion Charging scheme described above is an example of area-based pricing. In still another variation on HOT lanes, Box 13.2 considers how planners have used managed lanes to reward what they consider socially desirable behavior.

Parking

Alternatively, focusing on efforts to harness automobile travel, a logical strategy is to focus on space required for vehicle storage. After all, cars are parked 90 percent of the time, mostly in a garage or other parking space belonging to the vehicle owner. But the instant the car leaves home, it creates one of the most controversial, yet simple planning issues: the demand for parking. When citizens are assured of the availability of parking, either free or paid, they are significantly more likely to drive. When parking is scarce, the fuse is lit on

Legend

P^* = Optimal price with tolls Q^* = Amount of travel with tolls

P_o = Price without tolls Q_o = Amount of travel without tolls

◂ = Welfare loss without tolls

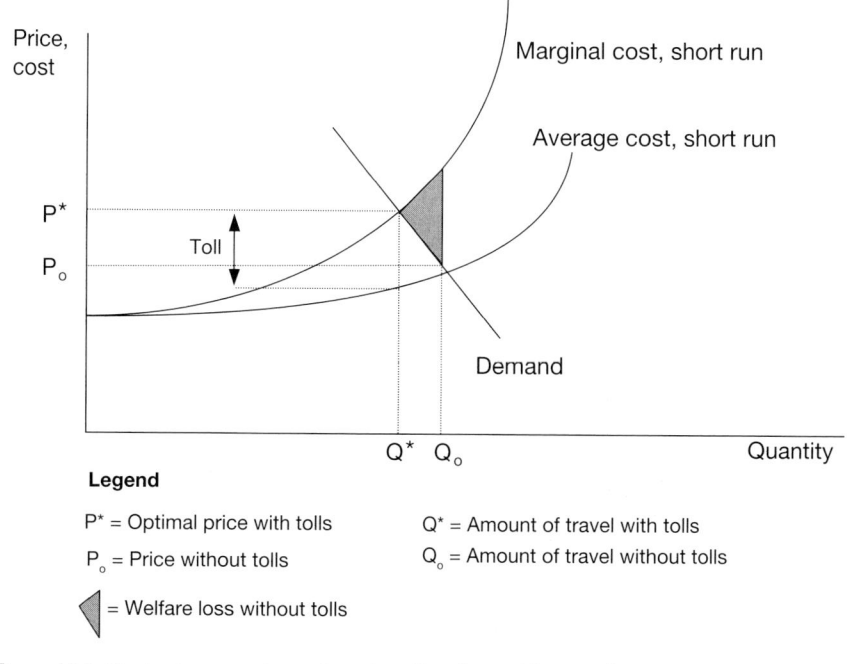

Figure 13.5 Optimal congestion toll, and welfare loss without toll

Box 13.2 Righteous lanes

In the beginning, engineers built the general-purpose lane, which allowed all persons to drive just about any vehicle on a first-come, first-served basis. And engineers looked it over and saw that it was good.

But congestion arose. Planners noted general-purpose lanes meant that cars carrying one person were on equal footing with cars carrying two or more persons as well as with buses. With gas shortages, air pollution, and congestion as problems, riding transit and carpooling came to be seen as socially preferred behaviors. But carpooling was hard; passengers had to coordinate their schedules, which cost time. So planners separated the light from the darkness and devised a "high occupancy vehicle" (HOV) lane, which only vehicles carrying more than one person could use. Thus, HOVs could avoid congestion. And planners saw that it was good.

But travelers noted that the HOV lanes were under-utilized, and travel analysts noted that most HOV passengers came from the same households and would have carpooled anyway. So planners thought hard and discovered that hybrid cars, which have both electric and gasoline powered engines and thus better gas mileage, were still few in number. It was argued that these hybrid cars, which were socially preferred because they pollute less, should be given preferential treatment by being allowed to use the HOV lanes. In effect, solo drivers who were willing to pay a fee—a fee in terms of purchasing a "more" environmentally benign mode of transportation—were allowed to compete on equal footing with single occupant vehicles. Virginia implemented such a policy, and soon had the second-largest number of hybrid cars in the nation. In 2000, Virginia had only 32 registered cars with "clean fuel" tags, but by the end of 2004, there were 6,800 registered hybrids with appropriate tags, comprising almost 20 percent of the traffic in the state's HOV lanes. Thus, hybrid vehicles could avoid congestion, and hybrid owners saw that it was good.

HOV drivers did not agree. They could see that an annually compounded tripling of hybrid vehicles could not compound for very long before the HOV lane capacity became filled. At the time of this writing, the Virginia Department of Transportation is recommending that the system be phased out, but a potent new lobby of righteous hybrid owners will want to maintain their right to save time as well as gas. Other states, including California, are trying variants of this policy [13] —some capping the number of permits to use the HOV lanes, others passing the benefit without having many HOV lanes in place, and are still awaiting Federal Highway Administration approval.

Allowing one group to use a lane because they have fewer social impacts (or because they pay more costs) is one means to achieve a social good, but certainly not the most efficient. Other ideas, such as allowing SOVs (any SOVs, not just hybrids) to buy into high-occupancy vehicle lanes (designated as High Occupancy Toll or HOT lanes), have been tried and shown to be effective at providing reliable transportation for anyone who values it sufficiently. Because there is very little excess capacity on the freeway system during peak periods, how it gets allocated is an important question, and policies that exhibit long-term sustainability should be preferred to short-term gimmicks. Policies that add to pollution by having non-polluting cars run at free-flow conditions while polluting cars idle in congestion have, at least, short-run costs that should be considered.

the transportation-land use planner's most secret weapon to foster the use of other modes: a lack of parking.

Parking poses an interesting set of problems with regards to ordinances that state and local governments apply to reduce the number of vehicle trips stemming from different land uses. As it currently stands, free parking is an untaxed benefit that many employers are able to provide to their employees. The internal costs to an employee for driving to work are therefore less than true costs, leaving aside the other externalities mentioned earlier. In theory, the pricing of parking should encourage the marginal single occupant vehicle (SOV) commuter to find some other means of getting to work. Similarly, reducing the quantity of parking provided should also encourage a certain percentage of the workforce to switch from driving alone to ridesharing, transit, or some other means of commuting.

Shoup, who has written the most comprehensive book on the parking issue, [14] suggests that communities charge fair-market prices for curb parking, return the revenue to neighborhoods, and remove all off-street parking requirements. He also suggests unbundling the charge for parking from charges for the other uses of land (turning the hedonic model introduced in Chapter 3 on its head). People bundle things all the time to reduce costs and increase convenience (e.g., the lot and the house are generally purchased together rather than in separate transactions). Bundling, however, puts the cost of parking into the cost of everything else we purchase at stores, or the cost of rent for offices. As an example of an unbundling strategy, Bellevue, Washington (a suburb of Seattle) requires that parking costs be listed as a separate line item in leases and that a minimum cost be imposed for parking. This aims to reduce use of inexpensive parking as an incentive by developers to attract tenants, and by employers to attract employees. A number of localities have also written regulations to permit a reduced minimum number of parking spaces per unit of floor area to help implement travel demand management strategies. However, many developers still choose to construct larger numbers of spaces to lure tenants. A ceiling on the number of parking spaces allowed may prove more effective. As an incentive to rideshare, the employer can provide guaranteed, preferential parking for high occupancy vehicles, thereby minimizing the walk from the parking lot to the front door.

Large parking lots often discourage walking or biking, while the positioning of those lots can make transit undesirable (acres of parking is not the ideal pedestrian-friendly environment). To alleviate these disadvantages, applying "good" site design techniques can make these other modes more desirable, thereby increasing their use and reducing the number of vehicle trips. "Good" site design from the trip reduction perspective would involve placing buildings in a reasonably dense pattern close to the main roads, sidewalks, and paths in order to encourage both pedestrian and bicycle movement, and to encourage transit ridership. Likewise, mixing retail and restaurant uses with offices makes having an automobile during the day less important.

The principal reason municipalities adopt minimum parking requirements is relatively simple: spillover into adjacent areas. The minimum parking requirement, however, remains a blunt instrument to solve a relatively simple problem. Minimum parking standards are usually constructed around peak demand for a single use, rather than taking into account the many uses that may be within close proximity to a certain area. As shown in Figure 13.6, different uses have different parking demands based on time of day. An optimum solution would be to contain the average cumulative number of spots for each geographic area. Unfortunately, it is not this simple. Thus, the most effective solution to the problem of minimum parking standards is to eliminate them (and regulate or charge for parking on the "commons," the public street). Business owners who feel parking is necessary will provide it; others will share on an as-needed basis. Using these strategies, parking becomes a matter that can be effectively harnessed through strategies that can be aligned by charging over the shorter term.

Queuing (long term)

How many home buyers would be interested in cheap houses without roads, water, sewers, schools, parks and other urban amenities? [15] Few. This is why many governments have implemented growth management programs to match the demand for places to live with the supply of services. But matching supply and demand often poses a problem for non-market services, foremost roads and schools, but water and sewer systems as well. Some infrastructure services (e.g., electricity, which is generally provided by a public utility or a private company providing a public service for a fee with monopoly privileges) do not seem to suffer from the same problems in accommodating development (though blackouts in the northeastern United States in August of 2003 suggest issues lie in that sector as well).

Growth management emerged in response to concerns about how much growth would be allowed, where and when it would be permitted, who would pay the bills. It has been formally defined as deliberate and integrated use of the planning, regulatory, and fiscal authority of state and local governments to influence the pattern of growth and development in order to meet projected needs. [16] Included in this definition are such tools as comprehensive planning, zoning, subdivision regulations, property taxes and development fees, infrastructure investments, and other policy instruments that significantly influence the development of land and the construction of housing. In a nutshell, growth management aims to affect the location, character, amount and timing of development.

Famous early growth management experiments included the development management system in Ramapo, New York, urban growth boundaries in Petaluma, California, and the Municipal Urban Services Area line in the Twin Cities of Minneapolis and Saint Paul, Minnesota. Growth management is fundamentally a rationing or queuing strategy, which says only a limited

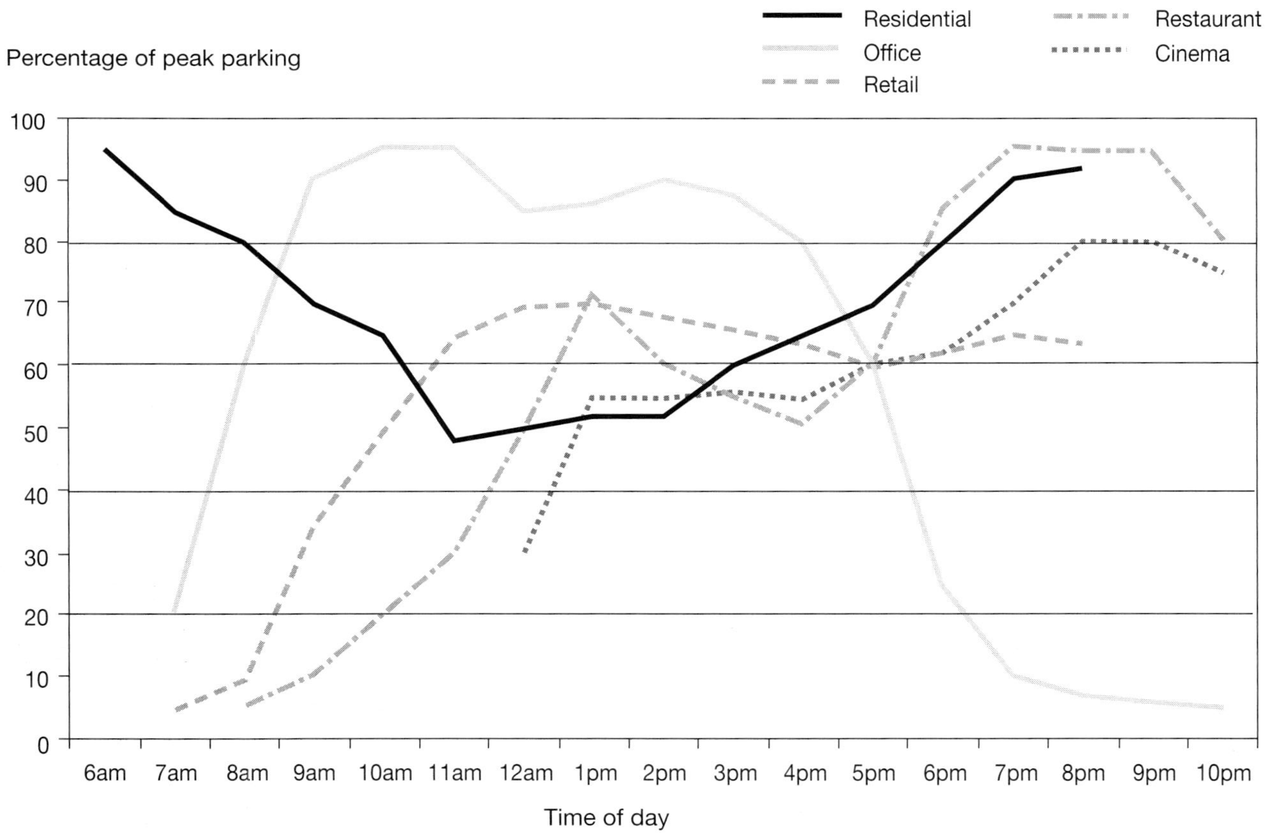

Figure 13.6 Parking demand, by time of day

amount of development can be approved at a given time (based on the capacity of infrastructure) and that all other development must wait until infrastructure is available. By constricting the inflow of people and firms, growth management schemes aim to synchronize expenditures on new capital facilities such as roads, schools, water, sewer, and parks with the development that requires them. Local government financing capacity may not be able to respond to demand for new infrastructure immediately, but can do so over the longer term.

Fast-growing communities have adopted a variety of growth management strategies, with varying degrees of success and numerous problems. [17–23] The rationing of new development may make existing properties more valuable, Katz and Rosen [24] and Pollakowski and Wachter [25] have found price premiums in areas with growth controls.

In addition to efficiently providing infrastructure consistent with land use, growth management may also carry with it additional baggage: affordable housing, environmental and open-space protection, jobs-housing balance, and financing. However, the more goals one attaches to a policy, the less effective the policy will be in achieving any one of them. Growth management coordinates land use control (and planning) and capital investment, which some would argue is what traditional plans do. However, growth management is distinguished from more traditional planning strategies by its intent and scope rather than by its implementation techniques.

A dark side of growth management arises when development spills over into neighboring, less regulated areas. These effects accord with theory, which suggests as a commodity is made scarce, its price rises and substitutes are sought. The exact amount and nature of different spillover effects is an open question, [26, 27] depending on the choices available to developers and consumers. Growth management is both a political and a pragmatic response to circumstances, but whether it is economically efficient locally and/or regionally depends on the program (e.g., Montgomery County's growth management system discussed in Box 13.3).

Charging (long term)

Transportation infrastructure is financed via a variety of sources. Major roads in the United States, for instance, are typically paid for by states with gas taxes, while medium-sized roads are financed by cities and counties with property taxes, and small roads by developers who dedicate the streets to local governments. What defines a major, medium, and minor road in this classification system can be location-specific, is not always clear, and is more fully discussed in Chapter 12. Clearly, roads that serve only the purpose of accessing a development are relatively minor compared with roads that connect multiple developments or various metropolitan areas. Most communities exact on-site improvements from developers. However, large developments would be at a disadvantage in such a system, as simply by subdividing, they could

Box 13.3 Growth management programs

Between 1986 and 2004, the Annual Growth Policy (AGP) of Montgomery County, Maryland coordinated the timing of development in accord with the provision of adequate transportation and other public facilities. Growth management in Montgomery County began in 1974, with the release of a report recommending the provision of adequate public facilities for new development, the enactment of development district legislation, and the implementation of a staging policy in each local area master plan. Through the mid-1970s, the theory of growth management was presented to the public, though no regulatory system was implemented. Briefly, the theory was built upon the idea that an area has a carrying capacity (only so much traffic can be tolerated), which depends upon the level of infrastructure (such as roadway capacity). [28] Because only a limited amount of infrastructure was actually deployed at any given time, only a limited amount of development could be permitted while maintaining adequate carrying capacity. The system was to be implemented with computerized models tracking development, demographics, traffic, and environmental impacts.

The method to regulate development established "staging ceilings" in each policy area in the County. The growth policy defined staging ceilings as the number of permitted jobs or housing units in that area. These staging ceilings were set to ensure the satisfaction of transportation level of service standards. Areas with too much traffic were placed in moratoria for new jobs, housing, or both; while areas with less congestion than their standard were allowed more development. Transportation, though nominally one of several public facilities considered for growth management, clearly became the critical constraint. The measurement and standards of congestion were critical issues.

The objective in setting staging ceilings was to produce a land use pattern that minimized the difference between the forecast traffic congestion and the traffic level of service standard, given the existing network plus roads that were fully funded within the first four years of the county and state capital improvement programs. The level of transit accessibility determined the standard in each policy area (areas with greater transit accessibility were permitted more congestion). Being too congested or too uncongested were equally bad in this framework, as the former condition implies excess delay (travel costs), and the latter implies excess investment in infrastructure (construction and operation costs) for the amount of permitted development. These development capacities were estimated by transportation modelers working for the Planning Department and recommended to the Planning Board, which adjusted and forwarded them to the County Council. The process was reminiscent of the "rational planning" model. Actually solving this problem exceeded the technical capabilities of the modelers, but additional constraints allowed an approximate solution to be presented. These constraints aimed to minimize disruption from existing public policies (and existing staging ceilings), and acted as a brake on the changes in staging ceilings that could be made from year to year.

In the terminology of the day, this approach relied on "police" powers to control private development rather than "purse" powers to provide public facilities. Apparently, and somewhat surprisingly, no written consideration was given to using taxing powers to raise revenue from private development to directly fund public facilities. Although the Planning Board did not have taxing powers, it did have regulatory powers.

The political structure of an independent planning commission and department, which shaped the historical path on which Montgomery County embarked, evolved from the 1920s "good government" movement. But putting taxing powers in the hands of the County Council and executive and regulatory powers over development in an independent Planning Commission resulted in growth policy decisions that did not even consider the taxation alternative. As a consequence, there were a hodgepodge of infrastructure financing systems being implemented by the Executive without a planning outlook and plans being created without financing mechanisms. The county has only taken baby steps in the direction of matching supply and demand with payments rather than queues.

Montgomery County has been firmly in the camp of proactive planning, attempting to comprehensively direct both the timing and placement of development. But such direction creates inefficiencies and inequities: a development trapped in a moratorium creates a dead-weight loss; it cannot proceed even if it is willing to pay some of the infrastructure costs. In contrast, non-moratorium areas have a surplus of infrastructure, indicating past bad investment decisions. The result is an infrastructure funding shortfall.

"Just-in-time" has become a watchword in manufacturing, and the idea underlying it should be considered in planning as well. Clearly, infrastructure planning, engineering, and construction occur within a time frame of years rather than the hours and days of manufacturing. To apply "just-in-time" does not mean collapsing the infrastructure cycle to something on the order of manufacturing, but in addition to shrinking that time, building in response to a demand that pays its full cost rather than (1) subsidizing transportation in advance of a speculated demand, or (2) building infrastructure long after congestion has become intolerable (and economically inefficient) and new development has been placed under a multi-year moratorium.

off-load road infrastructure costs onto the public (i.e., onto residents of other developments).

Broader funding schemes are demanded when infrastructure is inadequate and existing revenue sources have been tapped dry. One strategy is to tax the use of land—a relatively powerful mechanism, though not widely employed in the US. Where a land tax has been administered, it seems to have positive benefits. For example, since 1975, Harrisburg, Pennsylvania has taxed land at a rate six times that of improvements. Although there are probably a variety of things going on in Harrisburg, this policy is largely credited with reducing the number of vacant structures in downtown Harrisburg from about 4,200 in the early 1980s to less than 500 and has created additional revenue for the city. Box 13.4 thinks about trying to encourage property owners to build to the fullest using the Single Tax on Land, thereby increasing agglomeration. Another way of trying to raise funds to exploit the agglomeration of business is to impose a commuter tax, this is addressed in Box 13.5.

Charging developers for the impacts on infrastructure that their developments impose is a different strategy to refill proffers that have been tapped dry. [29] The incentive for the developer to participate in such a charging system is that the only alternative is not being permitted to develop at all (or, more precisely, not being able to develop until infrastructure is adequate, which may take considerable time). Theory predicts, and evidence corroborates, that imposing developer charges in only a limited area will distort the market, so there will be overdevelopment of the un(der)charged areas. There are a variety of such mechanisms available, including: impact fees (taxes), development districts, infrastructure proffers, tax increment financing (discussed in Chapter 8), and trip mitigation.

Impact fees (or taxes) are imposed on development based on its expected capital impact on infrastructure. There are a number of ways to determine impact fees. A top-down approach might conclude that when an area is built-out, there will be an additional 50,000 housing units and an additional 50 lane-km of major roadway. At build-out, it is concluded that the roadway level of service would be adequate. By linearly interpolating, each additional 1,000 housing units requires one lane-km of roadway. Calculations may suggest that one lane-km of roadway costs $10,000,000. Therefore, each housing unit would be charged $10,000 for its share of the necessary roads in order to cover the development's impact. This strategy is often considered an "average cost" approach.

Alternatively, a bottom-up approach might recognize that some infrastructure is less expensive than other infrastructure. It may be easiest to build the least expensive infrastructure first. There may also be some excess capacity at present, which new development could exploit. Thus, the impact of the first development might be less than the impact of later development. This could be measured with a more careful study to determine the *marginal cost* of a new development on infrastructure. The result might be that early development is subject to a much lower impact fee than later development. Although the first may be seen as more fair, the second is more economically efficient.

A variation on the impact fee is the development district. These enable development in a designated area to proceed after paying into a fund established to cover the construction of planned infrastructure. In a proffers system, a developer may voluntarily provide infrastructure to meet transportation level of service requirements in order to gain approval for their projects. Developers may band together to form "road clubs," based on a contract signed by a group of developers and the local government to collectively finance and build transportation infrastructure (roads). However, this option is only open to developers, or a coalition of developers, of sufficient size to be able to afford major infrastructure.

Box 13.4 Henry George's single (land) tax

> Men did not make the earth ... It is the value of the improvement only, and not the earth itself, that is individual property ... Every proprietor owes to the community a ground rent for the land which he holds.
>
> (Thomas Paine, Agrarian Justice, paragraphs 11 to 15)

> A tax upon ground-rents would not raise the rents of houses. It would fall altogether upon the owner of the ground-rent.
>
> (Adam Smith)

According to his granddaughter, actress and choreographer Agnes DeMille, Henry George was, at his death in 1897, the third most famous man in America, behind only Thomas Edison and Mark Twain. Over 100,000 people attended his funeral in New York. George was a political figure, a two-time mayoral candidate in New York (dying just four days before the election on his second bid), a newspaper publisher, and an economist. Born in 1839, he was in California during the gold rush and the railroad boom. He noted how railroads drove up land value and rents at a rate faster than wages.

He proposed in his best-selling economic tract *Progress and Poverty* that:

> We should satisfy the law of justice, we should meet all economic requirements, by at one stroke abolishing all private titles, declaring all land public property, and letting it out to the highest bidders in lots to suit, under such conditions as would sacredly guard the private right to improvements. Thus we should secure, in a more complex state of society, the same equality of rights that in a ruder state were secured by equal partitions of the soil and, by giving the use of the land to whoever could procure the most from it, we should secure the greatest production.
>
> (www.henrygeorge.org/chp15&16.htm)

These views do not make George a "communist," though some have dubbed his ideas "commonism" because the land is held in common. In modern language, his most famous proposal is that of a single tax on land. The idea is simple in its core, but is easily confused with other concepts due to the complexity of modern tax codes. First, it is a single tax, so no other tax would be required. Second, it is a tax on land, not property. So the question of "What is land?" should be answered. Land is, in short, nature's bounty; it includes geographic spaces, but also mineral deposits, natural resources, and the electromagnetic spectrum. It is what would exist without labor. The value of land, particularly the value of geographic spaces, does depend on labor and what is done with other geographic spaces. A square meter of land in downtown Tokyo may be worth a square kilometer (or more) in Alaska. Most of the value of that square meter of Tokyo, however, is due not to the improvements by its owner, but rather to the accessibility to the land, which is created by everyone else in society.

Taxing land based on the land value, rather than the property value, encourages full development of the land. The property tax discourages development of land, since all improvements are taxed. This helps result in an urban form of surface parking lots in big cities rather than developed land. The property tax also encourages leap-frog development in the suburbs. In contrast a land tax would apply the same tax to a parcel whether or not it were developed, thereby encouraging development to help pay the tax. This land value tax (LVT) is the current incarnation of the Georgist proposals. It is currently used in Singapore, Hong Kong, Estonia, and Taiwan, though not as the only tax.

This idea, however, is not as radical as it seems; four Nobel-laureate economists urged Mikhail Gorbachev to adopt the land tax in 1990 as the Soviet Union was turning away from communism. Modern Georgists generally favor movement towards a single tax, but recognize the political impossibility of an overnight change, especially one which would eliminate not only property taxes but also sales and income taxes. The idea is illustrated in Figure 13.7.

This policy would be in stark contrast to tax increment financing. Instead of subsidizing firms to develop fully, they would be taxed as if they were fully developed, and thus would it be more expensive if they don't.

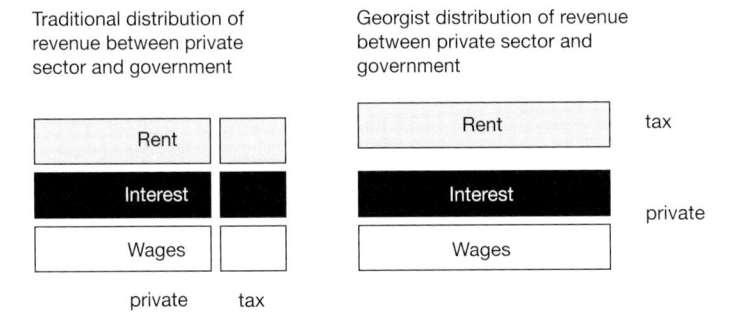

Figure 13.7 Traditional versus Georgist taxing proposals

Box 13.5 Commuter tax

A commuter tax is a tax charged on individuals who work in a particular place but live elsewhere. In some respects, many places already have a commuter tax. All states that have income taxes do tax non-resident income and give credit for any income paid to other states. A commuter tax is charged in some form in Birmingham (Alabama), Cincinnati (Ohio), Cleveland (Ohio), Detroit (Michigan), Louisville (Kentucky), Newark (New Jersey), and Philadelphia (Pennsylvania). Other cities have, or want, income tax revenue as well. The commuter tax has been proposed for a number of large cities, among them New York and Washington, DC, to capture additional revenue to pay for the costs of providing services for these commuters. New York City had this tax until 1999, when it was repealed by New York State to appeal to suburbanites, and there have been calls to reinstate it, by Mayor Michael Bloomberg among others.

Washington, DC, with its special legal status, is prohibited by Congress from imposing this tax. Fifteen states and the District of Columbia allow reciprocity, so that people can pay taxes to their state of residence instead of state of employment, which greatly simplifies tax collection and can be advantageous to both parties if cross-border flows are approximately equal. However, the District of Columbia imports far more workers than it exports, so it loses significant revenue to Maryland and Virginia as a result of the required reciprocity. The allocation of joint costs associated with public goods brings with it inequity. The commuter tax is one attempt at remediation.

By shifting additional tax-burden to non-city residents, taxes on city residents can be lowered. The thought is, by taxing workers of firms who benefit from economies of agglomeration, some of those positive externalities can be captured by the local municipality. The risk to the city is that offices will migrate to the suburbs as well, and the city loses more tax base, rather than gaining.

Alternatively, a developer may enter into a trip mitigation program in order to attain approval. These arrangements rarely impose direct monetary charges on the developer; rather, such an agreement might specify that development approval is "traded" for a reduction in vehicle trips. However, reducing vehicle trips costs money (otherwise we would have no congestion). Trip mitigation programs include ride-matching, shuttle services, construction of park-and-ride lots, transit subsidies, and other measures that are intended to get vehicles off roads. Their estimated cost is on the order of $500 per trip per year (somewhat less than $5,000 for a ten-year program). By mitigating peak-period, peak-direction trips, ideally the developer will eliminate the bulk of the traffic impact of the development. This is considered in more detail in Box 13.6, which describes the history of travel demand management.

Operating wrap up

Transportation networks and the land they serve are scarce resources whose use can be allocated in any number of ways. This chapter has examined various mechanisms for operating and allocating scarce resources: pricing and rationing in the long and short run. Queues, making people wait in line, are one, and prices are another mechanism to ration demand: prices ration based on ability and willingness to pay money, queues ration based on ability and willingness to wait. From an economic point of view, prices are more efficient than queues.

The ideas discussed in this chapter are less about providing new services than about ensuring an equilibrium between supply and demand for travel, using means having either a temporal component (e.g., congestion or growth management) or a financial dimension (e.g., congestion pricing or impact fees).

We confront the choice: *charges* or *congestion*. To date, with the exception of London, Singapore, and a few others, cities have largely chosen congestion. The costs of collecting tolls are one argument against using road pricing. When everyone has to stop at a tollbooth to pay, the delay may be worse than the charge. But collecting tolls electronically now helps avoid many such issues. A more fundamental question grows out of concerns over equity. Questions of what it is and its importance come to the fore. A central issue is that concerns over equity often get blown out of proportion.

Pricing strategies help create choices. In some cases, the pricing might be used to help place the auto on more equal footing with other modes. For example, William Vickrey, source of the quotation at the beginning of this chapter, leaned this way when he wrote that "even greater preference should be given to space economizing modes of transport than would be indicated by rent and tax levels. And our rubber-shod sacred cow is a ravenously space hungry, shall I say, monster?" Though, such a sentiment comes as little surprise from the Nobel laureate who roller-skated to class at Columbia University well into his seventies.[3]

In other cases, pricing creates a more efficient system. After all, a system that priced certain roads at higher rates to provide a higher level of service would create choices. Pay and drive fast, or do not pay and drive slowly, where presently there is only one option: to drive slowly. This differentiates people according to the value they place on their time, which might be deemed inequitable. On the other hand, people have different values of time at different times; even the poor may have a high value of time sometimes, and be willing to pay a premium to ensure they can reach their destination in a known and shorter time. How is it fair that people in different circumstances get the same level of service? Isn't it more fair to allow people to buy their way out of congestion sometimes, rather than to require everyone to be congested always. We contend that prices create choices, and choices are fair.

Box 13.6 History of travel demand management

In 1982, Placer County, in Northern California, enacted legislation requiring developers to reduce the number of vehicle trips to 80 percent of what would normally be expected from the Institute of Transportation Engineers' trip generation tables, creating what is widely considered to be the first Trip Reduction Ordinance (TRO). Trip reduction, reducing the number of peak vehicle trips coming out of developments, is one strategy for implementing Travel Demand Management (TDM). TDM comprises "programs designed to maximize the people moving capability of the transportation system by increasing the number of persons in a vehicle, or by influencing the time of, or need to travel." [30] We refer to the three following classes of TDM programs:

- alternative work schedules (e.g., staggered, flex time, four-day week, telecommuting);
- alternative means of travel (e.g., carpools, vanpools, subscription buses);
- parking management (e.g., preferential parking, parking pricing, parking ratios, park-and-rides).

Most TDM programs are applied to commute trips and other travel during peak periods, and are organized through employers, most of which are relatively large firms. At first, many of the attempts to change traveler behavior were voluntary, such as government encouragement of ridesharing and transit use through advertising and marketing campaigns. Later the formation of Transportation Management Organizations (TMOs) showed private-sector interest in dealing with these problems. But for many communities, volunteerism and the private-sector response were too little and too slow. There was no external enemy to align private desire with the public good. New and existing land use developments were simply generating more peak hour trips than the existing (unpriced) road capacity could handle, and supply side increases were, for various reasons, undesirable or not feasible. For instance, Montgomery County, Maryland requires employers in designated Transportation Management Districts (TMD) to join together in an association, thus bringing about involuntary cooperation.

Another way to encourage Transportation Demand Management (TDM) is to provide incentives for employers or developers who choose to apply certain techniques. One incentive that has been proffered is reducing parking required by the development. This idea has met with mixed success, as developers do not necessarily want to reduce parking, which can be considered a selling point. Similarly, employers view parking as a perk for their employees, and so desire it when searching for office space.

Communities can also choose to implement TDM techniques on a development-by-development basis. Conditional zoning is the granting of permission by the community for the construction of a development that is otherwise not allowed in an area if the developer meets certain conditions. The conditions lead to Negotiated Developer Agreements (NDAs), whereby the developer agrees to do certain things, such as expand road capacity or apply TDM trip reduction measures for rezoning. But negotiating with each developer can become inefficient and arbitrary, particularly in a larger community with many developers, leading to the enaction of trip reduction ordinances.

In order to establish a rational nexus between a law and its intended effects, many communities place a series of "Findings" at the beginning of an ordinance. Reviewing the findings included in multiple ordinances, one finds a clear pattern in the reasoning behind TROs: the community has experienced, and is experiencing, economic growth, resulting in new employment; this new employment produces additional peak hour trips, which produce noise, reduce air quality, impair traffic circulation, and increase energy consumption; the application of relatively simple, inexpensive and effective TDM trip reduction measures will reduce the number of peak-hour trips, and thus mitigate the negative impacts of the economic growth.

These ordinances generally have a stated goal of ensuring that developers and employers participate in a program to mitigate traffic impacts and the air quality impacts by using TDM trip reduction measures.

In general, the ordinances establish specific objectives, such a maximum number of employees who are commuting to work in a single-occupant vehicle, mandating the number of outbound employee vehicle-trips per area unit of rented space, or, as in the case of the South Coast Air Quality Management District's (SCAQMD) Regulation XV, which between 1987 and 1995 required that employers achieve a specified Average Vehicle Ridership (AVR). It was repealed in part because of its heavy administrative burden, though it did increase AVR by about 3.4 percent at firms employing more than 100 workers.

A TRO can apply to developers, new employers, and even to existing employers. Different communities have established different requirements. Generally, however, they apply only to larger developments and larger employers. Targeting larger businesses makes administration easier and increases the chances of success for the various trip reduction techniques. Enforcement takes several techniques, including the use of penalties and fines, the requirement that firms post bonds that are surrendered, and the issuance by the community of only a Temporary Certificate of Occupancy.

Ridesharing is one of the most common techniques promoted to achieve a reduction in peak-hour vehicle trips. In fact, the first TROs were ridesharing ordinances. TROs define three kinds of ridesharing: carpools, vanpools, and buspools. Carpools use the vehicles of the employees, and permit two to four persons to share a ride to work. Vanpools generally use employee-provided vehicles (owned or leased) with an employee/commuter serving as the driver, and can comfortably accommodate up to 15 persons. The third kind of ridesharing has been dubbed the buspool. In this case a group of 30 to 40 commuters ride in a chartered bus to a particular worksite or urban location. The driver is a professional and the bus is often provided by a private operator for a fee (for instance, the Columbia, Maryland system evolved from the mid-1970s and was ultimately taken over by the state).

As viewed by employees, one of the major drawbacks to ridesharing is the lack of a vehicle during the day. Two strategies that have been attempted to alleviate these fears are the guaranteed ride home and loaner vehicles. As its name implies, the guaranteed ride home is a regulation that requires employers have a system in place to hire a taxi to take workers home if they must leave earlier or stay later than their rideshare. The provision of loaner vehicles, or the subscription to carsharing services, has been suggested as a means to provide workers with mid-day mobility.

Promoting public transit is another technique for the reduction in peak-hour vehicle trips. Promotion techniques are similar to those used for ridesharing. An Employee Transportation Coordinator[a] may be required to post bus/train routes and schedules on a transportation information board. The employer or developer may be required or encouraged to build bus shelters in front of the building. Additionally, the employer may be required or encouraged to give transit fare subsidies to employees. Fare subsidies help to counter the allure of free parking, a tax-free subsidy that employers are permitted to provide. Again, guaranteed rides home and loaner cars make taking transit more palatable. Sacramento, California has encouraged developers to pay for all or part of the costs of bus and light rail transit stations. Shuttle buses connecting business areas with each other and with regional transit services are another option.[b]

While the TRO typically aims to limit the number of trips occurring during peak hours, one set of techniques is not to reduce the total number of trips, but rather to change their timing. These techniques are collectively called alternative work hours. There are several kinds of alternative works hours: compressed work weeks, staggered work hours, and flexible work hours. Companies are often reluctant to implement such policies, but with the right set of incentives or requirements, this technique can be used successfully. The premier example of alternative work hour strategies is the 1984 Summer Olympic Games in Los Angeles, when, for a limited time, many of the city's major employers implemented alternative work hours to great success. [31]

Another set of strategies for restructuring of the traditional work schedule to achieve a reduction in the number of vehicle trips is telecommuting and work at home. Although working only at home has major drawbacks, working at home several days a week is a realistic possibility for many employees that could go far towards relieving congestion problems. Including a "work at home" program may meet program requirements. The provision of bicycle lockers and showers at the work site are techniques obviously designed to encourage bicycling. There are certain climates where bicycles will be used more frequently than others, and certain topographies. Northern California seems particularly supportive of bicycle transportation.

In practice, every Transportation Management Program employs a different mix of trip reduction techniques. This mix depends on the following employer characteristics:

- Firm—type of business, work shifts, type of employees;
- Location—physical characteristics, parking availability and access, cost of parking, congestion of local highways and streets, location in a business or industrial park, transit access, and bike and pedestrian path availability;
- Density—employee population of the company, population of the surrounding business district;
- Budget—availability of money or incentives to implement trip reduction program;
- Management Support—level of support and commitment demonstrated in budget and policy decisions to suppport a trip reduction program.

On the whole, it can be said that compliance and enforcement provisions of most of the current TROs are weak. Results are often self-reported by the company, and it is only when the company does not even bother to file a report that any action is taken. There are typically several levels of warning and appeal before fines go into effect. However, just because mandatory compliance is not strongly enforced does not mean that companies are not complying. Particularly in areas where the private sector instigated the Ordinance, companies want to comply for their employees' benefit if not out of the kindness of their heart.

On the whole, studies appear to support the conclusion that a reduction of peak hour vehicle trips on the order of 15 to 25 percent is possible using conventional site-specific TDM programs. Larger impacts come at a larger price, most often in the form of incentives. The most successful programs, reviewed elsewhere, [32, 33] use financial incentives and disincentives to persuade commuters not to drive alone; these strategies include vanpool and transit subsidies, financial disincentives (e.g., parking charges), bicycle and walk programs/subsidies (e.g., bike loans), or parking supply management (e.g., limiting parking).

a Most TRO's require the appointment of an Employee Transportation Coordinator (ETC). This individual becomes responsible, on the employer's side, for the implementation of the Ordinance's requirements. The ETC often conducts a survey of employees' travel behavior, determining what their trips are and by what mode they are made. This information is important in designing the trip reduction program. The ETC generally must report the progress of the employer's program to the local government agency responsible for administering the program, and the Coordinator serves as a liaison between the local government and the employer. Finally, the Coordinator is given the task of promoting the various techniques of trip reduction as are described below. SCAQMD's Regulation XV required that Coordinator's go through a special training course to become Trained Transportation Coordinators. The TRO may specify certain methods for the promotion of ridesharing. These include posting of potential carpools/vanpools on a transportation information board. It may even be so involved as to provide for personalized computer matching of commuters with potential rideshare partners. As a further encouragement to ridesharing, preferential parking may be required for high occupancy vehicles. This parking, in addition to being guaranteed, may also be less expensive than other spaces and closer to the front door of the building. As a technical matter, some ordinances have gone so far as to require that there be adequate van height clearance on parking structures (Bellevue, Washington). The final major incentive sometimes required in TRO's is monetary. Some jurisdictions require that a small amount of money be provided to those employees who rideshare.

b At the Hacienda Business Park in Pleasanton, California, shuttle buses are provided that run within the development, to and from the local transit station, and to shopping. Much as good site design, shuttle buses both encourage riding transit and diminish the need for having a vehicle during the day.

Notes

a The input output diagram helps understand delay. Imagine, for example, that we assume vehicles are able to stack vertically. That is, queues take place at a point. This is of course wrong, but not too wrong. The resulting travel time we get is almost the same as if we were to measure the queue taking place over space, the difference is that the time required to cover distance is included when we make the better assumption—even under free-flow conditions it takes time to travel from the point where a vehicle entered the back of the queue to the point where it exits the front. We can make that correction, but when queuing is taking place, that time is often small compared to the time delayed by the queue. Another assumption we make for exposition is that this is a deterministic process, that is vehicles arrive in a regular fashion and depart in a regular fashion. However, sometimes vehicles bunch up (drivers are not uniform), which leads to stochastic arrivals and departures. This stochastic queuing will in general increase the measured delay.

The delay resulting from queuing can be approximated by a number of functions. The famous Bureau of Public Roads function is one example developed in the 1930s, and still used today in planning models. However, we don't need to use approximations like that any more.

b   Vehicles interacting would still lead to some congestion.
c   See, for example, www.dailyrepublican.com/nobelprizewinners.html.

References

[1] Taylor, B., 'Rethinking Traffic Congestion', *Access*, 2002, vol. 21 (Fall): 8–16.
[2] Small, K. and Chu, X., 'Hypercongestion', *Journal of Transport Economics and Policy*, 2003, vol. 37 (3): 319–352.
[3] Vickrey, W.S., 'Pricing in Urban and Suburban Transport', *The American Economic Review*, 1963, vol. 53 (2): 452–465.
[4] Wachs, M., 'Will Congestion Pricing Ever Be Adopted', *Access*, 1994, vol. 4 (Spring): 15–19.
[5] Giuliano, G., 'Equity and Fairness Considerations of Congestion Pricing', *Curbing Gridlock*, 1994, vol. 2: 250–279.
[6] Richardson, H.W. and Bae, C.H.C., 'The Equity Impacts of Road Congestion Pricing', *Road Pricing, Traffic Congestion and the Environment Issues of Efficiency and Social Feasibility*, 1998.
[7] Small, K.A., 'Using the Revenues from Congestion Pricing', *Transportation*, 1992, vol. 19 (4): 359–381.
[8] Giuliano, G., 'An Assessment of the Political Acceptability of Congestion Pricing', *Transportation*, 1992, vol. 19 (4): 335–358.
[9] Jones, P., 'Gaining Public Support for Road Pricing through a Package Approach', *Traffic Engineering and Control*, 1991, vol. 32 (4): 194–196.
[10] Meyer, M.D., Saben, L., Shephard, W., and Drake, D.E., 'Feasibility Assessment of Metropolitan High-Occupancy Toll Lane Network in Atlanta, Georgia', *Transportation Research Record*, 2006, vol. 1960: 159–167.
[11] Levine, J. and Garb, Y., 'Congestion Pricing's Conditional Promise: Promotion of Accessibility or Mobility?', *Transport Policy*, 2002, vol. 9: 179–188.
[12] Levinson, D. and Zhang, F., 'Ramp Meters on Trial: Evidence from the Twin Cities Metering Holiday', *Transportation Research, Part A: Policy and Practice*, 2006, vol. 40 (10): 810–828.
[13] Rogers, P., 'Hybrids Cleared for Carpool Lane', *The San Jose Mercury News*, August 11, 2005.
[14] Shoup, D.C., *The High Cost of Free Parking*, Chicago, IL: APA Planners Press, 2005.
[15] Godschalk, D.R., 'In Defense of Growth Management', *Journal of the American Planning Association*, 1992, vol. 58 (4): 422–424.
[16] Nelson, A.C., Pendall, R., Dawkins, C.J., and Knaap, G.J., *The Link Between Growth Management And Housing Affordability: The Academic Evidence*, The Brookings Institution Center on Urban and Metropolitan Policy, 2002.

[17] Chinitz, B., 'Growth Management Reconsidered: Good for the Town, Bad for the Nation?', *Journal of the American Planning Association*, 1991, vol. 56 (1): 3–8.
[18] Dalton, L., 'The Limits of Regulation', *Journal of the American Planning Association*, 1989, vol. 52 (2): 151–168.
[19] Downs, A., 'Growth Management: Satan or Savior 1: Regulatory Barriers to Affordable Housing', *Journal of the American Planning Association*, 1992, vol. 58 (4): 419–422.
[20] Fischel, W., 'Growth Management Reconsidered: Good for the Town, Bad for the Nation? A Comment', *Journal of the American Planning Association*, 1991, vol. 57: 341–344.
[21] Landis, J.D., 'Do Growth Controls Work? A New Assessment', *Journal of the American Planning Association*, 1992, vol. 58 (4): 489–508.
[22] Pendall, R., 'Do Land-use Controls Cause Sprawl?' *Environment and Planning B*, 1999, vol. 26: 555–571.
[23] Rodriguez, D.A., Targa, F., and Aytur, S.A., 'Transport Implications of Urban Containment Policies: A Study of the Largest Twenty-five US Metropolitan Areas', *Urban Studies*, 2006, vol. 43 (10): 1879–1897.
[24] Katz, L. and Rosen, K.T., 'The Interjurisdictional Effects of Growth Controls on Housing Prices', *Journal of Law and Economics*, 1987, vol. 30 (1): 149–160.
[25] Pollakowski, H.O. and Wachter, S.M., 'The Effects of Land-use Constraints on Housing Prices', *Land Economics*, 1990, vol. 63 (3): 315–324.
[26] Chinitz, B., 'Growth Management: Good for the Town, Bad for the Nation', *Journal of the American Planning Association*, 1990, vol. 56 (1): 3–8.
[27] Fischel, W.A., *Do Growth Controls Matter? A Review of the Empirical Evidence on the Effectiveness and Efficiency of Local Government and Land Regulation*, Cambridge, MA: The Lincoln Institute of Land Policy, 1990.
[28] Schneider, D.M., Godschalk, D.R., and Axler, N., *The Carrying Capacity Concept as Planning Tool*, Planning Advisory Service Report No. 338, Chicago, IL: American Planning Association, 1978.
[29] Altshuler, A.A. and Gómez-Ibáñez, J.A., *Regulation for Revenue: The Political Economy of Land Use Exactions*, Washington, DC: Brookings Institution Press, 1993.
[30] US Bureau of Transportation Statistics, *Overview of Travel Demand Management Measures: Final Report* [DOT-T-94-11], 1994. Available at: http://ntl.bts.gov/DOCS/273.html.
[31] Giuliano, G., Haboian, K., Rutherford, K., Prashker, J., and Recker, W., *Evaluation of 1984 Los Angeles Summer Olympics Traffic Management* [UCI-ITS-WP-87-8], Irvine, CA: University of California, 1987.
[32] Ferguson, E., *Transportation Demand Management*, Chicago, IL: American Planning Association, 1998.
[33] Ferguson, E., *Travel Demand Management and Public Policy*, Aldershot: Ashgate, 2001.

Chapter 14

Drawing the curtain

"Let's consider a reevaluation of the situation in which we assume that the stuckness now occurring, the zero of consciousness, isn't the worst of all possible situations, but the best possible situation you could be in. After all, it's exactly this stuckness that Zen Buddhists go to so much trouble to induce..."

Robert Pirsig from *Zen and the Art of Motorcycle Maintenance*

We began this book by suggesting that urban areas were at a crossroads. Traffic congestion has existed for centuries and current observations suggest that it will not relent in the foreseeable future; this state of affairs suggests no visible crossroads. Notwithstanding relatively new urban developments (both greenfield and brownfield), cities' core transportation networks (i.e., roads) and accompanying property lines and buildings have been established for many years. These facilities, legal constructs, and structures have long lives; this, too, suggests no imminent crossroads.

The degree to which a crossroads truly exists arises from the confluence of need and opportunity. Need for a new path emerges from recognizing the environmental and ethical costs incurred by procuring and burning oil and by paving the landscape. Need for change arises from the imminent retirement of an influential population cohort, the "Baby Boom" generation, which will place new demands on the built environment. Need also rains from the sheer level of wasted time in congestion experienced by the *average* commuter, which perhaps only now is reaching levels that are unbearable for most.

Opportunity comes from new forms of cleaner transportation—either in the form of new mass transit systems, or of energy efficient cars. New information technologies lessen the need for temporal and spatial coordination, and enable remote and effective asynchronous work for many, while retirees have a great deal of locational flexibility (and all the time in the world). Other new information technologies enable strategies such as road pricing to reduce congestion by effectively (and electronically) signaling to travelers the truer cost of their trip, without making them stop at a tollbooth to pay that price. The ability of society to recognize these needs and opportunities depends on changes in the political climate, the availability of new and better solutions, and shifts in socio-demographic and lifestyle trends.

Chris Nelson, although not explicitly mentioning a crossroads, argues that immense changes lie ahead for cities in the United States. [1] He projects that the composition of American households in the future will differ markedly from the past; that the market for traditional suburban housing is waning; that in 2025, an emerging labor force will demand twice as much non-residential space as existed in 2000; and that new residential construction may equal half of all residential units that existed in 2000. Nelson suggests that "now is the time for planners to craft a new template that meets the challenges of the next planning era."

A crossroads permits travelers to turn left or right, make a U-turn, or continue heading in the same direction. If travelers (or cities, for that matter) prefer the status quo, they can proceed forward on the current path. But a turn left or right implies a change in how communities plan place and plexus, and a corresponding change in expectations that will result.

This book has offered new perspectives—presented in the form of five "Diamonds," of Choice, Exchange, Design, Assembly, and Operation—to understand the behavior of individuals, firms, and governments. We, the authors, have also offered strategies to think about evaluating the actions of government agencies (e.g., the Diamond of Evaluation). Throughout, we have suggested that there are many alternatives that individuals, firms, and governments could pursue; it is just that the alternatives that will make a difference exceed the scope that the polity has been willing to accept. Thus, our core message has been that continuing on the current trajectory, capitalizing on an incrementalist fix where it seems appropriate, will only move communities incrementally in the direction they want to go.

Changes and perspectives for future planning

Few communities would reject the goals of reduced pollution, increased preservation of natural resources, and better quality of life. But the existing land use-transportation environment in many central cities, most inner-ring suburbs, and almost all outer-ring and exurban settings requires significant changes to be able to make progress towards any one of these goals.

Some may consider a major change refreshing. It would certainly be consistent with the most popular quote in the history of city planning. In 1909, the co-author of the historic Chicago Plan, Daniel Burnham, wrote:

> Make no little plans. They have no magic to stir men's blood and probably themselves will not be realized. Make big plans; aim high in hope and

work, remembering that a noble, logical diagram once recorded will never die, but long after we are gone will be a living thing, asserting itself with ever-growing insistency. Remember that our sons and grandsons are going to do things that would stagger us. Let your watchword be order and your beacon beauty. Think big.

Executing a big plan may be exactly what most cities need at this point. However, such exuberance needs to be tempered. Despite individuals getting excited about the idea of change, collectively people are incrementalists. They are mostly adverse to change when it comes to dramatically affecting their behavior. A collection of eight incrementalist strategies, offered by Giuliano and Hanson, [2] to produce better urban transportation and more livable communities illustrates the conventional professional wisdom.[a] Furthermore, people prefer to have changes prove themselves as "superior" before permitting the next change. This might be why the incrementalist policies and behavior we have witnessed time again over the past decades is probably best phrased as "Make no big plans, they will stir men's blood."

Such a dichotomy poses a dilemma. The current land use-transportation environments need change; however, people are generally cautious or adverse to it, yielding the following situation:

- Cities have big problems;
- Small changes cannot address big problems;
- To be consistent with most planning visions and to solve big problems, metropolitan regions need big changes;
- The populace resists big changes.

Like the recurring tag line by Ulysses Everett McGill (George Clooney) in the film O Brother, Where Art Thou?, "Damn, we're in a tight spot."

Developing strategies to solve the slate of land use-transportation problems requires an ambitious plan. This plan must demonstrate a comprehensive understanding of how these systems operate, and propose strategies for future action. The preceding chapters help provide this understanding. However, these chapters have not specified which changes could and should take place. Nor have we furnished a multi-point plan to do so. Sorry to disappoint the reader, but we do not claim to have all the answers. After years of researching land use-transportation interactions, we contend no one does.

What we can offer, however, are eight suggestions to better position many of the policies and actions debated in planning circles. We suggest planners utilize four primary methods and undertake four challenging strategies for change:

Methods

1   First, do no harm;
2   Prize evidence-based practices;

3 Scrutinize the merits of claims;
4 Let a hundred flowers bloom, but cull the laggards.

Strategies for change

5 Recognize the confines of mature systems;
6 Relocate intelligence and incentive;
7 Rewire the plexus;
8 Reinvent the city.

Each suggestion charts a new direction - a direction for land use-transportation issues that will result in a greater likelihood of succeeding. We briefly describe each below.

1 First, do no harm

The disastrous urban renewal programs of the 1950s and 1960s are folkloric for today's planners. The history books describe how declaring healthy communities blighted and subsequently replacing them with a tangle of superhighways, surface parking lots, subsidized housing, and shopping malls seems like heresy against the backdrop of contemporary planning initiatives. Yet policies deemed "progress" at the time have set back by decades many vibrant neighborhoods. Some cities have yet to recover. Replacing functioning (though officially "blighted") neighborhoods with high-rise towers of subsidized housing hardly serves the previous residents.

Medical doctors generally view their task as improving the health of the patient, not worsening it. Even worse, urban renewal that destroys parts of cities with no clear plans for what will replace them is analogous to first-year medical students cutting open a body and leaving it there while they complete medical school. Certain aspects of the urban planning profession have failed to heed the advice "first, do no harm."[b]

The seemingly predominant dogma in urban planning is that policy intervention is best. However, the planning profession is now realizing the error of its ways and the limitations of this dogma. Echoing Chapter 11, "we have met the enemy and [it] is us."[3] On a daily basis, several other policies, considerably more mundane than urban renewal, interfere with the healthy functioning of networks and neighborhoods. Many have equally detrimental effects. A number of problems are created by society itself, abetted by planners. These policies include, but are certainly not limited to:

- land use regulations that reduce development densities and lead to more exclusive suburban communities than would otherwise arise; [3–7]
- a zoning monoculture fostering sterile "garagescapes" in many suburban environments; [8]

- minimum parking requirements designed to satisfy demand on one or two days a year, leaving acres of blacktop to absorb the sun and produce polluted runoff the rest of the year; [9]
- transportation standards that mandate wide streets and a sweeping roadway geometry; [10]
- transit systems that run nearly empty buses on suburban routes to ensure the political capital required to fund nearly full buses in the heart of the city (transit systems which are designed to serve not their customers, but their potential customers). [11]

Working toward many of the suggestions offered in this book does not always require directed planning intervention. In some cases, it requires the absence of directed planning intervention. Prior to prescribing the next generation of design mandates to heal cities' place and plexus, one must recognize that the existing pattern of development is governed by a web of municipal regulations. [12]

2 Prize evidence-based practices

Continuing with the medical theme and borrowing from a burgeoning medical movement initiated in Britain could also be of benefit. The movement, termed evidence-based practice, claims that current medical prescriptions based on intuition, unsystematic clinical experience, and/or pathophysiologic rationale is insufficient. This so-called paradigm shift urges that decisions about the care of individual patients be informed by conscientious, explicit, and judicious use of current best evidence. The practice aims to integrate individual clinical expertise with the best available external clinical evidence from systematic research. [13] Under such guidelines, little attention is paid to the "best practices" for population-based (public) health.

The above may come as a surprise to many, as it did to the authors; we certainly thought our doctors already prescribed treatments based upon documented success rather than upon intuition or simple emulation. Perhaps we were over-assuming. Evidence-based approaches are gaining steam in other disciplines as well, including public health [14] and business. For example, two Harvard Business School professors question the wisdom of managers making decisions based on the obsolete knowledge they picked up in school, long-standing but never-proven traditions, patterns gleaned from experience, methods they happen to be skilled in applying, and information from vendors. [15, 16] The alternative, under evidence-based paradigms, is to base decisions on facts and logic, not ideology, hunches, fads, or poorly understood experience. Several principles of evidence-based management include:

- facing the hard facts, and building a culture in which people are encouraged to tell the truth, even if it is unpleasant;
- committing to "fact based" decision making—which means being committed to getting the best evidence and using it to guide actions;

- treating your organization as an unfinished prototype; encourage experimentation and learning by doing;
- looking for the risks and drawbacks in what people recommend; even the best medicine has side effects;
- avoid basing decisions on untested but strongly held beliefs, what you have done in the past, or on uncritical "benchmarking" of what winners do.

If it is good enough for the medicine (the body) and business (the firm), we ask: why not urban planning (the city)? Should we not expose urban environments to the same standards? Most urban planners are trained as social scientists, after all. They should therefore appreciate policies that are carefully conducted and objectively evaluated.

3 Scrutinize the merits of claims

Evidence-based practice is based on research. However, the above recommendation, although clean in theory, needs to be tempered by an understanding that the urban planning profession is less amenable to some of the "cleaner" research available in other fields such as medicine or business. For example, after nearly three decades of increasingly sophisticated research using ever improved datasets, statistical methods, and techniques for geographic analysis, Boarnet and Crane [17] state the following "our conclusion is not that urban design and transportation behavior are not linked, or that urban design should never be used as transportation policy. Rather, we conclude that we know too little about the transportation aspects of the built environment." This is a rather unsettling conclusion, especially given the reams of research devoted to the topic.

Research examining land use and transportation is troubled by a number of issues, including:

- the relatively long lag times required for initiatives to take effect (some policies may take years to come to fruition);[d]
- the difficulty of controlling for confounding effects that determine behavior (there are often many different reasons for individual decision-making);
- the general impossibility of implementing rigorous experimental designs—under which a randomly selected control group would live and travel in controlled environs, and an experimental group might be exposed to various treatments— in free and democratic environments where people decide where to live and work;
- the inability to monitor treatment and develop rigorously controlled studies (it is impossible to tell people where to live);
- the marginal effect of many policies in a free society (there are larger economic forces influencing human behavior, such as the low costs of automobile travel); and

- the lag time between gathering data and subsequent analysis (a matter that is undoubtedly improving but still an issue).

Such issues create a difficult role for research. How does one know which research to believe, and when? In some cases, the research supports an intervention. In other cases, it supports a "do nothing" approach. In many contexts, it shifts the terms of the debate.

Consider the raving excitement around the New Urbanist movement over the past two decades. In part, the excitement initially stemmed from the claimed ability of land use design to reduce traffic congestion and private vehicle travel. Years of research has now peeled back the multiple layers of this onion, however. Communities of researchers—and practicing planners, for that matter—are learning that such initiatives have not satisfied the original claims. Instead, the primary benefit of such designs may lie in making transit, walking, and cycling more attractive and thereby improving overall quality of life for those who reside in such neighborhoods.

A similar initiative grew out of arguments that advocates of non-motorized transportation often put forward to advance their cause. They claim that investments in non-motorized infrastructure will reduce congestion, increase physical activity, decrease natural resource consumption, increase livability, and decrease smog. Can these arguments reliably be supported with evidence? Several leading practitioners and academics suggest that many of the benefits touted for walking and cycling facilities—decreased congestion, decreased consumption of natural resources, and even overall increases in physical activity—are *not* the benefits that will ultimately come to fruition. A close review of research to date indicates that it is easier to get people out of their cars than onto their feet. [18, 19] In addition, rates of bicycling are currently so low that even a quadrupling of the number of people in the United States who bike to work would lessen environmental and other harms from motorized vehicles to an immeasurably small degree.

Upon a close examination of the literature, a few select benefits related to the relatively ambiguous goals of "livability" appear to hold more hope for meeting their expectations. From a policy perspective, the subject of non-motorized transportation presents a bit of a dilemma. Statistics are spotty and the literature appears to be heavily laced with advocacy. Thus, the overarching policy questions are whether non-motorized transportation, in fact, is a transportation services issue or a lifestyle issue. Giuliano and Hanson suggest, "building communities with abundant walking and biking opportunities may be more about livability than solving transportation problems."

These examples suggest a highly nuanced role for research. Where available, research should certainly guide decision-making about applicable policies (consistent with proposition number two above). Where the research is not available (either due to scale, timing or some other limitation), transportationists must inform the discussion as best they can. We suggest there might be merit in heeding the advice of the late Mel Weber, who reportedly taught his students to be skeptics: "if an idea was crystal-clear and simple, he suspected it was wrong and asked question after question to peel back the layers of complexity that lay behind most matters of public policy in the city."[e] Research can help inform the debate, which in turn can help scrutinize the merits of different claims.

4 Let a hundred flowers bloom, but cull the laggards

Using evidence to make decisions is good where an idea has already been implemented. But there are new ideas that have never been tested. Well-constructed policy experiments (testing hypotheses) are an essential part of the scientific method. We enthusiastically support letting a hundred flowers bloom.[f] These opportunities provide refreshing glimpses into ways of changing behaviors and operating systems that are often assumed to be immutable.

The key lesson is to distinguish between policies that work (i.e., that achieve their desired outcome) and those that do not. Any good planning initiative has objectives against which to measure progress. It is important to advance those policies or initiatives that are empirically demonstrated to work.

The problem comes when policy experiments are, despite evidence to the contrary, declared successes and become institutionalized. We believe that hard choices (like abandoning someone's pet project) must be made in order for progress to occur. Rigorous benefit-cost analysis that includes both the real benefits and the full costs is one way to discern whether or not a project serves its purpose in improving efficiency. On the other hand, many projects fail to advance economic efficiency, but rather aim to ensure that all people have some opportunities (equity), to improve the environment, or to create a better experience for travelers and residents. Those goals are all fine, but in an era of scarce resources, some projects achieve them better than others. Comparative evaluation must be conducted. Continuation of failed experiments diverts resources from alternative uses. For every decision there is an opportunity cost.

5 Recognize the confines of mature systems

If planners adopt the four methods set forth in this chapter, we believe plans will only get better. But there are some specific strategies that can also help. To understand these specific strategies, planners must first understand the reasons underlying the current state of "stuckness" in which many cities find themselves mired.

Most of the major towns in today's Europe were founded prior to 1300; [20] development of most US cities was initiated prior to 1900. Although metropolitan regions are certainly growing in size, few are being developed anymore with a new, robust economic center. As each new city was established, the number of remaining good sites for new cities diminished. When new territories were discovered by European explorers, (the Americas, Australia), additional new towns were formed, but the same process of site-elimination resumed. As the planet has been settled, the number of new places left to discover is minimal.[f] Metropolitan regions, considered as entities in and of themselves, are subject to the same cycles represented by the traditional growth and development process: birth, growth, maturity, and decline. An important caveat is that metropolitan areas are characterized by a particularly long phase of maturity; their decline is not imminent. These regions are long-lived mature systems.

The reason they are long-lived is that urban areas are formed by laying roads, assigning property rights, and building structures (though not necessarily in that order). The longest-lasting imprint for any city is the road system. The more mature it becomes, the harder it is to change. The transportation story in the United States (and almost all other developed countries) shows few recent advances. Ever since the Interstate Highway System was completed in the US, the entire transportation network has grown minimally (Figure 14.1),[8] suggesting an extremely mature system.

Looking at the places to which people travel along the roadways system—the buildings and other structures—tells a similar story, though buildings are not quite as long-lasting. Figures 14.2 and 14.3 show the year of construction for residential and non-residential uses in the US, respectively. Looking at residential uses, we see that many housing units are durable and long-lived. Almost 10 percent of all units were built before 1920; more than one-quarter are at least a half-century old. For non-residential uses, more than 70 percent of buildings and total floorspace in 1995 were constructed prior to 1980, and more than 50 percent of buildings and floorspace were constructed prior to 1970. Based on data displayed in the figures we calculate the (weighted) average age of the current residential and non-residential building stock to be 36 and

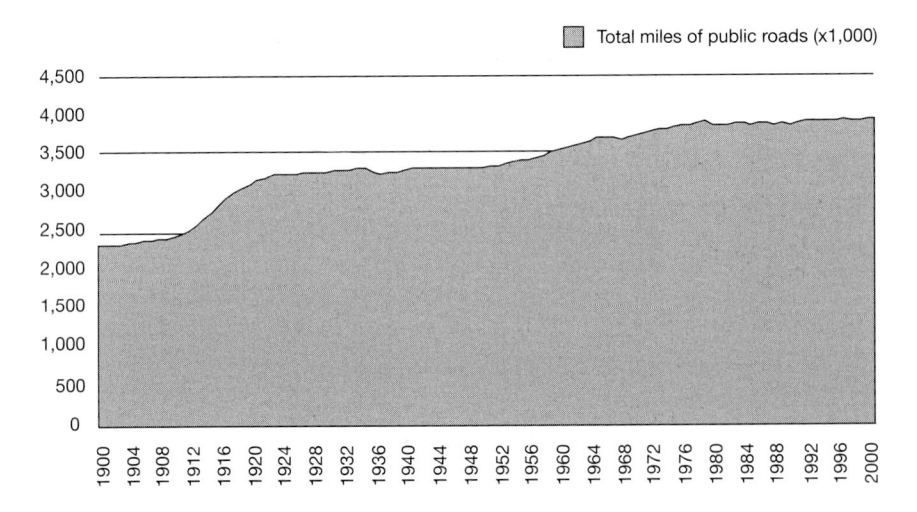

Figure 14.1 Miles of public roads in the United States

37 years, respectively. The existing building stock, although certainly not as stable as the rights-of-ways of roads, is not exactly volatile either. Once structures are built, their use may change with some frequency,[i] though they stay around for a long time, also suggesting a relatively mature system.

If one recognizes that urban areas are mature systems, one must accept a central property of such systems: additional improvements to existing systems have only marginal effects. For example, each additional network segment has a smaller effect in reducing travel time. This is because as the world gets more connected, the time saved by any additional segment is less than the time saved by the previous segment (this is also known as the law of diminishing marginal returns). At market saturation, traditional investments have little marginal impact. Why? Because they are unlikely to move the accessibility gradient. free-flow times cannot be changed much on a mature network. Similarly, slower, or lower-quality investments will not significantly change accessibility. These themes have been examined throughout this book. For example, we demonstrated how:

- New bus service on city streets will minimally reduce transit travel time, and it will still be slower than auto time. Unfortunately, the evidence is similar for a single grade-separated transit line (e.g., rail). Although it will

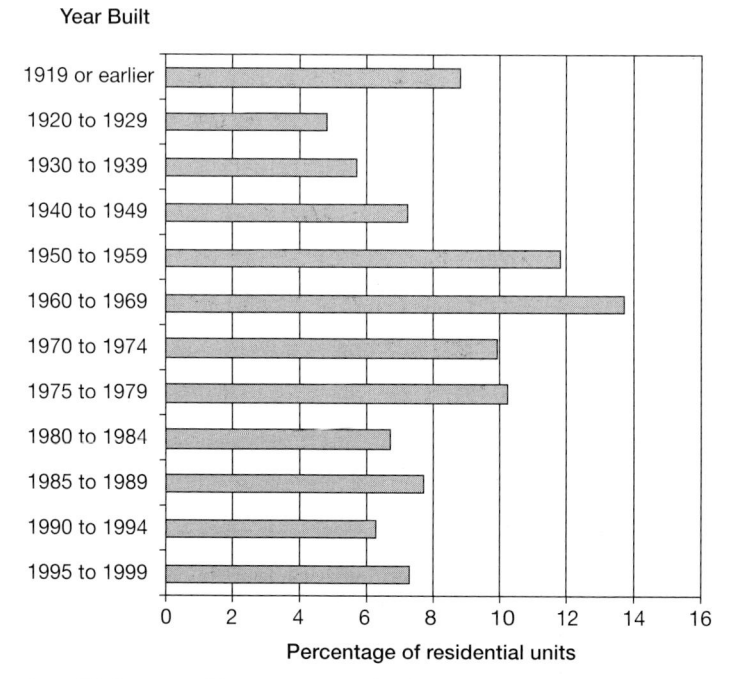

Figure 14.2 Year of construction of existing residential properties

Figure 14.3 Year of construction of existing commercial properties

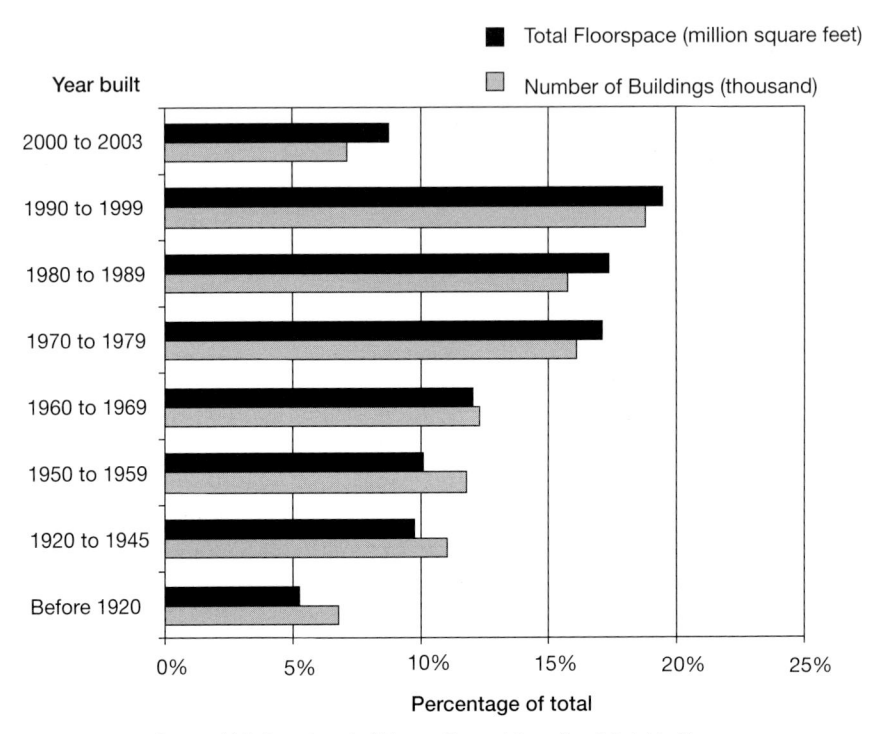

serve people in a specific corridor going from origins along the line to destinations along the line, that number is relatively small compared to the travel market in general;
- A new roadway is a relatively small addition to the existing system of roadways in most metropolitan areas;
- Physically locating jobs closer to housing is a strategy that proves relatively weak compared to the social networking approaches most people employ to find work;
- Even quadrupling the number of people in the United States who walk or ride bicycles to work would not come close to measurably lessening environmental and other harms from motorized vehicles.

In mature land use-transportation systems in metropolitan areas, incrementalist policies will have even more incremental effects. This is not to say that the benefits of incremental changes do not exceed their costs. It is just that incrementalism fails to solve urban transportation problems and needs to be understood in this context. There should be no expectation that incrementalist policies will dramatically alter the course of events.

6 Relocate intelligence and incentive

This book has been constructed from the perspective of agents: individuals, firms, governments. Each agent makes decisions, and each has unique perspectives and local knowledge. Agents behave according to rules and an important rule is to respond to incentives. A relatively recent fashion in planning circles urges "Smart Growth," a series of principles offered to guide the next generation of planning policy. Smart Growth Online[10] lists the following principles:

1. Mix Land Uses;
2. Take Advantage of Compact Building Design;
3. Create Range of Housing Opportunities and Choices;
4. Create Walkable Neighborhoods;
5. Foster Distinctive, Attractive Communities with a Strong Sense of Place;
6. Preserve Open Space, Farmland, Natural Beauty and Critical Environmental Areas;
7. Strengthen and Direct Development Towards Existing Communities;
8. Make Development Decisions Predictable, Fair and Cost Effective;
9. Provide a Variety of Transportation Choices;
10. Encourage Community and Stakeholder Collaboration.

Some of these would be pursued by the private sector if operating in laissez-faire conditions; others would not. An examination of cities (for example, London) prior to the advent of planning regulation suggests they were mixed use, high density, multiple income, walkable, and distinctive. By concentrating development, less land was consumed, and existing places became more populous. The first seven development principles are in line with market forces *circa* 1900. Principle 9 was as well, as privately operated buses, streetcars, surface rail, and underground all competed with the personal horse and carriage and later motorcar. Principles 8 and 10 are artifacts of the modern regulated world. The problems faced by London at the dawn of the twentieth century are the same problems we opened with in Chapter 1—congestion, crowding, pollution—and are in many ways consequences of the very urban characteristics the advocates of Smart Growth now seek to re-establish.

But today Smart Growth, however it is defined, does not happen where the planners want it. It happens where the developers want to build (supposedly within the confines of development regulations). Developers have a relatively straightforward objective: maximizing profit. If developers can "do well by doing good," they are happy to do so (and claim credit in their marketing materials). But with this objective, they behave as dumb growers; they follow relatively basic incentives, first and foremost the lure of money.

Fortunately, a few government planners cannot better respond to individual wants and desires than individuals themselves. Instead of proactively overplanning, we suggest it is more effective to manage the costs of development on publicly provided infrastructure. In other words, charge development based on the costs it imposes on society, and if necessary provide discounts (or subsidies) for better quality (e.g., developments that provide better user experiences) and use the resulting funds to build infrastructure and remedy any damage done. If the Smart Growth advocates are correct, the costs will be higher where new infrastructure is necessary, and lower where old but underutilized infrastructure can be effectively and inexpensively exploited. The prices need to be smart, but the development process need not be excessively fettered by complex regulations. In policy terms, this means impact fees, taxes, and other similarly administered policies with low transaction costs. In short, we need "smart prices" to steer "dumb growth."

Smart prices should also be applied to travelers using congested roads, through tools such as electronic tolling. Like developers, travelers are self-interested and generally wish to reduce their own travel times. If this imposes costs on others, so be it—this is, in the poetic words of Roughgarden [21] "the Price of Anarchy."[k] But travelers who were charged a toll based on how much they inconvenienced others would then consider that in their calculations, and decide, to the benefit of society, whether or not to travel on that route at that time.

7 Rewire the plexus

A major under-recognized issue affecting place and plexus is the hierarchy that organizes it. Whatever the scale, all areas have central places (e.g., downtowns, suburban activity centers, regional centers, town centers, neighborhood centers). Connecting these places requires hierarchically organized transportation networks: the airline hub-and-spoke system, ports connected to intercity highways and railroads, major urban highways linking signalized arterials, collectors, and distributors.

Is the existing layout optimal? Almost certainly not. The layout for many of these environments was developed during a time in which different economic structures and more primitive technologies held sway (centuries or half-centuries ago). We would be misguided to think that the current layout could be made optimal. Does a grid street network result in too much traffic in front of people's homes? Does the hierarchical limited-access suburban collector distributor systems concentrate too much traffic on too few links, leaving traffic flow vulnerable to small perturbations and resulting in congestion, with no good alternatives? Do rail corridors serve current demand patterns? Any restructuring of the plexus inevitably restructures the place that accompanies it; these decisions cannot be taken lightly. The current hierarchy of roads is the result of design guides such as the AASHTO Green Book. Using cookbook-like design guides produced by professional associations helps low-skilled staff mass-produce elements of complex systems. But should complex systems be mass produced?

Context matters. Solutions that work in some environments do not work in others.

A new transit network becomes more valuable with more segments (the more places that are connected); yet, the first segments (or first five) rarely justify their cost. Sometimes, substantial capital and long-term planning must be committed in order to build a network for the long term. On the other hand, the network may never pay off if it comes after existing land use patterns have been established and if overall growth is small. The decision makers who ignored the early losses end up throwing good money after bad in the futile hope of achieving possible (prospective?) future gains. It is considerably less risky to invest in a known, largely built-out network than in one whose success depends on speculative forecasts. In particular, it may help to know if the new network is technologically superior to what surrounds it. If a travel mode is faster and cheaper and connects places people want to go, it is a reasonably good bet people will ride, but if it remains slower or pricier, the logic falls short. When everyone is traveling on surface streets at 50 km/h (30 mph), a 100 km/h (60 mph) freeway is a big deal. When everyone is traveling at 100 km/h, a 50 km/h light rail line is not.

Over time, as cities become larger and transportation networks become faster, origins and destinations tend to become more diffuse (less concentrated). In most cities, there is no dominant "central business district" (CBD) which concentrates the majority (or even a strong plurality) of employment. Places with sufficiently concentrated development, in sheer numbers if not market share, such as Manhattan, Hong Kong, Tokyo, and London, are well suited for rail transit and usually have it. But those places have interlinked rail networks, not individual lines, nor do they rely on a simple hub-and-spoke system. In short, trying to return to the rail-based, monocentric CBD, walking city model, although it has nostalgic appeal, could be considered analogous to tilting at windmills.[1] One is basically trying to recreate the city of 1910 while neglecting the changes that have occurred since then.

Each city is a living organism that, like the brain, needs to continuously rewire itself to take into account changed circumstances. People don't stop learning once they leave school, and cities similarly need to reallocate resources and forge new networks to compete in a world economy. Those investments must be chosen carefully, however, with full attention paid to the context in which the city finds itself.

8 Reinvent the city

The previous point considered rewiring the plexus, a step most cities think they need to pursue in order to be competitive. However, cities and the infrastructure technologies that serve them are not local. A good technology—a superior transportation infrastructure—is almost always replicated, and subsequently customized for more specific purposes.

The concept of S-curves helps demonstrate this point. First popularized by Everett Rogers in the *Diffusion of Innovations*, S-curves suggest that good ideas are diffused through a process where an innovation is communicated through certain channels over time and members of a social system. [22] The life cycles of technologies (birth, growth, maturity, and ultimately decline) are driven by the constraints imposed by old technologies and the opportunities created by new ones. Figure 14.4 shows how such ideas have played out for transportation infrastructure over time.

The idea of progress itself is embedded in a capitalistic market system: old technologies cease to have special profits (profits in excess of normal rates of return); new technologies promise extraordinary profits, and attract investment. Capitalists, as profit-seeking (and risk-taking) investors, chase the potential for special profits associated with the new technologies while milking the old.

Local governments behave similarly, as bodies with mercantilist interests to promote their own communities. Furthermore, the transportation sector has evidence of the new technology proving itself superior to the old and then the old technology being abandoned. The turnpikes and canals in the late 1800s, and passenger railroads and streetcars in the mid 1900s are prime examples.

In Chapter 12 we discussed land use feedbacks by visiting George Lucas's planet of Coruscant. Returning to Coruscant, we witness spaceships and flying cars. Clearly, the technology there is more advanced than that of early twenty-first-century Earth. If an English-speaking Earthling asserts today: "I commute by spaceship," the statement would be understood and grammatically correct, but absurd; of course, few take a spaceship to work (unless they work at the International Space Station). It is not technologically impossible to take a spaceship, even though manned spacecraft have been in existence for over forty years. It is too expensive for just about everyone to commute by spaceship on a regular basis. Some people (wealthy space tourists) could do it once in a while, others do it with large subsidy (government employees working for NASA).[m]

This "of course" relies on a tacit understanding of what is feasible. What is feasible today is very different than what was feasible a hundred or a thousand years ago. Once, travelers were unable to take a car or a train or an airplane to work either. Why? These modes did not exist. The set of available choices was (and is) constrained by technology. It is not only the future that is absurd. I could say, "I commute by horse." Although also grammatically correct and understandable, it is not quite as absurd as the sentence "I commute by spaceship," but still seems highly unlikely for contemporary citizens. The issue is not that no one commutes by horse (many of us had ancestors who almost rode the back of a horse frequently). It is just that the era of the horse has passed. Horses are inefficient compared with the alternatives. This was shown in the 1800s with the rise of the railroad and the electric streetcar, and the dominance of the "iron horse" over the steed of flesh and bone.

S-curves, such as those presented in Figure 14.4, suggest that inventions related to the same category (in this case, transportation) often come in waves. That is, first came canals, then rail, then streetcars, followed by the interstate highway system. What is the next wave? It is difficult to tell. Some have speculated on automated highway systems or intelligent vehicles. Other

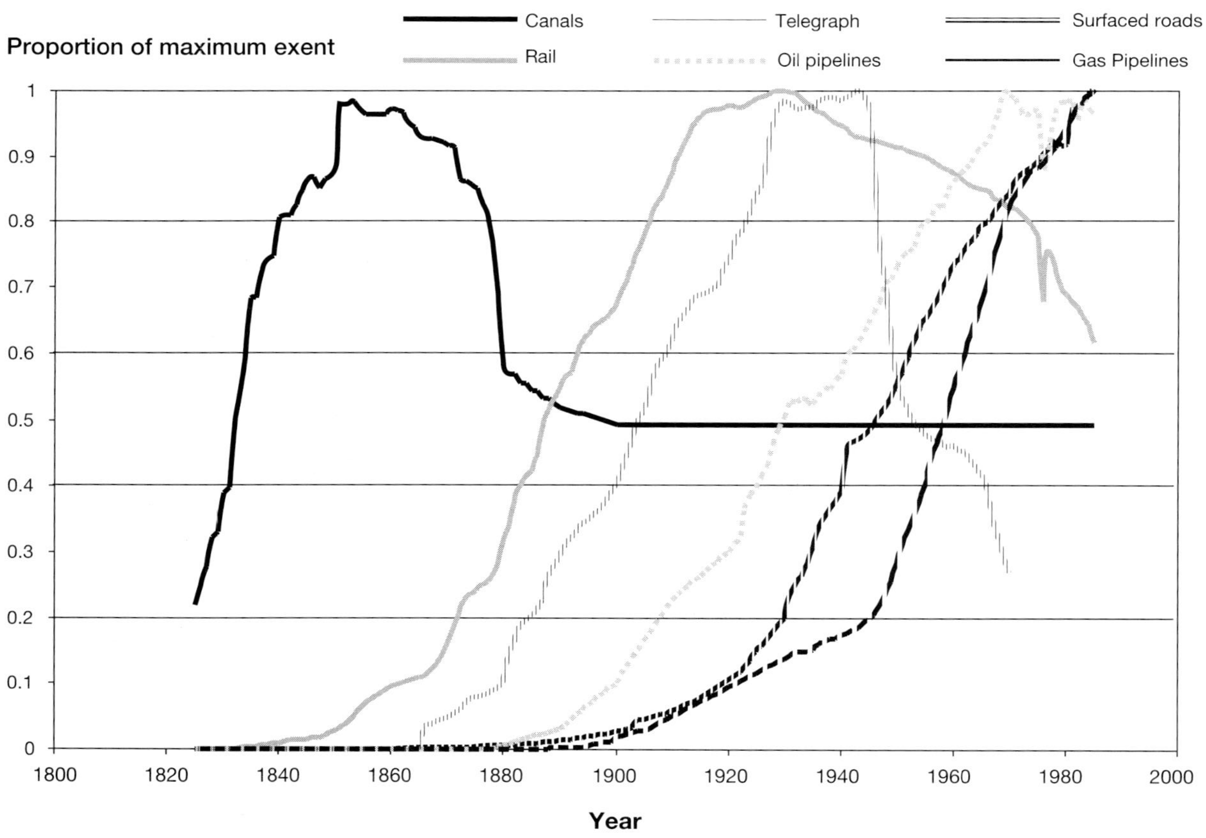

Figure 14.4 Size of networks as a proportion of their maximum extent

possible futures include flying cars, or a bit more prosaically, an airplane or helicopter in every driveway. Maybe small or narrow vehicles will more efficiently move people. Perhaps existing (large) vehicles will move more people. Possibly, people will live in small places without much need for movement. A great deal of investment is still being made in fixed-rail transportation. These futures are still speculation. No transition will be easy or without costs, a factor those leading the change must be aware of.

Wrap up for place and plexus

Which came first—the place, or the plexus? Does place lead or lag behind plexus? A great deal of debate centers on these questions, which we answer with a resounding: Yes! Academics (including ourselves) continue to attempt to tear this relationship apart employing time series with lead-lag variables or other econometric gymnastics. Transportation investments both serve existing markets and create new ones; some projects lean more towards one than the other, and by doing so recreate accessibility patterns which in turn change future development patterns.

The outstanding question relates to the degree to which transportation investments can be used to create desired land use patterns, or whether they should merely respond to past unmet demands. If so, how?

These are lofty issues that beg even loftier questions such as:

1   What is the appropriate role of government?
2   Is there consensus on what constitutes a "desired" land use pattern? and
3   Can government action (and the actions of planners, architects, engineers) effectively shape such an outcome?

In several instances throughout this book, we have suggested that the planning profession has long wrestled with such questions. Because planners are wrestling in an arena (i.e., the city) that is "mature," however, their policy interventions have served only to tweak things here or there. Such tweaking of itself is not necessarily a bad thing; it is just that after 40 or so years—after 10,000 blows of the axe—the tree is still not cut down.

Our claim throughout this work has been that tweaking the land use and transportation system here and there may pose a distraction from real policy shifts that might make a difference. There is an opportunity cost faced by the body politic. Society, and planners for that matter, must recognize differences between what is necessary and what is sufficient for change. Many things are necessary, but alone are not sufficient. Society must also recognize that there are influences on land use and transportation that are well outside the control of transportationists or urbanists.

By moving away from the normative world-view that dominates planning today (to eliminate congestion, save the environment, and interact with their fellow humans, people "should" take transit) to one that is based on empirical evidence (people will be more likely to take transit when the following conditions are in place . . . , transit agencies will provide more services when . . .), society can move toward actual solutions instead of the self-satisfying soliloquies that fill far too many plans. The physical nature and composition of communities is not a topic short on description. New Urbanists are often criticized for behaving as though they have all the answers, for telling others how to live (and how not to live), and for operating in a normative world of "what should be." Because the authors of this book are positivists before we are normativists, we first want to know what is—how the world actually operates.

What seems to be lacking is a strong tie between the types of places designers and other urban philosophers think people should live and the types of places in which most people reside. We advise the reader to take heed of the warning posed by Governor Kathleen Babineaux Blanco of Louisiana, speaking about the evacuation strategies employed during the 2005 New Orleans Hurricane Katrina tragedy, "Sometimes ideas that make sense to planners do not make sense to people." [23]

Planning proposes a course of action to achieve a particular end. In particular, land use and transportation planning aims to achieve desirable places and plexuses. Too much attention has been given to the question of what ends are desirable, rather than to identifying the course of action (and the participants who must engage in that action) necessary to achieve any particular end. Visions of end-states are fine, and may be necessary motivators, but a path from the present-state to such a Vision involves interactions between many different agents with diverse motives.

Aligning the motives of individuals with the aims of the group (the Vision) has been a missing element from planning. Rethinking the world from a bottom-up, agent-centered perspective rather than simply proposing top-down Visionary fantasies will ultimately move society closer to reaching its Vision.

Notes

a   The eight strategies they recommend are: (1) *selectively* implement pricing strategies to reduce problems of auto use, (2) *selectively* increase and improve public transit service, (3) make walking and biking safer, (4) take advantage of new technology, (5) remove barriers to flexible use of the transportation system, (6) *selectively* increase and improve highway capacity, (7) promote more flexible land development and redevelopment, and (8) promote reinforcing suites of strategies that are appropriate for local conditions.
b   The phrase "first, do no harm" in fact does not appear in the original Hippocratic oath, though a similar sentiment does.
c   The phrase comes from a twist on Oliver Hazard Perry's words after a naval battle: "We have met the enemy, and they are ours." The updated version was first used in a poster featuring comic strip "Pogo," by Walt Kelly, on Earth Day in 1970 and referred to pollution, earlier versions of the quote by Kelly referred to McCarthyism.
d   This is particularly problematic in a US context, where there is a two or four-year timeframe between elections; it therefore takes "courageous" politicians to be willing to engage ideas that might take longer. The problem of risk aversion among

politicians is reflected in this quote "If you wish to describe a proposal in a way that guarantees that a Minister will reject it, describe it as courageous." (Jonathan Lynn and Antony Jay, *Yes Minister* vol. 1 (1981)).

e Martin Wachs describing Melvin Weber in his obituary, www.berkeley.edu/news/berkeleyan/2006/12/07_Webber.shtml, accessed December 5, 2006.

f The expression, "let a thousand flowers bloom" is an adaptation of Chairman Mao Zedong's "let a hundred flowers bloom; let a hundred schools of thought contend." The slogan was used in the summer of 1957 when the Chinese intelligentsia was invited to criticize the political system then maintained in Communist China.

g Extra-terrestrial (or sub-terranean or oceanic) development would certainly pose as much of a shock as the discovery of new continents in previous centuries, and would change the nature of the discussion from working within mature systems to working with new systems.

h However, we warn against suggesting that new road improvements are not a big deal to residents in countless communities who have devoted decades to blocking new road developments.

i As mentioned in Chapter 11, flexibility in the changing use of existing buildings is actually a goal of many current planning initiatives such as form-based codes.

j smartgrowth.org, accessed December 14, 2006.

k In more technical language, the price of anarchy is the ratio of the user equilibrium and system optimal outcomes.

l The phrase "tilting at windmills" comes from an episode in *Don Quixote* by Cervantes in which the hero attacks windmills as he is under the illusion that they are giants. The expression therefore means to take on and fight an imaginary wrong, evil or opponent.

m Although it is too expensive now, perhaps in a hundred years it will not be. (Or in George Lucas's universe, maybe it wasn't too expensive a long time ago in a galaxy far far away.) Much science fiction revolves around relatively routine space travel. It's just that space travel isn't suited for daily use yet; it is still costly and risky. Moreover, given that most of us live and work here on earth, and it takes month of preparation for even Space Shuttle flights, the amount of waiting time makes it slower than more conventional transportation.

References

[1] Nelson, A., 'Leadership in a New Era', *Journal of the American Planning Association*, 2006, vol. 72 (4): 393–407.

[2] Giuliano, G. and Hanson, S., 'Managing the Auto', in G. Giuliano and S. Hanson (eds), *The Geography of Urban Transportation*, New York: Guilford Press, 2004, pp. 382–404.

[3] Fischel, W.A., 'Does the American Way of Zoning Cause the Suburbs of US Metropolitan Areas to be Too Spread Out?', *Governance and Opportunity in Metropolitan Areas*, Washington, DC: National Academy Press, 1999.

[4] Moss, W.G., 'Large Lot Zoning, Property Taxes, and Metropolitan Area', *Journal of Urban Economics*, 1977, vol. 4: 408–427.

[5] Peiser, R., 'Density and Urban Sprawl', *Land Economics*, 1999, vol. 65 (3): 194–204.

[6] Pendall, R., 'Do Land Use Controls Cause Sprawl?' *Environment and Planning B: Planning and Design*, 1999, vol. 26: 555–571.

[7] Thorson, J., 'Zoning Public Changes and the Urban Fringe Land Market', *Journal of the American Real Estate and Urban Economics Association*, 1994, vol. 22: 527–538.
[8] Southworth, M., 'Walkable Suburbs? An Evaluation of Neotraditional Communities at the Urban Edge', *Journal of the American Planning Association*, 1997, vol. 63 (1): 28–44.
[9] Shoup, D.C., *The High Cost of Free Parking*, Chicago, IL: APA Planners Press, 2005.
[10] Southworth, M. and Ben-Joseph, E., *Streets and the Shaping of Towns and Cities*, New York: McGraw Hill, 1997.
[11] Winston, C., 'Government Failure in Urban Transportation', *Fiscal Studies*, 2000, vol. 21 (4): 403.
[12] Levine, J., *Zoned Out: Regulation, Markets, and Choices in Transportation and Metropolitan Land-use*, Washington, DC: Resources for the Future, 2005.
[13] Sackett, D.L., Rosenberg, W.M.C., Gray, J.A.M., Haynes, R.B., and Richardson, W.S., 'Evidence-based Medicine: What it is and What it isn't', *British Medical Journal*, 1996, vol. 312: 71–72.
[14] Brownson, R.C., Baker, E.A., Leet, T.L., and Gillespie, K.N., *Evidence-based Public Health*, New York: Oxford University Press, 2003.
[15] Pfeffer, J. and Sutton, R.I., 'Evidence-based Management', *Harvard Business Review*, vol 84 (7/8) (July/August 2006): 13.
[16] Pfeffer, J. and Sutton, R.I., *Hard Facts, Dangerous Half-truths and Total Nonsense: Profiting from Evidence-based Management*, Cambridge, MA: Harvard Business School Press, 2006.
[17] Boarnet, M.G. and Crane, R., *Travel by Design: The Influence of Urban Form on Travel*, New York: Oxford University Press, 2001.
[18] Forsyth, A., Oakes, J.M., and Schmitz, K.H., 'Does Residential Density Increase Walking and Other Physical Activity?' *Urban Studies*, 2007, vol. 44 (4): 679–697.
[19] Oakes, J.M., Forsyth, A., and Schmitz, K.H., 'The Effect of Neighborhood Density and Street Connectivity on Walking Behavior: The Twin Cities Walking Study', under review.
[20] Hohenberg, P.M. and Lees, L.H., *The Making of Urban Europe 1000–1994*, Cambridge, MA: Harvard University Press, 1995.
[21] Roughgarden, T., *Selfish Routing and the Price of Anarchy*, Cambridge, MA: MIT Press, 1995.
[22] Rogers, E., *The Diffusion of Innovation*, New York: The Free Press, 1962.
[23] Goodman, B., 'Forum for Evacuees in Atlanta Draws Small, Unhappy Crowd', *The New York Times*, January 22, 2006: 1.23.

Index

accessibility 11, 22, 44–46, 48, 50–53, 55, 62–63, 66, 91–92, 141, 161, 198, 200–202, 212–213, 261–263, 271–272
  local 53, 55
  measures 48, 50–51, 53, 196, 198
  measuring 48
  regional 53, 55, 63
  relationship to congestion 9
  relative 50, 261
affordable housing, *see* housing, affordable
African-Americans, *see* race and ethnicity
agencies, public 213–214, 266
agent nodes 150–151
agglomeration, economic 13, 155–159, 165, 295, 298
agriculture 5, 42–43, 159
Alexander, Christopher 232
Alonso, William 43, 163
American Dream 5–7, 40, 61
Amsterdam 112–113, 248, 254
architecture 211, 214, 218, 223, 237, 241–242, 246, 248
arms race 12, 108–109, 111–112
auto travel 8, 199
  time 117
automobile mode share 107–108, 117
automobiles 2–5, 7, 9, 12, 25, 40–41, 53, 97–98, 107, 112, 116, 143, 185, 201, 208–209, 245–246
  hybrid 288
average
  delay 281, 284

  household size 27, 178
  travel time 79, 285

balance, jobs-housing, *see* jobs-housing balance
ballot initiatives 251–252
bid-rent 43–44, 63, 163
  theory 43–44, 63, 163
bottleneck 282–284
boundary effects 161
Brazil 119
bridges 3, 95, 151, 171, 191, 253, 269, 271, 274, 286
built environment 2, 3, 7, 11, 59, 237, 306, 311, 313
buses 20, 41, 94–96, 98–101, 107, 114, 116–117, 195, 197, 238, 278, 288, 301, 302–303, 310
  mode share 107, 117
  service 96, 101, 107–108, 114, 266
  strike 96
  travel times 107, 115, 117
business
  districts, central 43, 63, 157, 159, 162–163, 169, 251, 264, 319
  owners 191, 290
businesses, *see* firms, commercial

canals 36, 320–321
capacity 13, 65, 91, 102–103, 130, 184, 196, 219, 225, 229, 259, 261–264, 282–283, 292–293
  roadway 82, 107, 264, 293
capital 31, 141, 150, 197–198, 215, 271, 319

carpools and vanpools 77, 95,
    106–107, 300–301, 303
cars, cable 254–255
central place theory 163, 185,
    224–226, 228
chances, *see* opportunities
Chicago Area Transportation Study
    20
choices, mode, *see* mode choices
Christaller, Walter 163, 224–226
cities 41–43, 50–52, 82, 109–111,
    154–155, 157–160, 162–165,
    218–219, 225–228, 247–249,
    254–255, 273–275, 297–298,
    306–311, 313–314, 319
  central 4, 5, 42, 159, 164, 199,
    226, 242, 265, 307
  edge 159, 163, 227
  network 226, 228
  systems of 224–227
city planning, *see* urban planning
Cleveland Regional Area Traffic
    Study 124
clusters 64, 163, 165, 186
collision 113, 229, 282–283
community design 59, 60, 209
commuting time 75, 81–83
competition 11–12, 19, 24, 28–29,
    52–53, 87–88, 102–103,
    109, 113, 135, 144–148,
    152, 161–162, 165, 169,
    181–182
complementary 88, 145, 147, 150,
    182
complementors 11–12, 19, 24,
    28–29, 36, 98–99, 102, 130,
    135, 142, 145–146, 148,
    152, 155, 157, 162
congestion 1–4, 9, 12–14, 40–41,
    81–83, 99, 193, 200–201,
    256, 260, 262–263,
    278–284, 288, 293–294,
    299, 303
  costs 103
  index 81–82
  pricing 101, 151, 278–279,
    284, 286, 298
  problems 195, 302
constraints 11–12, 14, 18–19,
    23–24, 26, 28–30, 32, 73,
    79, 90, 121, 128, 130, 135,
    239, 293
consumer cooperatives 31

consumers 20, 30–31, 45, 58,
    111, 146, 159, 169, 179,
    183–186, 196–197, 212,
    229, 256–257, 259,
    272–273
cooperation 29, 31, 105, 144
cooperatives 31, 144–145
Coruscant 271, 320
costs, average 285, 287
crime 4, 6, 22, 56, 61, 104, 117,
    168
cross-elasticity 147
cycling, *see* bicycling

demand curves 76, 181, 216,
    256–257, 259–260, 264,
    272–273
development 39, 153–154, 223,
    226, 231, 240–242, 245,
    261, 271–272, 292–295,
    297, 299, 317
  charges 206
  districts 295
  effect on accessibility 51
  factors affecting 140, 262–263,
    272
  financing 153
  high-density 317
  impact fees for 295
  induced 261–263
  infrastructure required by 7
  in Japan 240
  models of 226
  networks constructed for 232,
    241
  New Urbanist approach to
    231
  open-source model 143
  parking required by 300
  regulation of 2, 7, 215, 245,
    280, 290, 292–294, 310
  role of design in 241–242
  rural areas 5, 191
  suburban 3, 39, 40, 209, 242
  temporal sequence of 225, 279
  transportation as factor in 66,
    261, 270–271
Diamond
  of Action 11–12, 17–18, 20,
    22–24, 26, 30, 32, 36, 38,
    147–148
  of Assembly 215, 262, 271
  of Design 214, 222–223, 246

of Evaluation 11, 13, 190,
192–194, 196, 198, 200,
202, 204, 206, 208, 210,
212, 214, 216, 247
of Exchange 11–12, 138, 140, 142,
144, 146, 148, 150, 152
District of Columbia 19, 24, 36,
122, 298
Downs, Anthony 7, 8, 260
drivers 99, 103, 111–112, 138,
203, 237–238, 242, 244–245,
260, 278, 281–282, 284–286,
288, 301–302

economic geography 158–159
economies of scale 28, 142,
144–145, 159, 165, 186,
224, 228, 233
edge cities 159, 163, 227
Edge City Dynamics 159–160
efficiency 2, 13, 25, 193–197, 199,
201, 208, 211, 213–214,
233, 247, 285
elasticity 147, 263–265
electricity 165, 270–271, 290
elephant, parable of 190
employees 53, 88, 94, 158, 162,
188, 197, 289, 300–301,
302–303
employers 33, 70–72, 79, 88, 95,
158, 163–164, 266, 289,
300–301, 302–303
employment 12–13, 23–26, 48,
50–53, 65, 70–73, 75–77,
79–81, 87–92, 95–96,
133, 135, 157–164, 188,
195–196, 301
  constraint on time 121
  distance to 128, 160–161
  finding 316
  location 52, 65, 72, 75, 88, 90,
95, 106, 311
  related travel 73, 76–77, 81,
91, 96–97, 121, 126, 130,
132, 161–162, 185, 289,
297, 301, 312, 320
  seeking 11, 70–71, 73, 78, 82,
84, 86, 88, 90
  telecommuting 302
environment 13, 193–194, 203,
247, 318, 322
  built, *see* built environment
  physical 21, 59, 60

environmental impacts 160, 186,
205–206, 208, 211, 213,
293
equity, social 203
expediency 13, 193–194, 202,
210–211, 247
experience, quality of 29

factories 139, 150, 153
factors, push 35, 55–56, 58
fairness 199, 202, 204
farming, *see* agriculture
Federal-Aid Highway Act of 1944
124
firms, commercial xiii, 10–13,
52–53, 88, 139, 141–142,
145–148, 150, 152, 155–158,
160, 162–163, 165, 169, 261,
307
Flyvbjerg, Bent 265, 276
forecasting 197, 264–266
forecasts 194, 204, 264–266, 273
Francisco 37
freeways 63, 99, 163, 211, 222–223,
225, 228–230, 240, 253, 255,
319
Fresno, California 246

game 28–29, 35, 45, 70, 88,
104–105, 139, 173, 246
  non-zero-sum 29, 30
  one-time 104–105
  theory i, 35, 88, 98, 104,
107–108, 110
Gans, Herbert 40
gender 123, 204
geography 25, 53, 140, 184, 225
  business location theory 155
  gravity models in 79
  movement of goods and services
147
  relation to social networks 162
  social networks 86
  spatial mismatch 91
Giuliano 162
Granovetter, Mark 12, 79
gravity models 12, 63, 71, 73–79,
90–91
grid 24, 228, 233, 246, 270
growth
  controls 279–280, 292
  management 290, 292–293,
299

Hagerstrand, Torsten 127
hedonic model 43, 60, 62, 64–65, 89, 216, 289
hierarchy 214, 222–223, 225–226, 229–230, 237, 246, 267, 318
  of places 13, 163, 223
  of plexus 225
  of roads ix, 13, 223, 228–229, 246, 267, 318
Highway Act of 1944, see Federal-Aid Highway Act of 1944
highways, interstate 41, 103, 191, 225, 230, 233, 314, 320
home
  location 25–26, 41, 52, 268
  price 63, 65
  sale price 62–63
  sales 62
  value 62–63
homebuying 11, 39, 40, 42, 44, 46, 48, 50, 52, 56, 58, 60–62, 64, 66, 68
Homestead Act 6
Hong Kong 178, 200, 271, 297, 319
household size 26, 89, 123
housing xi, 10–11, 32, 39, 40, 52, 58, 64, 66, 79, 88, 140, 157–160, 178, 227, 292–293, 315–316
  affordable 40, 60, 64, 292
  competition with businesses 157
  prices 61
  single-family 40, 61, 236
HOV lanes 288
human behavior 21–22, 192, 197, 311

impact fees 295
induced
  demand 256, 258–259, 263–264
  traffic 258, 263
industrial location theory 156
inelastic demand 92, 263–265
information networks 247
infrastructure 1, 13, 153, 197, 214–215, 223, 252, 254, 262–264, 267, 269, 271, 279–280, 284, 292–295, 317–318
  cost of 3

investments 52, 101, 253, 255, 262, 268, 290
  planning 265, 294
Internet xiv, 9, 25, 177, 180, 183–184
Interstate Highway Act of 1956 5
interstate highways, see highways, interstate
intervening opportunities model 91

Jacobs, Jane 232, 265–266
Japan 120, 139, 145, 206, 230, 240, 255
job
  location 72–73
  markets 71, 87–88
jobs, see employment

Krugman, Paul 159–160

land
  prices 242, 262–263
  tax 294, 297
  use-transportation
    planning 158, 196
    policy 58, 105–106, 233
    systems 12–13, 194, 205, 214, 237
landlords 105
Landlord's Game 46–47
law 49, 74, 112–113, 163, 208, 215, 227–228, 296, 300, 315
Levine, Robert 119
Levittown 5, 39–41
light rail 210, 229, 251–253, 255–256, 274–275
links 24, 46, 48, 73, 85–86, 102, 113, 148, 150–151, 195–196, 228–229, 238–239, 246, 260–261, 266–267, 283
  connection 150–151
  upstream 283
livability 209–210, 312
location decisions 41, 66, 71, 135, 152
  business 140, 155–156, 163
  residential 11, 22, 33–34, 52–53, 60, 65, 71–72, 79
locational triangle 156
locators 12, 140, 151
London 114, 319
Losch, August 163, 185, 224–226

malls, *see* shopping centers
manufacturer 138–139, 179, 197
market area 75, 181, 186, 224–225
markets 13, 40–43, 48, 50, 60, 88, 138–145, 148–152, 157, 162–163, 171, 180–181, 186, 199, 200, 227, 261
  input 149–151
  output 149–151
Marshall, Alfred 159
maximum travel costs 181
migration 74
Milgram, Stanley 84
Minneapolis 62, 64, 109–111, 168, 178, 227, 261
mobility
  income as factor in 89
  measures of 82, 198
  relationship
    to accessibility 198, 200–201, 284
    to congestion 191, 193
modal competition model 108, 116–117
mode
  choice models 20
  choices 12, 97–98, 104, 124
  share 96–97, 117
models 10, 20–21, 23, 35–36, 42, 57–58, 62, 73, 76–77, 81, 116–117, 125–126, 158–160, 224–226, 237–238, 272–273
  behavioral 52
  land use 48
modes 15, 19, 20, 35, 53, 97–98, 104, 113, 116, 136, 200–201, 209, 215–216, 245, 263, 289, 299, 303
MOEs 193, 195, 198
Mokhtarian, Patricia 37, 68, 90, 92, 133, 136–137
Monopoly (board game) x, 44–45, 110
monorails 251–253
Morill Act 6
morphology 13, 214, 222–223, 230–233, 236–237, 246–247

Nash Equilibrium 104–105, 107
National Household Travel Survey 128–129, 187, 209
neighborhood design 55, 59, 60

neighborhoods 21, 55–61, 63, 66, 105, 145, 172, 183–184, 186, 188, 196, 202, 208, 236, 242, 309
Netherlands 185, 242–244
network
  analysis 35, 46, 148, 267, 277
  effects 98, 101, 108
  expansion 99, 271
  externalities 165, 183, 272
  growth processes 267
  model 149
  size 49, 50, 66, 321
  system 226, 228
networks xii, xiii, 1, 47–49, 98–99, 147–148, 150–151, 155, 159–160, 183, 226–227, 232–233, 238–239, 246–248, 266–267, 270–273, 319
New Economic Geography 159
New Urbanism 178, 223, 231–233, 241–242, 312, 323
New York City 94–95, 99, 120, 222–223, 226, 233, 298, 319
nodes 46, 48–50, 85, 148, 150, 226, 230, 238–239, 267
non-motorized transportation 210, 217, 247, 312
non-work travel 25, 52, 67, 90, 121, 132–133

Olympic Games 246–247, 255, 302
opportunities 11, 24, 29, 30, 43, 88, 127–8, 135, 145, 147, 268, 301
ordinances 289, 301, 303

parking 90, 106, 140, 169, 172, 182, 186, 215, 219, 287, 289–290, 300, 303
parks 55, 140, 167, 240, 290, 292
A Pattern Language 232
peak hour 96–97, 302
pedestrians 95, 169, 173, 203, 222, 243, 245–246, 280, 289
Phoenix 255
physical activity 36, 216, 312
planners 41, 59, 60, 76, 98, 116, 139, 145, 191, 193, 196–198,

241, 265–266, 287–289, 307–309, 312–313, 322–323
planning viii, xiii, 7, 10, 109, 194–203, 240, 246, 284, 290, 292, 294, 307, 322–323
  applications 198, 200–201
Poisson model 82
pollution 6, 63, 205, 270, 272, 288, 317, 323
positive feedback loops 99, 101
preferences 7
pricing 13, 215, 278–280, 284, 298
prism, space-time, *see* space-time prism
prisoner's dilemma 12, 104–108
production 148–150, 157–158, 237, 261
productivity 193, 196–198, 212–213
property taxes 94, 153–154, 290, 292, 297
psychology 20–21, 97
public
  roads 117, 124, 240, 314
  transit, *see* transit
Putnam, Robert 86, 87

queues 280–283, 294, 298, 299
queuing 91, 275, 279–281, 283, 290, 299

race and ethnicity 5, 12, 22, 31, 56–58, 61, 65–66, 199, 202, 204, 245
railroads 6, 36, 43, 205, 230, 270, 296, 318, 320
ramp metering 269, 281–282, 284–286
Ravenstein, Ernest 74, 92
Raymond, Eric 143, 152, 226
Reagan, Ronald viii, 17–18, 36, 85
relocation costs 88, 154–155
rents 43–45, 58, 64, 105, 176, 219, 289, 296, 299
  bid-rent curves, *see* bid-rent theory
residences, *see* housing
residential location decisions, *see* location decisions, residential

retail 11, 139–140, 142, 144, 151, 163, 168–169, 178–179, 185–186
roads 12–13, 19, 99, 101–103, 215–216, 222–223, 228–230, 232–233, 237–247, 259–261, 269–270, 282–284, 289–290, 292–293, 295, 314–316
  building 5, 6, 201
  network 19, 223, 229–230, 232, 246
  pricing 151, 284, 298, 307, 318
Rome, ancient, congestion in 3
route choice 102, 136
routes 19, 28, 32, 41, 77, 99, 102, 114, 127, 136, 230, 238, 252, 260, 266, 318

San Francisco 82, 99, 114, 191, 226–227, 239, 248, 254–255, 287
Santayana, George 7
scale 3, 28, 40, 53, 55, 142, 144–145, 172, 186, 195, 211, 222, 224, 228–229, 233, 263
scarcity 90, 106, 206
scheduling time 120–121, 123–135, 137
Schelling, Thomas 110
schools xiii, 7, 22, 24, 54–55, 59, 61, 77, 126, 129–131, 208, 236, 290, 292, 310, 319
S-curves 319
Seattle 82, 111, 124, 187, 227, 243, 251–252, 273–274, 289
segregation 56, 66
shopping centers 58, 110, 125–126, 155, 167–168, 170, 172–174, 179, 185–186, 219, 246
signs 6, 7, 91, 237, 244–245, 264
siting 11, 153–156, 158, 160, 162–164, 166, 252
Smart Growth 317
social
  capital 86–87, 95, 143, 209
  networks xiii, 1, 12, 71, 79, 84–87, 91, 162, 183, 188, 247
space 43, 60
  distribution of activities across 1, 21, 23
  preference for 90

space-time prism 127–128
spatial mismatch 91, 163–164
sprawl 4, 8, 14, 113, 250
staging ceilings 293
strategies, female 71, 84
streetcars 25, 39, 114, 116, 229, 270, 317, 320
streets, local 225, 228, 230, 242, 246
subsidies 3, 5, 7, 64, 113, 118, 155, 303
  business location 115, 163, 245
  development 3, 7, 318
  transit 298, 302–303
  utility service 6
suburbs 5, 39–41, 50, 55–56, 59, 89, 90, 139, 153, 164, 172, 199, 230, 255, 289, 298
supermarkets 55, 174–176, 180, 182
suppliers 12–13, 140–141, 144, 146, 148, 150, 152, 155, 162–163, 169
supply chains 139, 155, 159, 247
Surowiecki, James 110
surplus 161, 196–197, 212, 257, 259, 294
sustainability 215, 278
systems, open 142–143

taxes 6, 7, 58, 153–154, 163, 176, 279–280, 294–298, 318
  commuter 295, 297–298
telecommuting 300, 302
Thünen, von, *see* von Thünen, Johann Heinrich
time
  budget 30
  choices vary by 32–33, 41
  congestion costs 4
  constraint on travel 18, 20–21, 23–25
  horizons 33
Tokyo 319
tolls 284–287, 299, 318
  congestion, *see* congestion pricing
topology 232–233, 246
traffic
  calming 223, 242–243, 246
  congestion, *see* congestion
  transit 2, 4, 12, 20, 61, 77, 95, 97, 99, 107, 112, 116, 288–289, 302–303, 310, 322–323
  accessibility 53, 116, 164, 215, 229, 293
  agencies 94, 124, 197, 323
  demand 20, 95, 99, 114, 117, 263–264, 274
  financing 4, 101, 115, 251
  forecasting 266
  infrastructure for 13
  level of service 41, 96, 197, 201, 210
  mode preferences 12, 32, 91, 98, 289
  ridership 99, 101, 108, 114, 117–118
  strikes 94–95
  subsidies 118, 303
  wait time 99, 100, 108–109
transport
  costs 25–26, 42, 146, 156, 159, 163, 169–170, 195, 224
  planning i, 76, 216
transportation
  infrastructure 10–11, 65, 91, 254–256, 261, 263, 273, 279–280, 292, 295, 320
  modes, *see* modes
  networks viii, 4, 13, 18, 79, 84, 98, 110, 130, 151, 201, 215, 225, 232, 239, 267
  planning models 75
  projects 202–204, 213
Transportation Demand Management 269, 300
travel
  behavior 9, 14, 20, 32, 72, 120, 126–127, 133, 231, 268, 273, 303
  theory 20
  complexity of 126–127
  cost 41, 43, 48, 50, 73, 261, 284, 293
  decisions 41, 55, 113, 130, 132
  delay 285–286
  demand 55, 92, 158, 256, 263–265, 268, 271, 279, 283
    management xiii, 32, 289, 298–299
  diaries 123, 125–126
  distance 89, 121, 129, 136, 162, 176, 184–185

leisure 133–134
markets 46, 209, 316
modes 20–21, 32–33, 58, 126, 132–133, 136, 210, 213, 319
positive utility of 90, 133–134
surveys 120, 123–126, 187
time 9, 12, 18, 51, 75–82, 89, 91–92, 102–103, 107–108, 115–117, 120–121, 123–124, 126, 134, 260–263, 285–286
  budgets 25, 120–121, 136
  estimates 82, 117
  reducing 256, 262, 315
  variability 229
trips
  classification 132
  linked 130, 132
  non-work 26, 89, 130, 185
  peak hour 300–301
  purpose 76, 126, 129, 131–132, 135
  sequence of 126
Twin Cities 191
  of Minneapolis 168, 290

United States Department of Transportation 187
urban
  form, influence on travel 68, 325
  planning xii, 3, 4, 10, 109, 178, 222, 230–231, 248, 307, 309, 311
  transportation planning, *see* transport planning

utility theory 20, 22–23, 30, 43, 61, 64, 133–134, 139, 197, 212, 237, 270

Vance, James 225
vanpools, *see* carpools and vanpools
Varignon frame ix, 156–157
von Thünen, Johann Heinrich 41–43, 163

walking 2, 25, 53–55, 59, 95, 97, 119, 125, 133, 146, 201, 209–210, 219, 231, 245, 312
Washington (state) 124, 200, 225, 243, 289, 303
Washington, DC, *see* District of Columbia
Web, World Wide, *see* Internet
Weber, Alfred 156, 165
work
  see employment
  trips 52, 91, 97, 130
  while traveling 124
workers 3, 12, 25, 40, 50, 53, 73, 76–77, 88–89, 91, 94, 106, 121, 141, 158–164, 301–302
workforce 66, 140–141, 155–156, 158, 162–163, 179, 289
World War II, carsharing encouraged 113

Zahavi, Yacov 120

eBooks – at www.eBookstore.tandf.co.uk

A library at your fingertips!

eBooks are electronic versions of printed books. You can store them on your PC/laptop or browse them online.

They have advantages for anyone needing rapid access to a wide variety of published, copyright information.

eBooks can help your research by enabling you to bookmark chapters, annotate text and use instant searches to find specific words or phrases. Several eBook files would fit on even a small laptop or PDA.

NEW: Save money by eSubscribing: cheap, online access to any eBook for as long as you need it.

Annual subscription packages

We now offer special low-cost bulk subscriptions to packages of eBooks in certain subject areas. These are available to libraries or to individuals.

For more information please contact webmaster.ebooks@tandf.co.uk

We're continually developing the eBook concept, so keep up to date by visiting the website.

www.eBookstore.tandf.co.uk